Hamidreza Kazempour

The Evolution of Muqarnas

in Iran

ii

Hamidreza Kazempour
Ph.D., Traditional Architecture of Iran
D.A., Restoration of Historical Buildings

School of Architecture and Environmental Design,
Iran University of Science and Technology,
Narmak, 1684613114 Tehran, Iran

ISBN: 978-1-939-12352-7
ISBN-10: 1939123526

Library of Congress Control Number: 2016908726

Publisher: Supreme Century, Reseda, LA, USA

Prepared for Publication by Asan Nashr

Preface

Muqarnas has always been one of the most complex decorative elements of world's monumental architecture. In muqarnas, niche–like components are combined together and arranged in successive tiers to produce a three–dimensional geometric shape, enclosing and embellishing features such as ceiling, soffit, portal, and vault. This unique structure has been intensely studied from various aspects by many scholars. Nevertheless, there is still lack of clarification about the structure's origin and path of evolution. There are some theories indicating that the structure is originated from squinches in Iran, but no further explanation is provided to fill the huge gap between the two, i.e. muqarnas and squinch, and to clarify the quality of the gradual development.

In this manuscript, the missing link between muqarnas and squinch is introduced that is in fact, another undefined form in traditional architecture of Iran, named *patkaneh*. Hence, the main effort is to investigate the possibility of differentiating this ornament from muqarnas and to introduce it as a different structure, which is the missing link between squinch and muqarnas. For that, a qualitative approach was employed that strives to demonstrate the steps of gradual deformation of muqarnas from squinch by defining the characteristics of the linking ornament, using an inductive approach. From among the diverse samples recorded for the purpose of this investigation on the origin of muqarnas in Iran, some critical samples of muqarnas and pseudo-muqarnas, as they are named before being identified, were selected and used as guides towards finding the gradual development of muqarnas.

In addition, I appreciate the moment to express my sincere gratitude to my precious supervisors, Associated Professors Dr. Gholamhossein Memarian and Syed Ahmad Iskandar bin Syed Ariffin, and my distinguished teacher of Islamic art, Maestro Aliasghar Sha'rbaf, for their kind encouragements and guidance. I am very much grateful to my darling wife, Dr. Moones Rahmandoust, for her kind and never–ending motivations and encouragements; without her support and patience, I would not have been able to dedicate my time to my research and to make my path toward greater success. I also admire and thank my respected parents, Mr. Abbas Kazempourfard and Ms. Malak Ghaderi; my distinguished parents-in-law, Dr. Mostafa Rahmandoust and Ms. Azar Rezaee; I owe all the nice and valuable moments of my life to them. Further, I appreciate my father as my first and best teacher of traditional art, without whom, I would not have the chance to understand the beauty of Islamic art and architecture, to this extent.

Many of my friends are also worthy to be very much appreciated here: Dr. Seyed Javad Asadpour, Dr. Reza Hashemi, Hadi Safaeipour, Ala Amirfazli, Dr. Feryal RezaeeMood, Shima Taslim, Dr. Nima Norouzi, Dr. Mohammad Ghomeishi, Dr. Hesam Sotoudeh, Dr Nima Moinzadeh, Dr. Hassan Chizari, Dr. Alireza Daneshpour and Dr. Hadi Ebadi for their friendly participation in our scientific discussions and sharing their views and tips to achieve better and more reliable results. I'm grateful to Rokhsareh Mobarhan, Mojib Majidi, Fahimeh Malekinezhad, Morteza Firouzi and Masoumeh Zibarzani as well, for their kind assistance and friendly help at various occasions.

Finally, I strongly hope that the book provides valuable knowledge for those who are interested in the beauty of muqarnas and the traditional art and architecture of Iran.

Hamidreza Kazempour

Contents

Abbreviations

C.E.	–	Current Era
B.C.	–	Before Christ
A.H.	–	After Hijra
ca.	–	circa, means about
i.e.	–	That is
e.g.	–	For example
2DPP	–	Two-Dimensional Pattern Plan
ETP	–	Edge to Point
ETE	–	Edge to Edge
PTE	–	Point to Edge
Msq.	–	Mosque

Glossary

Muqarnas   –   The most advanced ornament in Islamic architecture invented in early 14th century. This structure is purely decorative, adding dead load to the building. The constituent elements of *muqarnas* are confined to the eight basic elements, namely *shamseh, taseh, toranj, shaparak, parak, tee, espar* and *takht.*

Patkaneh   –   A set of niche–like components arranged in successive tiers in a definite geometrical regime, in order to cover vaults or ceilings. *Patkaneh* is a load-bearing transition structure, dating back to the 10th century. The structure consists of two main parts, the ribs that provide the structural role for the ornament, and the *tasehs*, built between these ribs, in two to maximum five successive projecting tiers. Plumb *paraks* and *shaparaks* are the intermediate elements observed in *patkanehs*, although exceptional samples had been recorded with three-dimensional *paraks* or *shaparaks* in late 12th century.

Squinch   –   It is the oldest known transition technique, which can be considered a single *taseh*. It consists of two load-bearing concave vault–sections intersecting each other in a line at right angle.

Shamseh   –   The central medallion, a three dimensional star.

Taseh   –   A symmetric edge to point *muqarnas* component consisting of one to eight concaved triangular segments. Though, a single-segment *taseh* may be also two-dimensional.

Shaparak   –   A flexible point to edge intermediate element, consisting of two attached concave triangles, capable of covering various angles efficiently.

Parak   –   A point to edge intermediate element, consisting of a single concave triangle.

Toranj   –   A tetrahedral element, with axial symmetry, which may be compressed or elongated based on its position in the ornament with respect to its neighbor elements.

Tee   –   An edge to edge plumb intermediate element that is in fact a long rectangle which is normally attached to a *takht* element on its lower edge and has two flanking *paraks* as its lateral neighbours, making the whole combination similar to letter T.

Espar   –   *Espar* is an edge to edge plumb element created in 12th century, with the lower edge bigger than the top one. The edges are connected to each other by means of two curved lines. The word is the Persian equivalent of "entablature" as an architectural term. *Espar* can be observed in the first few tiers of *muqarnas* on the walls and in the space between two *shaparaks*.

Takht   –   *Takht* elements are the only horizontal elements which are specific to *muqarnas* and cannot be seen in any pseudo-*muqarnas* structures, before 14th century. They may have regular or irregular geometric shapes. Regular *takhts* are star shaped flat elements, like three-, four-, five-, six-, seven- and eight-pointed stars and other shapes are defined as irregular *takhts*. The existence of any *takht* element in an ornament shows that the ornament is undoubtedly a *muqarnas*.

Tier   –   Each of the successive levels in which the constituent elements of *muqarnas* and pseudo-*muqarnas* ornament are arranged.

CHAPTER 1

Introduction

1.1   Chapter Opening

Muqarnas is one of the most complex ornaments of Islamic architecture also used in many non–Islamic buildings, such as cathedrals and palaces. The geometrical complexity of this structure has attracted the attention of many scholars and researchers in many famous and high ranking universities of the world.

Muqarnas is sometimes termed as a *stalactite* or *honeycomb* as well. Figure 1.1 shows a *muqarnas* dome built by the author in 2005, in Varamin, Iran. The pattern of this *muqarnas* veneer of designed by Maestro Sha'rbaf, who is a world famous traditional mason and designer. In her book, Necipoglu refers to him as the only reference for all the issues related to traditional techniques of designing and constructing *muqarnas* in Iran (Necipoglu, 1995). The author is proud to be mentored by Maestro Sha'rbaf for more than 10 years and to learn from him many subtle details and characteristics of *muqarnas* through working with him and constructing many *muqarnas* samples under his supervision in Iran.

As a result of the existence of *shamseh*, i.e. the central sunburst medallion at the apex of the work, as well as the star-shaped *takht* elements, dispersed over other tiers of the ornament in a radial scheme, the appearance of the structure induces the pattern of the sky in the eyes of the observer, in a way that since 12th century, many Persian poets used the term *muqarnas* in their poems as a metaphor for sky.

Figure 1.1   *Muqarnas* dome designed and built by author, Jame' of Varamin

In spite of the great amount of research carried out about this art, there are debates about the chronological or geographical origins of it. But what is for certain is that by the end of the 12th century, *muqarnas* was not only widely used all over the Muslim world, but it was in many regions mixed and developed by local traditional ornaments as well. However, some scholars believe that *muqarnas* originated from either Iran or Syria in the late 8th and early 9th centuries and some others trace its roots in northern parts of Africa or Baghdad in the 11th century (Al-Asad, 1995).

Among the available studies, there is a reliable theory by Michel Écochard introducing *squinch* as the functional origin of *muqarnas* in Iran, in the 10th century, but he does not explain the path of this gradual change (Écochard, 1977). Stierlin also believes that the origin of *muqarnas* goes back to the squinch. He explains his theory by mentioning that *muqarnas* structures are in fact several little squinches (Stierlin, 1976). Later on, Harb tried to explain the connection between *muqarnas* and squinch by means of Roman pendentives (Harb, 1978). To the author's best knowledge, these are the only available sources trying to explain how Iranian *muqarnas* structure was developed into its current state (Harb, 1978).

Practically, today's geometrically complex *muqarnas* originated from a simpler pseudo–*muqarnas* structure named *patkaneh* which should be defined and explained in detail in order to enlighten the evolution of *muqarnas* through passage of time, in terms of changes in its form and structure, as well as the applied construction techniques. *Patkaneh* is a set of niche–like components arranged in successive tiers in a definite geometrical regime, in order to cover vaults or ceilings. Hence, introducing *patkaneh* more specifically as the practical origin of *muqarnas* and as the missing link between squinch and *muqarnas* is one of the goals of this research.

By studying the earlier samples of *muqarnas*, one can assert that the structure was completely structural at the beginning, playing the role of a smooth transition from square base of the wall to the circular base of domes, vaults and ceilings, but later it was transformed into a purely decorative structure leading to today's ultimate form of *muqarnas*.

1.2   Evolution of *muqarnas*

Although in traditional Islamic architecture, there are usually divine concepts attributed to the appearance of forms and structures, they are simultaneously in complete conformity with construction rules. In other words, the architects and masons were capable of explaining the soaring ideas through the knowledge of structure and material.

There seems to be a general practice in Islamic architecture, that is, forms and figures of elements, which are constructed for structural purposes in a building, are gradually transformed into non–load–bearing decorative elements. Similar process of transformation may be observed in other architectural schools, but what makes Islamic architecture different is that the architects and masons were mixing the new elements with an Islamic theme (Edwards & Edwards, 1999). Another specific characterization of Islamic architecture is that it makes the observer confused of whether the specific form was supposed to be structural or decorative. Though, there are many samples in which a complete combination of both purposes can be seen.

Muqarnas is a unique example of a form in Islamic architecture that possesses the path of evolution and maturation from a pure structural role to a purely decorative one. This research intends to demonstrate the path of development of *muqarnas* in light of three factors, hence time, form and structure, and construction technique, as the factors that played the key role in the process of this reformation, based on the existing samples in Iran.

1.2.1 Chronology

After Holy Prophet of Islam passed away in 632 C.E., Islamic world was governed by Umayyad (661–750) and Abbasid (750–1258), expanding to Near East, Middle East, North Africa and a great part of Spain. In the 9th century, the capital of the Caliphate at Baghdad lost control over the eastern provinces and from early 11th century Seljuks established an independent Islamic sultanate expanding from Greater Iran to Near East. At that time, Egypt became the capital of the Caliphate of Fatimids (969–1171). Figure 1.2 (a) shows the governing borders of Seljuk dynasty.

Starting with the 13th century, after frequent invasions by the Mongols, Islamic government of the Middle East lost its power, and in 1256, Baghdad, the capital, was completely overran by the Mongols and the Ilkhanid dynasty was established. In 1295, the Ilkhanids embraced Islam and devoted their efforts towards expanding the Islamic architecture in Iran then after, by building mosques and schools and mausoleums. Figure 1.2(b) illustrates the borders of Iran during Ilkhanid dynasty. Later on, Timurids (1370–1506) conquered the eastern parts of their territories and after them Safavids (1502–1736) achieved the control over the East and the Ottomans (1300–1924) over the Western parts (Edwards & Edwards, 1999).

Based on the explanations above, the main focus of the current study about discovering the evolutionary path of *muqarnas* in Iran, includes mainly the Seljuk (1038–1194) and Ilkhanid (1256–1353) dynasties, although examples from periods, before and after the mentioned dynasties, were also considered, when necessary (Wilber D. N., The Architecture of Islamic Iran: The Il-Khanid Period, 1955; Schroeder, The Seljuk Period, 1977).

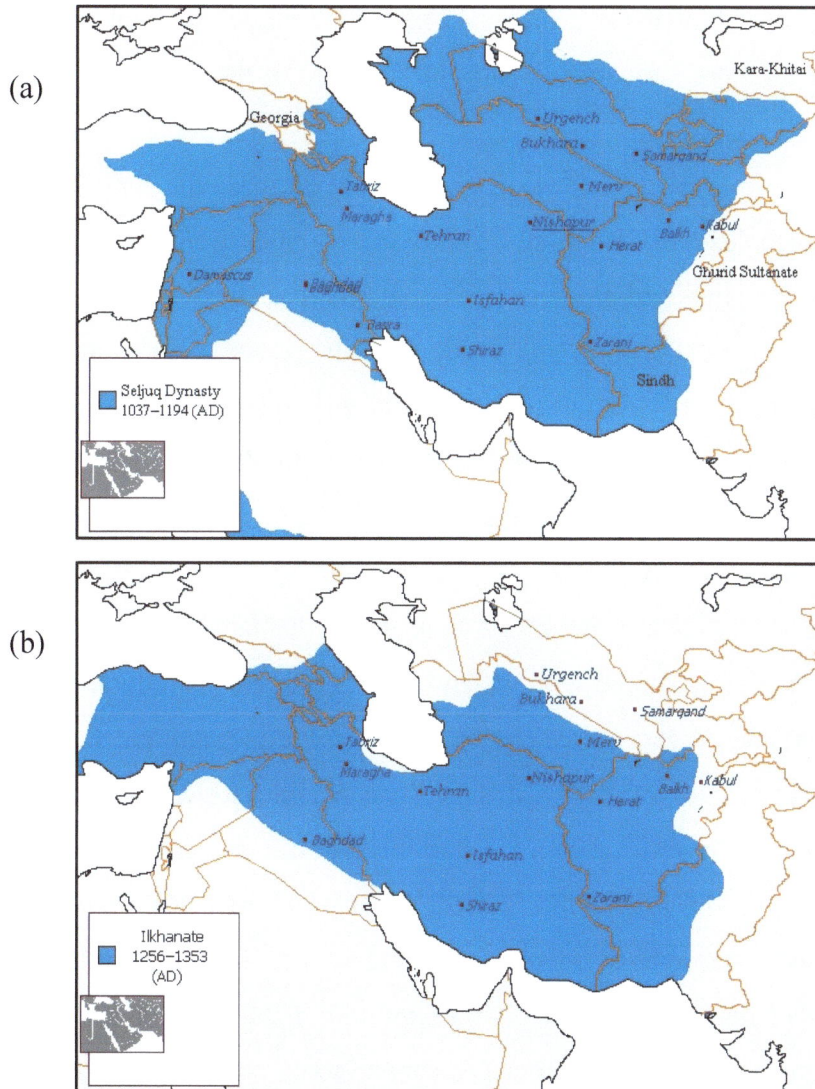

Figure 1.2   Map of Iran during (a) Seljuk and (b) Ilkhanid periods (Source: Wikipedia).

1.2.2   Form and Structure

In order to discover the progress of the evolution from squinch to *patkaneh* and *muqarnas*, existing samples of all mentioned structures were recorded and analysed in both form and structural aspects. By *form*, the appearance of the elements is intended, i.e. whatever the observer learns by looking at the *muqarnas*, including the finishing material, the number of tiers or rows, the size, and colour of elements of *muqarnas*. By *structure*, the material used to build the base of the studied ornament is targeted, as well as the geometrical shape of the constituent elements. In fact, there are many examples in which the material used for *muqarnas* is different from that of a building, e.g. masonry buildings with stucco *muqarnas*. Therefore, it is necessary to consider the structure of *muqarnas*. The two–and three–dimensional pattern of the *muqarnas* samples were also created and analysed to obtain more knowledge about the structure of the constituent elements of each sample.

Hence, with the purpose of achieving a substantial database for discovering the evolution of *muqarnas*, 100 samples were recorded on–site. Photographs were taken and measurements done. Furthermore, the two–and three–dimensional plans and patterns were illustrated and gathered for further studies. The form and structure of the constituent elements of the recorded samples were investigated, analysed and compared with each other in detail, to make it possible to differentiate and define the characteristics for each studied sample accurately.

1.2.3 Construction Techniques

Construction technique is another important concept in distinguishing the different types of *muqarnas*. The method and material with which the underlying construction of the *muqarnas* is pursued, the way the mason carried out the *muqarnas* from its plan, and finally if the studied *muqarnas* is a load–bearing structure or if it is merely a decorative element added to the building, are the items that were discussed and concluded in this part of research. There are samples which force us to go beyond its appearance and find out about its nature from what is hidden behind it and how it is attached to the wall or ceiling of the building, i.e. the hidden part of *muqarnas*.

Lack of knowledge among scholars about the traditional methods of drawing *muqarnas* plans and building them by experimental craftsmen and traditional masons has produced a large amount of questions in understanding and interpreting available *muqarnas* patterns and plans.

One of the most remarkable distinctions of the current study, with respect to other existing studies, is the author's wide knowledge on the traditional methods of drawing *muqarnas* plans and methods of building it. This rare and useful knowledge is obtained from a family with rich background in traditional architecture, as well as being mentored and having constructed many *muqarnas* structures for many years, under the supervision of world–famous masters of Iranian traditional architecture, namely, Maestros Sha'rbaf and Maheronnaghsh. This can be considered a privilege as it gives the author a better insight in understanding the phenomena and its complex nature.

1.3   Statement of the Problem

Muqarnas has always been one of the most complex decorative elements of the world's monumental architectures across history, in which niche–like components are combined together and arranged in successive rows to produce a three–dimensional geometrical surface, enclosing a ceiling, soffit, portal, vault, etc. This unique structure has been studied from different aspects by many scholars in many high–ranking universities such as Massachusetts Institute of Technology (MIT), and Harvard University in the United States, Tama Art University of Japan, Heidelberg University of Germany, King Saud University of Saudi Arabia, and Iran University of Science and Technology (IUST) and by many world–famous architecture historians and archaeologists (Necipoglu, 1995; Takahashi, 1973; Harmsen, 2006; Dold-Samplonius, 1996; Memarian, 2012; Yaghan & Hideki, 1995; Yaghan, 2000). Despite, there is still debate about the origin of the structure and about its definition.

There are theories asserting that squinch, which is considered an invention of Iranian architecture, is the predecessor of *muqarnas* (Écochard, 1977; Stierlin, 1976; Harb, 1978). Though, there is no explanation on the quality of gradual evolution. The problem arises from a very important misunderstanding about the definition of this structure and how it is different from other pseudo-*muqarnas* structures.

Muqarnas is in fact an advanced and famous form in Islamic architecture, which is rooted from a simpler unknown form in traditional architecture of Iran, named *patkaneh*. This structure, which will be clearly introduced in this research, is in fact the missing link between squinch and *muqarnas* that should be introduced to the body of knowledge.

In spite of the fact that the word *muqarnas* refers to a unique world-known structure, which is considered as the signature ornament of Islamic architecture, depending on the perception of the scholar who tried to define the characteristics of the structure and his available resources, as well as the location of the *muqarnas*, which results in a variety of characteristics and dimensions, vast and diverse interpretations of the phenomenon is available.

Although many researchers devoted their efforts to clarify different aspects of this structure, different concepts out of the same context are introduced and still there is debate about the very basic aspects of *muqarnas*, in terms of concept and basic physical characteristics. Even in Iran, which is known as one of the major candidates as the structure's place of origin, despite the diversity of *muqarnas* types, still little attention is paid to clarifying the concept as well as the physical characteristics of the structure and its different types, in association with its role in a building.

Squinches are introduced in some reliable theories as the functional origin of *muqarnas* in Iran. Though, the huge gap between the two structures has been ignored. Little attention is paid to demonstrating the gradual development of squinch towards *muqarnas*. Lack of clarification on categorization of the concept in association with the characteristics of its components, detracts the scientific understanding of these unique structures.

In other words, to differentiate the concept of *muqarnas* from its other similar structures, it is necessary to distinguish the actual structure from those, which will be referred as pseudo-*muqarnas* structures. This differentiation enables the researcher to explicitly recognize these different structures, which in turn assists in identifying each category based on its particular characteristics. The clarification of characteristics and architectural attributes of different types of the structure would be useful in classifying the achieved findings towards a systematic expansion of the knowledge. This knowledge will help in protecting and codifying these valuable historical ornaments not only by providing better perception of their details but also by making a scientific documentation of them to be transferred to the next generations.

1.4   Research Aim and Objectives

Based on the explanations above, this investigation strives to clarify the definition of *muqarnas* by demonstrating the evolution of this ornamental structure in Iran, from 10th to 14th century and by seeking the missing link between *muqarnas* and squinch.

The stated aim of this research is expected to be achieved through the following objectives:

(1) To seek the time period in which *muqarnas* structure is fully developed to its optimum form by recording existing *muqarnas* and pseudo-*muqarnas* samples in Iran.

(2) To investigate the constituent elements of *muqarnas*, with the purpose of developing a minimized but general set of basic constituent elements that could cover all elements of all *muqarnas* structures of all times.

(3) To clarify the similarities and differences between *muqarnas* and other pseudo-*muqarnas* structures; and finally,

(4) To define each ornament clearly, i.e. squinch, *patkaneh*, decorative *patkaneh* and *muqarnas*, from three aspects, namely chronology, form and structure, construction techniques.

The abovementioned objectives should be accomplished in a sequential order. In other words, the results from each objective will be used as a tool to achieve the next objective.

1.5 Research Assumptions

Muqarnas is an ornamental structure which is considered to have originated from another simple structure known as squinch. Though, the connecting link between the two completely different structures is unknown. The link is in fact another ornament which is simpler than *muqarnas* but more complex than squinch. The connection, that is assumed to be a structure named *patkaneh*, should have similarities and differences with both limiting cases. *Patkaneh* is made up of several ribs, containing small squinches inside them. This structure which is generally made of the same material as the building is built from bottom to top, in a way that each tier is supported by its lower one, transferring its weight through that or the hidden ribs to the walls of the building, making the combination load-bearing.

Muqarnas, on the other hand, is built on suspended layers, which does not necessarily have the same material of the building. These layers are prepared separately on the ground and later they are attached to the wall or ceiling. Finally, after installing the suspended layers, the space between them is filled with ornamental constituent elements. Having at least three tiers, *Muqarnas* should include horizontal *takht* elements. The load of the decorative is transferred to the supporting ceiling or walls by tensile elements, such as rope and timber. Hence *muqarnas* is considered a purely decorative structure. The first recorded sample of *patkaneh* was constructed in mid 10th century in Iran, whereas mature *muqarnas* was developed in the early 14th century.

1.6 Significance of Research

This research intends to establish the evolution of the structure known as *muqarnas* from its functional origins, i.e. squinch. Although squinches are accepted as the predecessors of *muqarnas* in Iran, but the gap between these two structures is very big and there has been no explanation about the process of evolution of the structure. This big gap which has never been bridged since

1977 will be clarified based on the results of this research. As mentioned before, the reason behind not finding the missing link between *muqarnas* and squinch is in fact the uncertainty about the definition of the structures and their characteristics. Hence, misinterpretation of several pseudo-*muqarnas* structures as *muqarnas* is observed in many scientific publications.

The connecting link that is called *patkaneh* has to be introduced to the body of knowledge and studying the gradual evolution of *muqarnas* from squinch is only possible by detailed review of these structures. To the author's best knowledge, this is one of the first attempts in studying *muqarnas* that reviews all the details of the structure, including its constituent elements, as well as the structure's other specifications, namely, chronology, form and structure, and construction techniques.

1.7   Research Scopes

The effective historical period in the scope of this research is from pre-Seljuk to Ilkhanid, i.e. from 10th to 14th century. This period was chosen for two main reasons. With reference to standard contemporary *muqarnas* patterns of Morocco, Necipoglu mentions that although efforts have been made to add innovative designs, one can conclude that since the 14th century the rules governing the elements and details of designing Moroccan *muqarnas* have not changed and creativity can only be observed in applying novel proportions of elements and tiers in the structure (Necipoglu, 1995). Iranian *muqarnas* and Moroccan *muqarnas* have similar basic characteristics, with *taseh* and *shaparak* as their main constituent element. However, they are built using different construction technique and material. Based on the axial coding of the gathered database of 100 *muqarnas* and pseudo-*muqarnas* structures in Iran, the author concluded that Necipoglu's theory about Moroccan *muqarnas* can be extended to Iranian *muqarnas* as well. In other words, the forms in Iranian *muqarnas* have been also fully developed by the 14th century and then after only minor innovations are observed in the composition of elements with respect to each other.

Studied samples are selected from hundreds of available *muqarnas* and pseudo-*muqarnas* structures inside the geographical region of Iran. These 100 samples were intentionally chosen to from specifically important buildings, in terms of architectural history, based on the opinions and publications of world-famous archaeologists, and architectural historians who studied this region (Pope, 1965; Blair & Bloom, 1994; Godard, 1965). Finally, this region had been selected for four main reasons. First, there are reliable theories that seek the origins of *muqarnas* in Iran. Second, there is a rather rich literature about the architecture of Iran during Seljuk and Ilkhanid periods. Third, the significant role of Seljuk architecture in Iran, on the expansion of these forms and structures to other parts of the world should not be neglected (Edwards & Edwards, 1999; Bloom, 1988), and finally, the importance of the author's practical and academic knowledge about the traditional architecture of Iran, which is obtained by working with great masters of traditional architecture, as well as being involved directly with constructing *muqarnas* for many years.

1.8   Research Limitations

There have been great limitations on the study of *patkaneh*, as the topic is newly introduced as a scientific issue, and hence, one can hardly find any available literature on the topic. The unknown and confusing concepts and key words required to define the structure were also influencing the research. There is a wide variation in the structural characteristics, applications, and dimensions of the studied samples. In many cases, in order to find out more about the structure's construction techniques and doing some measurements, it was necessary but impossible to access the hidden parts of it. Furthermore, many samples are now located within the political borders of other neighbour countries, such as Iraq, Afghanistan, Syria, and etc., which were not accessible due to political instability and continuous civil unrest, except the one in Samarqand, Uzbekistan.

1.9   Summary

This manuscript is organized in six main chapters. Chapter 1, the current chapter, looks mainly on the significance of the research topic and provides general information about the major concepts dealing with the research. The scopes and final contributions of the study are also explained and elaborated in this chapter. Chapter 2 is arranged to provide a comprehensive review of the available literature about *muqarnas* and its path of development and evolution as well as the definitions offered so far for *muqarnas* and its constituent elements, where as the methodology of the research is completely explained in the 3rd Chapter. Chapters 4 and 5 are where the gathered and recorded research data are presented, depicted and analysed in detail. As the predicted results of the first objective were necessary to proceed with the investigation, it was studied, concluded and validated in Chapter 4, and then the results were used to proceed with the study in the next chapter. Finally, a conclusion of the whole research and the achieved goal is introduced in Chapter 6. A list of cited references is provided after the last chapter and there is a complete set of appendices including all research data, for any further inquiries.

As the scope of this research on tracing the evolutionary path of *muqarnas*, the structure and its filiations are sought and their detailed characteristics are clearly studied, to resolve the misunderstanding about the definition of the structure, using a qualitative approach. A database of available *muqarnas* and pseudo-*muqarnas* samples in Iran is collected and the three–dimensional patterns and two–dimensional plans of the studied structures are created to facilitate the understanding of the similarities and differences of the aforementioned structures and their architectural characteristics, i.e. their structural role, as being load–bearing or merely decorative, their construction technique and the complexity of their geometry through time, using an inductive method. Based on the outcome of this research, after clear definition of *muqarnas* and its constituent elements, *patkaneh* will be introduced as the precedent phenomenon leading to the creation of *muqarnas*. Referring to the achieved definitions, the gradual development of squinch toward *muqarnas* will be then demonstrated in detail.

CHAPTER 2

Physical and Theoretical Concept of Muqarnas

2.1   Introduction

Islamic art is not independent in epistemology, and results undoubtedly from a combination of the traditions of East and West. Islamic architecture was able to use pre–existing arts from the conquered lands, namely Persia and the Byzantine, to create its own artistic tradition. It has borrowed domes over transition zones, barrel vaulting, and glazed tiles from Persia, and stone arches, and pendentives from the Byzantium (Varjavand, 2009). The probable non–Islamic origin of structures is a vital technical issue that should be considered when investigating any topic about development of an Islamic architectural form and element.

In addition, a comprehensive understanding of Islamic art will only be possible if one realizes the reflection of the divine revelation in it, as manifested in the exegesis of Islam (Nasr, 1987). Burckhardt claims that Islamic art has pursued the mystical end as its objective, and to this end, architecture finds a pivotal role in the school with all other crafts including calligraphy adjusting to its demands (Burckhardt, 2009). Therefore, there are many efforts that try to interpret the soaring ideas of architects and craftsmen, based on their masterpieces. Though, after a detailed exegetical analysis of available mystical interpretations of Islamic designs, Chorbachi comes to the conclusion that no evidence exists on whether any of these interpretations were intended by the craftsmen who actually produced these patterns centuries ago (Chorbachi, 1989). This is while Grabar and Gombrich refute the symbolic interpretation of geometrical designs considering them as pure decorations. Grabar's hypothesis of intermediation posits that through the process of abstract mental inferences, the viewer develops an interchange with the message of the abstract art–piece that bridges the boundaries of decoration and reality (Gombrich, 1994; Grabar, 1988). Hence, other than some exceptional interpretations of the concepts behind the studied architectural form as explained below, the focus of this investigation is mainly on the structural characteristics of *muqarnas* and its historical origins.

2.2   Concept of *Muqarnas*

In spite of the significance of the phenomenon known as *muqarnas* in Islamic architecture, and various efforts to study the topic thoroughly, discussions on its origins and development into the current state are still subject to debate. The analytical studies on the physical and technical attributes of the structure have remained inconclusive and scattered among various works. Indeed, even the roots of the word are obscure and the reason behind all the uncertainties may lie in the lack of classical literature on the phenomenon. When one encounters the term *muqarnas*, different aspects of the word may highlight in his mind as the concept behind the word, including the etymology of the word, its definition, as well as different types of ornaments that are referred to as *muqarnas*. These different aspects that create the concept of *muqarnas* as a whole are studied here as follows.

2.2.1   Etymology of *Muqarnas*

The term *muqarnas*, which may have also been written as *mukarnas* in English (Behrens-Abuseif, 1993), is spelled as مقرنس in Persian and المقرنص in Arabic. Furthermore, the term has a European counterpart as *mukarbas*, taken from the Spanish equivalent *mocárabe* (Fernández-Puertas, 1993) and is mostly referred to as stalactite or honeycomb, i.e. *alvéolus* in French, in many European scholarly texts (Tabba, 2009).

2.2.1.1   Persian and Arabic Origins

The Arabic language is certainly an important candidate for the roots of the word *muqarnas* (Dold-Samplonius, 1996). Tabba believes that the term *muqarnas* develops in 12th century Arabic, though related nouns are used a century before to describe what he calls a "deeply carved and moulded stucco ornament" (Tabba, 2009) Yet, although there is no specific explanation in any of the medieval Arabic dictionaries for *muqarnas*, to be associated with its architectural function, the term is in near exclusive use by masons.

Based on the investigation of the quadroglyphic root of *muqarnas*, i.e. قرنص or QRNS, in medieval Arabic wordbooks, such as *Lisan Al–Arab* (Manzur, 1984) and *Qamus Al–Mohit* (Firuzabadi, 1983), the meaning of the word has nothing to do with construction, and this can be explained by the fact that these encyclopaedias attempt to record the Arabic language as spoken at the advent of Islam, and a period when it would have been devoid of technical construction terms, as used in the centres of civilization, outside the Arabian peninsula (Amin & Ibrahim, 1990; Al-Zabidi, 2008). The meanings enlisted for QRNS under the named sources include:

Qarnas   A term used in falconry; like a rooster flapping wings and running from another rooster;

Muqarnas   A corrugated sword; and

Qernās   Protruding rock formation.

It is speculated that the latter two may hint to the use of the word in the architectural sense, since they imply protrusion and corrugation. Further, the encyclopaedia of *Kitab Al–Ein*, ترتيب كتاب العين, which is one the oldest available Arabic wordbooks, describes *qernās* as: "a rocky protrusion like the protrusion of the nose on the face" (Ahmad, 2004).

When one looks up the Arabic term المقرنص on the web, other than the conventional meaning of the word, which is the stalactite vault, one other distinctive result calls the attention, and that is that it also implies *concrete pavement blocks* in Iraqi Arabic. The reason for such nomination is probably that the edges of the blocks are corrugated. Figure 2.1 shows an image sample of the abovementioned search result.

Figure 2.1   *Muqarnas* may mean concrete pavement blocks in Iraqi Arabic (Source: creativit.montadarabi.com).

Zamani, in his Persian paper titled "Decorative *Muqarnas* in Islamic Monuments of Iran" analyzes *muqarnas* and claims that the word could be derived from any of the following Persian roots (Zamani, 1972):

Qor   concave;

Qernas   the summit of a mountain; and

Qarniz   i.e. Cornice, refers to a soffit or projection over an awning.

On the other hand, the major Persian encyclopaedias such as Dehkhoda and Borhan have accepted that the word *muqarnas* is not Persian in origin (Dehkhoda, 1997; Tabrizi, 2003). Dehkhoda speaks of the link between the roots of the words *qernas* and *muqarnas*. *Qernas* in Arabic represents protrusion and subsequently, *muqarnas* refers to a structure with protrusions. The Encyclopaedia of Dehkhoda speaks of a rather common composition, i.e. سیف مقرنس [*Saif–e muqarnas*] as an example for the usage of the word, which translates into "ladder–like sword". Similarly, Mofid and Ra'iszadeh admit that *muqarnas* is a decorative structure in the form of a ladder or steps which appears like hanging stalactites of the caves, and the word refers to rock projections where layers of soil have been washed from underneath the sedimentary formations (Ra'iszadeh & Mofid, 2010). Dehkhoda further explains that the Persian equivalent for *muqarnas* is آهوپا [*Ahou–pa*] which transliterates into "deer limb" (Dehkhoda, 1997). Memarian, on the other hand believes that *ahou–pa* is in fact the hanging portion of *muqarnas* with a three– or four–pointed flat–star *takht* element at its end (Memarian, 2008).

Clearly allegorizing the sky and the universe that is constantly in change and capricious as a *muqarnas* vault, must have entered Persian literature at a time when people would have been familiar with the decorative phenomenon. Numerous occasions are met in the Persian literature, since early 12th century, where mention is made of the word *muqarnas* as a metaphor for sky, the oldest of which belonging to Jamal Al–din Isfahani (ca. 1100), and Am'aq (ca. 1130). A number of such examples are listed in chronological order, as follows:

(1) Jamal Al–din Mohammad Abd Al–Razaq Isfahani (ca. 1100) (Isfahani, 2000):

از بر این خاک توده یک تن آسوده نیست

زیر این سقف مقرنس یک دل خرم نماند

There is no comfortable body on this mass of soil

There is no happy soul under this *muqarnas* ceiling (sky)

(2) Shahab Al–Din Am'aq Bokharayi (ca. 1130) (Bokharayi, 2010):

زمین گردد از نعل اسبان مقرنس

هوا گردد از گرد میدان مغبر

The Earth has become like *muqarnas* (sky) from the horseshoes

And the sky is cloudy from the dust

(3) Khaghani (1120–1190) (Khaghani, 1996):

این هفت تاب خانه مشبک شد از دعا

تا شاه در مقرنس ایوان نو نشست

The seven face house (the world) was reticulated by prayers

Until the king was finally seated under the new *muqarnas* porch (sky)

(A lot of prayers were done for the victory of the new king)

On the other hand, the same metaphor can also be found in the texts by different historians. The Mogul historian, Sharaf Al–Din Ali Yazdi allegorizes the ribbed vault at the portal of the edifice at Baaq–e Delgosha (Garden of Felicity), in Samarkand as "the *muqarnas* vault of the heaven"; as do other contemporaries of his, such as Abd Al–Razaq Samarqandi (1413–1482), and Khandmir ibn Ala Al–Dowla Samarqandi (–1534) (Thackeston, 1989).

2.2.1.2 English and Spanish Origins

Other than Arabic, another probable origin of the word *muqarnas* is considered to be Greek. Doris Behrens–Abuseif argues that the word *muqarnas* otherwise written as *mukarnas* originates from the Greek word κορωνίσ, i.e. *coronis* in Latin, *corniche* in French, and *cornice* in English. She claims this arguing that "transitional zones" are the most common locations where *muqarnas* is found (Behrens-Abuseif, 1993), however the aforementioned derivation is not approved by any Persian or Arabic source (Tabba, 2009). In addition, she refers to the word *mukarbas*, i.e. *mocárabe* in Spanish, to describe North African and Spanish *muqarnas* examples. Although the two names are considered to be the same, Fernández–Puertas believes the word *mukarbas* has a different Greek origin, i.e. *karbas* or *karbasa*, which means base or plinth, respectively and belongs to the Western Islamic world (Fernández-Puertas, 1993).

Similarly, Ibn Jubayr, in his book of travels when describing the Jame' of Allepo in 1180, refers to a *karbasa* in wood and ivory, i.e. a *muqarnas* decoration. Using the term *karbasa* in 1180s to describe *muqarnas*, suggests the root of the word *muqarnas* to have been the word *muqarbas*, which itself originates from the Spanish term *mocárabe*. Further, Ibn Jubayr proceeds in describing the phenomenon of *karbasa* as that with stars and polygons, prior to its stalactite–like attributes, which complicates the etymological research (Ibn-Jabayr, 1907).

Elmo is of the opinion that the word *mocárabe*, used by de Arenas to describe stalactites is indeed the corruption of the word *muqarnas* (Necipoglu, 1995). De Arenas' stalactites comprise of four elements in the form of prisms, where the first form comes with a rectangular profile and the other three with isosceles triangular bases of 45°, 90° and 135° apexes (Arenas, 1633).

Finally, it is important to refer to Fernández–Puertas's lengthy and valuable geometrical and structural analysis. He concludes that *mukarbas* is actually the North African and Spanish type of *muqarnas*. He adds that the structure is different from its Eastern version in terms of proportions and cutting of its pieces, as well as its etymology (Fernández-Puertas, 1993).

2.2.2 Definition of *Muqarnas*

Although a concise and technical definition of *muqarnas* and its structural analysis, as an architectural element, is necessary, it raises more questions too. Focusing on details of the proposed definitions for *muqarnas* leads to the necessity of clarifying its definition and differentiating it from other structurally different pseudo–*muqarnas* structures which are relatively unexplored. What remains is only the difficulty of contributing a reliable and technical definition; a goal that will be followed in the next chapters of this manuscript. There are various definitions proposed for the word *muqarnas*. Scholars look at the issue from a variety of perspectives and discuss its etymology accordingly.

Although Al–Kashi's vocabulary is vague and at times completely incomprehensible, this source is still the oldest and one of the most valuable ones available. In fact, Al–Kashi's calculations pertain to construction estimates for *muqarnas* (Al-Asad, 1995), as a known architectural element of his time, allowing him to do away with the explanation of the nature of *muqarnas* in his book and focusing on the arithmetic aspects of the structure.

In any way, the first definition presented here for *muqarnas* is Al–Kashi's description. He explains *muqarnas* as a structure that comprises of tiers, that are in turn divided into cells, in round, stair–like arrangements on the vault, where each cell connects to the next with a predetermined angle (Al-Kashi, 1427). By Al–Kashi's description, the dimensions of the largest side on the base of a *muqarnas* element determine the scale of the work. Al–Kashi provides profile designs of arches and domes in addition to a *muqarnas* module that he attributes to the builders of his time (Jazbi & Al-Kashi, 1987).

Golombek and Wilber describe *muqarnas* as a geometrical network of space–filling repetitive and symmetrical concave cells that connect equally–divided tiers of space (Golombek & Wilber, 1988). This description matches that of Al–Kashi, which considers *muqarnas* as a set of cells arranged in several tiers. Hillenbrand also describes *muqarnas* as "honeycomb or stalactite vaulting made up of individual cells or small niches; often used as a bridging element" (Hillenbrand, 2004).

Doris Behrens–Abuseif defines *muqarnas* as a structure "composed of a series of niches embedded within an architectural frame, geometrically connected and forming a three–dimensional composition around a few basic axes of symmetry" (Behrens-Abuseif, 1993); while Sha'rbaf, who is himself a traditional master in the field states that *muqarnas* is a decorative veneer in spatial form, that comprises of a number of elements such as what is traditionally called as *Shaparak*, which literally means butterfly, *taseh*, which literally means scooper, *toranj*, which literally means bergamot, *tanoureh*, which literally means traditional furnace, and star–shaped *takhts* that are only arranged horizontally between the aforementioned unit cells. For the tiers he uses the term قطار [*qataar*] (Sha'rbaf, 2006). Hence, in some texts, *muqarnas* is referred to as *qataar–bandi*, meaning tier installation or tier work (Ra'iszadeh & Mofid, 2010).

Hoag introduces *muqarnas* as the most prominent feature of Islamic architecture, which resembles the sedimentary stalactites naturally occurring in caves. He adds that the simplest form of this decoration comprises of seven prismatic elements, mounted over each other. Based on Hoag's definition, stalactites are decorative and non–load–bearing, usually resting over some structural support such as column, arch, or squinch. Although these decorations are made at times in stone, such as in Syria, and at times in baked clay, i.e. tiles, such as in Iran or Turkmenistan, the decorative structure, as shown in Figure 2.2, is mostly built in plaster. Hoag adds that the forms of stalactites do vary according to the construction tradition to which they belong (Hoag, 2004).

Figure 2.2   A plaster *muqarnas* of a school in Nimavard, Isfahan (Source: Author).

Similarly, Vincenza Garofalo describes *muqarnas* decorations as three–dimensional forms created by assembling simple prismatic elements through various compositions. These elements are similar to portions of vaults composed based on precise rules, in overlapping corbelled levels. She completes her definition by adding that the composition, size, shape, and form of the elements not only depend on geographical regions and historical periods of the buildings, but they also depend on the location of the *muqarnas* in the structure as well as the construction materials (Garofalo, 2010).

Dold–Samplonius also looks at the phenomenon from the inside–out. She explains, as illustrated in Figure 2.3, that *muqarnas* comprises of roofs and facets, as

"*Muqarnas* is a ceiling like a staircase with *facets* and *roofs*. Every facet intersects the adjacent one at either a right angle, or half a right angle, or their sum, or another combination of these two. The two facets can be thought of as standing on a plane parallel to the horizon. Above them is built either a flat surface, not parallel to the horizon, or two surfaces, either flat or curved, that constitute their roof. Both facets together with their roof are called one *cell*. Adjacent cells, aligned on the same horizontal surface, are considered one *tier*."

(Dold-Samplonius, 1992)

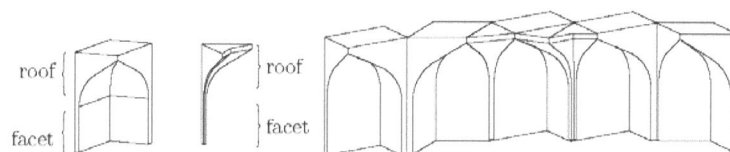

Figure 2.3   Basic cell and intermediate elements of *muqarnas* and their combination (Dold-Samplonius et al., 2002)

In the Grove encyclopaedia of Islamic art and architecture, Tabba describes *muqarnas* as "A three–dimensional decorative device used widely in Islamic architecture, in which tiers of individual elements, including niche–like cells, brackets and pendants are projected over those below."

(Tabba, 2009)

Necipoglu looks at the structure from a different point of view. She is of the opinion that *muqarnas* is a three–dimensional interpretation of the two–dimensional knot–work or گره [*gereh*], all of which is draw from similar base patterns. She argues that the side–by–side appearance of both phenomena in the scrolls –as would eventually be on the structures– testifies to the complimentary role assumed of them by the builders, such that the three–dimensional *muqarnas* would spatially fulfil what role *gereh* plays on the surface (Necipoglu, 1995).

Al–Asad believes that the *muqarnas* vault is in fact a secondary ceiling, and thus inherently only manifest at the interior view of a vault; however by his claim, at occasions such as the domes over Imam Dur mausoleum in Samarra, Iraq, and Nur Al–Din Hospital in Damascus, Syria, the *muqarnas* finds external manifestation as well. These two examples display tiered conical structures that result from the convexities at the back (exterior) of an otherwise internal *muqarnas* vault (Al-Asad, 1995). Figure 2.4 shows Imam Dur mausoleum, as an example of such external manifestation. Memarian agrees with Al–Asad's definition by explaining that *muqarnas* is mainly a secondary ceiling that hangs from the primary roof structure, which comprises of a set of geometrical elements in concave and convex form (Memarian, 2008).

Figure 2.4   General view of Imam Dur Mausoleum, Iraq (Tabba, 1980).

2.2.3   Categorizations of *Muqarnas*

Muqarnas structures had been categorized from various groups, considering variety of aspects, by different scholars. Al–Kashi divided *muqarnas* into four main groups, namely:

(1) Simple, ساذج ,ساده [*sadeh*],

(2) Clay–Plastered, مطين ,گل پوش [*Gel–poosh* or *Motayyen*],

(3) Arched or curved, قوس [*Qaws*]; and

(4) Shirazi.

Of the four types, *sadeh* only comprises of straight lines, having triangular and rectangular prisms, in constant tier height. *Gel–poosh* is similar to *sadeh*, with the only difference being that the height of the tiers are not uniform. *Qaws muqarnas* includes arched profiles, and finally, *Shirazi* is the most complicated of the four. Al–Kashi describes the *Shirazi muqarnas* as the most complex of the four types known to him, explaining that it has radial arrangement, whereas the other three forms comprise of rectangular plans (Al-Kashi, 1427). Furthermore, the *sadeh* and the *Gel–poosh muqarnas* types have plane facets and roofs, whereas in the *qaws* and *Shirazi* types, the roofs of the cells and the intermediate elements are curved (Dold-Samplonius, 1996) and that the variety of elements is greater in *Shirazi* style (Hoeven & Veen, 2010).

Golombek and Wilber also mention the radial symmetry of *Shirazi muqarnas* as its identical characterization (Golombek & Wilber, 1988). As a result, the two–dimensional pattern of *Shirazi muqarnas* incorporates different polygons such as pentagons, hexagons, octagons and multi–pointed stars, unlike the other three styles, where the top views consist of only triangles and quadrilaterals (Hoeven & Veen, 2010). Necipoglu believes that the dissemination of the *Shirazi muqarnas* has been from Iran towards Turan, i.e. Eastern Iran and Central Asia (Necipoglu, 1995). Furthermore, in reference to the *muqarnas* plans in the Topkapi Scroll, She divides *muqarnas*, from which she means the *Shirazi* type, into two categories on the basis of the central medallion or *shamseh*, i.e. radial *muqarnas* with fan–like medallions and radial *muqarnas* with oyster–like medallions (Necipoglu, 1995).

Another categorization of *muqarnas* is introduced by Shiro Takahashi. He classifies *muqarnas* into three main groups, i.e. square lattice style, pole table style and other style, which comprises of any other type that does not belong to the first two groups (Takahashi, 1973). Figure 2.5 shows the *muqarnas* groups defined by Takahashi and Figure 2.6 shows their distribution in the Middle East, Europe and North Africa. However, more details on the definition of Takahashi plan types are provided in Chapter 4.

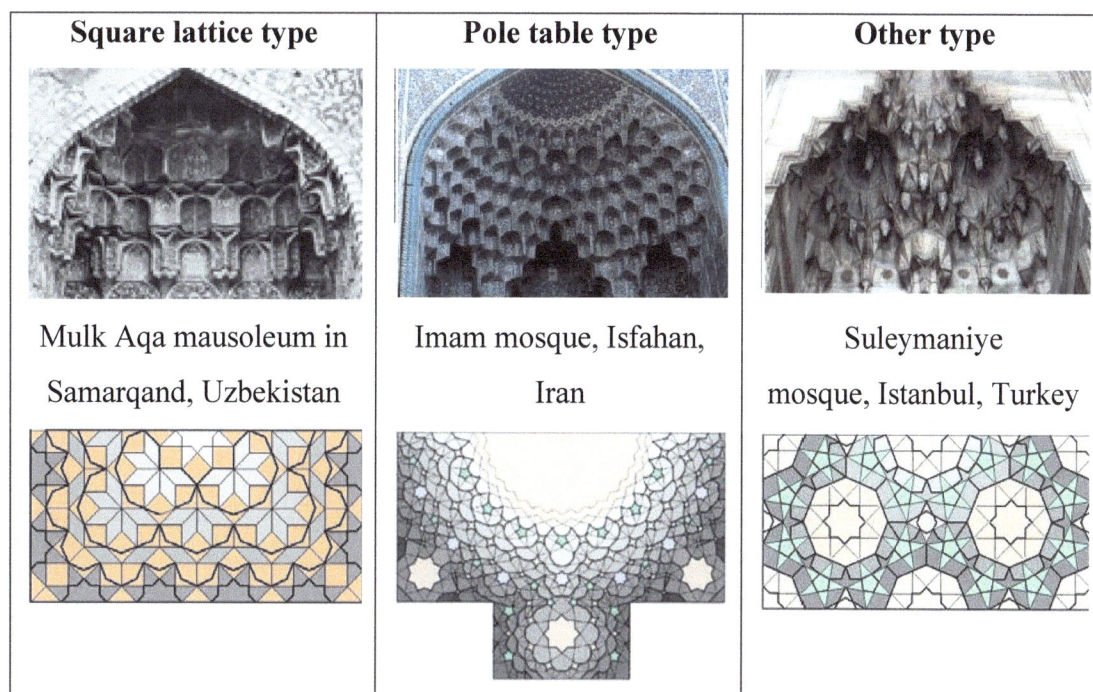

Square lattice type	Pole table type	Other type
Mulk Aqa mausoleum in Samarqand, Uzbekistan	Imam mosque, Isfahan, Iran	Suleymaniye mosque, Istanbul, Turkey

Figure 2.5   Shiro Takahashi *muqarnas* types (Takahashi, 1973).

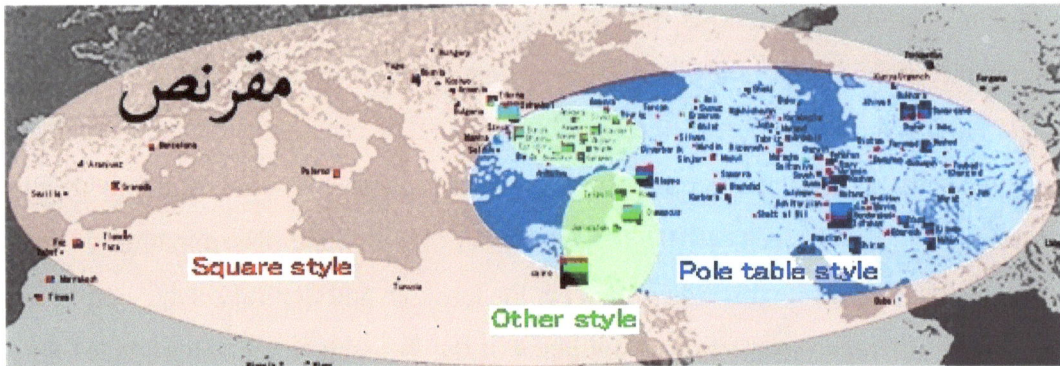

Figure 2.6 Distribution of different *muqarnas* types in the Middle East, Europe and North Africa (Takahashi, 1973).

As Hoeven believes, the most important characteristic of the square lattice *muqarnas* is that it has four-fold symmetry and the plan of the *muqarnas* has squares and 45° rhombuses. Square lattice *muqarnas* structures were developed in the 11th century C.E. and then spread throughout the Islamic world (Krömker, 2007; Hoeven & Veen, 2010); examples include the stalactite decoration of Alhambra, Granada, and Mulk Aqa mausoleum in Samarqand. This type is the equivalent to Al–Kashi's curved *muqarnas* (Post et al., 2009). The apex of the pole table *muqarnas*, on the other hand, is usually divided into 4 to maximum 7 segments and sometimes 11 segments radially, making the centre star, or *shamseh*, i.e. a 4–to 11–pointed star element (Hoeven & Veen, 2010). The pole table type was developed during Ilkhanid period in Western Asia and Iran. This type of *muqarnas* does not play any structural role in the building and is considered an independent structure. In the 17th century the pole table type reached its peak in Imam Mosque, Isfahan.

The type of material used for constructing *muqarnas*, plays an important role in its structure. Using bricks, most of the times with a gypsum mortar was very common in Iran (Edwards & Edwards, 1999), whereas other materials such as stone or wood were used more frequently in Turkey, Syria, Egypt, Spain and Italy (Castri & Campisi, 2007).

In the discussion of the mechanics of *muqarnas*, Paccard categorizes the existing samples based on their construction materials. Plaster *muqarnas* structures, mostly found in the area of Greater Iran, fundamentally differ from the wooden and stone types, which are the carved stalactites, common in North Africa, Morocco, Syria and Spain, in that the latter are based on grids of 45° and 90°, and may be arranged according to wider angles, whereas in a plaster *muqarnas* the mason decides the angles and curves more freely. Paccard's documentation of the technical vocabulary of the Moroccan builders is also of interest since in his words, the circumscribing circle of the design is called the "mother", the subsidiary circles the "sisters", and the lines and contours connecting these are called the "alleys" of the work (Paccard, 1980).

In confirmation of with Paccard's findings, Necipoglu speaks of a scroll attributed to a Moroccan builder named El–Bouri, displaying plans of wooden *muqarnas* the components of which are juxtaposed by angular arrangements of 45°, 90°, and 135° degrees (Necipoglu, 1995). The simplicity of El–Bouri's rectangular module, shown in Figure 2.7, is explained by the fact that wooden stalactites demand a more rigid arrangement than the plaster equivalents that are based on concentric modules. It is for this reason that the plaster stalactites of Morocco, as inspired by those in the Orient, do not necessarily abide with the rigid angular arrangements of 45°, 90° and 135°.

Figure 2.7   El–Bouri's *muqarnas* module (Necipoglu, 1995).

The use of materials led to the creation and evolution of many *muqarnas* type that cannot be easily categorized into the abovementioned groups. Takahasi defines a third type which he generally clusters together as the *other* style to cover all these distinguishable methods that do not follow the rules of the first two groups properly. Almost simultaneous with Takahashi, Zamani, in his research paper, divides *muqarnas* into four categories defined as:

(1) Projected *muqarnas* which is usually composed of the same material used on the exterior façades, and very durable in construction,

(2) Tiered *muqarnas* that is typically made of the same material as the rest of the building; mostly composed in bricks, stones and plaster, on both interiors as well as exteriors, and enjoys medium strength,

(3) Buoyant *muqarnas*, what is referred to as *stalactite*, designed to hang from interiors of ceilings, composed of plaster or ceramic tiles mounted over concave surfaces, with minimal strength, a sample of which is available in Figure 2.8; and

(4) Honeycomb *muqarnas*, which is very similar to the buoyant type and its Western equivalent, is known as *alvéolus*. Examples include the Palatine Chapel at Palermo, shown in Figure 2.9 (1132–1143) (Castri & Campisi, 2007). This form that appears in a rather honeycomb profile could hardly be met in Iran (Zamani, 1972).

In many texts, especially when dealing with details of different *muqarnas* structures, scholars refer to the *muqarnas* vaults common in North Africa and Europe, as well as Spanish, Moroccan or Western *muqarnas*, as stalactites (Garofalo, 2010; Arenas, 1633; Bloom, 1988; Castri & Campisi, 2007; Dold-Samplonius & Harmsen, 2004). As mentioned before, this type of *muqarnas* is different from the Eastern one in terms of proportions and cutting of its pieces, elements and construction materials (Fernández-Puertas, 1993).

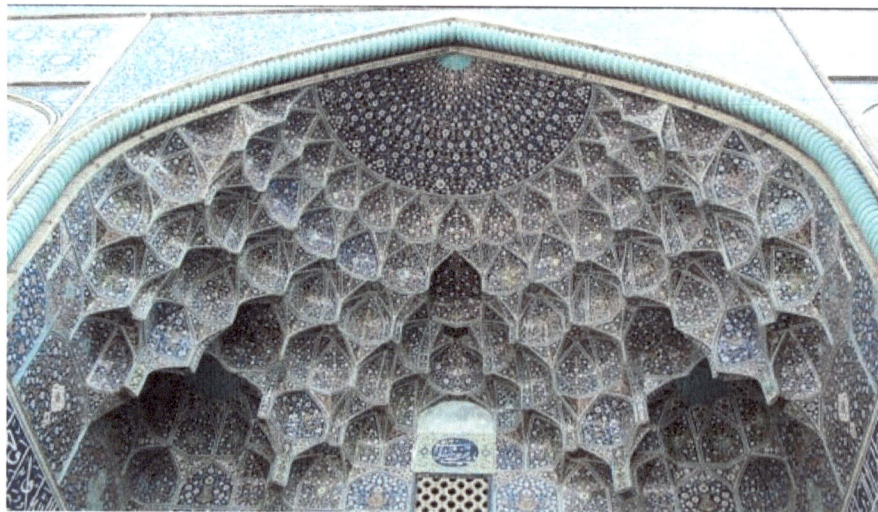

Figure 2.8   Tile *muqarnas*, Sheikh Lotfollah Mosque, Isfahan (Source: Author).

Figure 2.9   Wooden *alvéolus* of Palatine Chapel, Palermo (Source: flickr.com).

2.3   Theories and Models on *Muqarnas*

As mentioned before, in spite of the significance of the phenomenon known as *muqarnas* in Islamic architecture, and various efforts to study the topic thoroughly, discussions on its origins and development into the current state are still subject to debate. The analytical studies on the physical and technical attributes of the structure have remained inconclusive and scattered among various works. Indeed, even the roots of the word are obscure and the reason behind all the uncertainties may lie in the lack of classical literature on the phenomenon.

2.3.1   Pioneers of Recordings *Muqarnas*

The oldest existing text on the subject comes in the passages of مفتاح الحساب [*Miftah Al–Hisab*], meaning *Key of Arithmetic*, by the 15th century mathematician and astronomer Giyath Al–Din Jamshid bin Mas'ud Al–Kashi (Dold-Samplonius, 1996; Al-Kashi, 1954; Jazbi & Al-Kashi, 1987; Al-Kashi, 1427). In the last chapter of his book, Al–Kashi elaborates *muqarnas* from a practical perspective, by means of geometry, calculating the surface areas of the *muqarnas* (Dold-Samplonius & Harmsen, 2004). Other available evidence on the subject come in the form of ancient drawings, the oldest of which belongs to the 14th century, as shown in Figure 2.10; the 50cm^2 plaster plate unearthed at the excavations at Takht–i Sulaiman (Ghazarian & Ousterhout, 2001; Harb, 1978; Dold-Samplonius & Harmsen, 2005; Harmsen, 2006), showing a quarter of a *muqarnas* plan (Harmsen et al., 2007). There have been other drawings found in the Topkapi scroll, drawn by a designer from Bukhara, in the 16th century and a much later scroll, belonging to Mirza Akbar, from the 19th century, preserved at the Victoria and Albert Museum (Necipoglu, 1995).

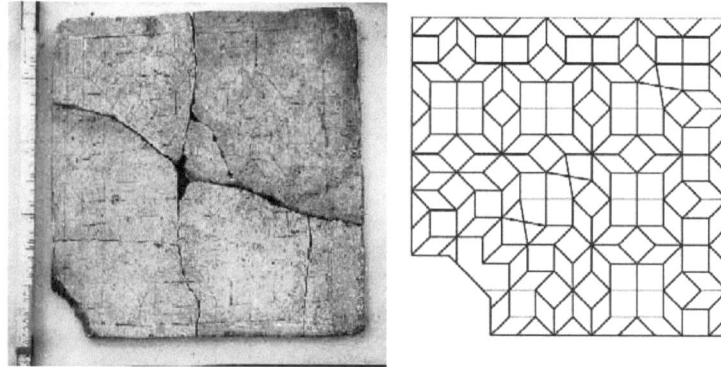

Figure 2.10   The plaster plate found at Takht–i Sulaiman (Dold-Samplonius & Harmsen, 2005).

In Europe, the first book to contain any technical drawing on stalactite vaults, as *muqarnas* is called in English, is that of Diego Lopez de Arenas, who wrote his treatise for carpenters and architects in a summary of a more elaborate book published in 1619 (Arenas, 1633). The book contains plans of two–dimensional geometrical ornaments with some simple designs for *mocárabe*, i.e. stalactite vaults, drawn in compass and straightedge. It is Noteworthy that the term *mocárabe* is the Spanish translation of stalactite that is in turn an equivalent for *muqarnas* in many European books.

Although de Arenas was writing 200 years after Al–Kashi, his work is still considered an attempt to document the techniques of an otherwise vanishing trade. Figure 2.11 shows a simple *muqarnas* pattern drawn by de Arenas and Figure 2.12 presents the pioneers of documenting Eastern and Western *muqarnas*, as well as the first pages of their books. *Miftah Al–Hisab* was hand–written by Al–Kashi in Samarkand and a copy of the book is available in the central library of the University of Tehran (Azarian, 1998). Clearly a research into the history of traditional and Islamic arts comes handy in the identification of the various types of elements in traditional and Islamic architecture, which at the inception of Islam dominated the entirety of the three continents of Asia, Africa, and Europe.

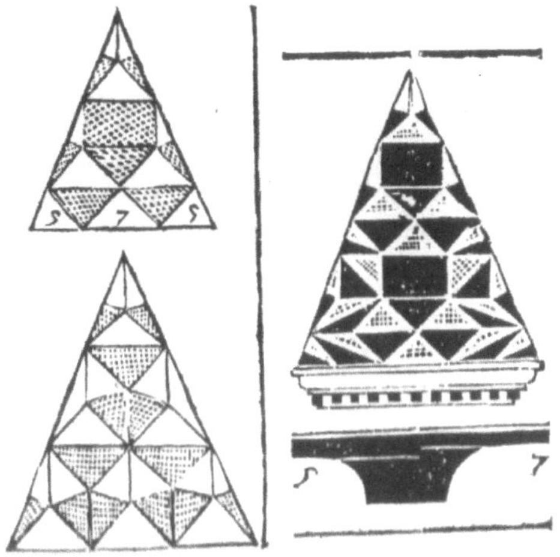

Figure 2.11 Triangular elements of a wooden *muqarnas* (Arenas, 1633).

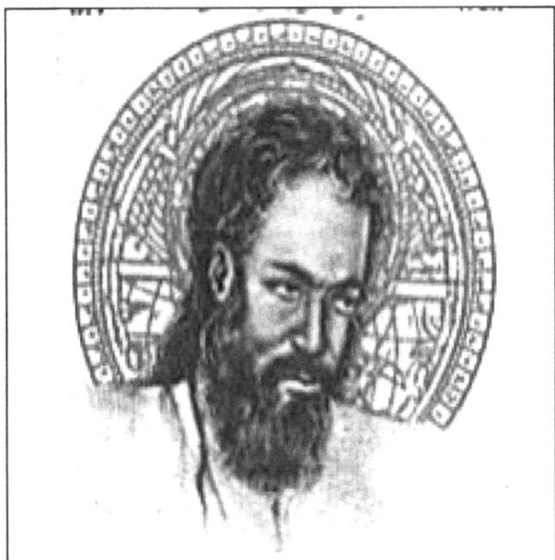

Giyath Al–Din Al–Kashi

(1380–1429)

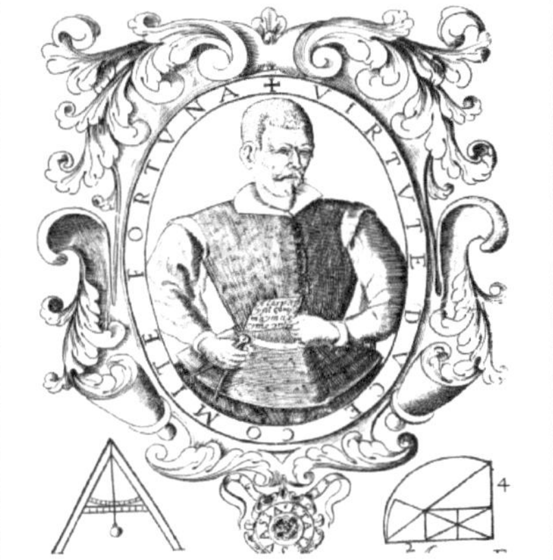

Diego Lopez de Arenas

(–ca. 1650)

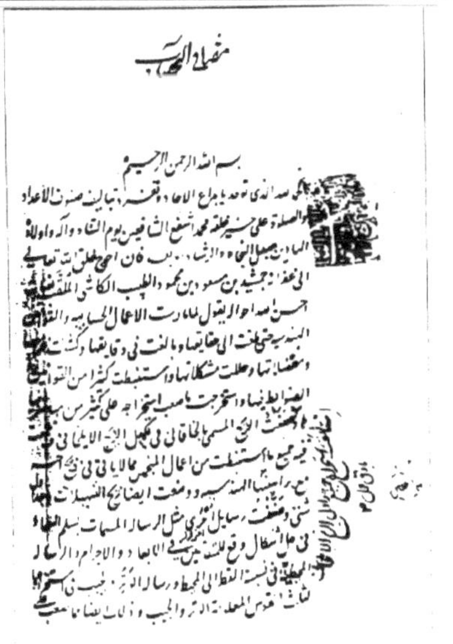

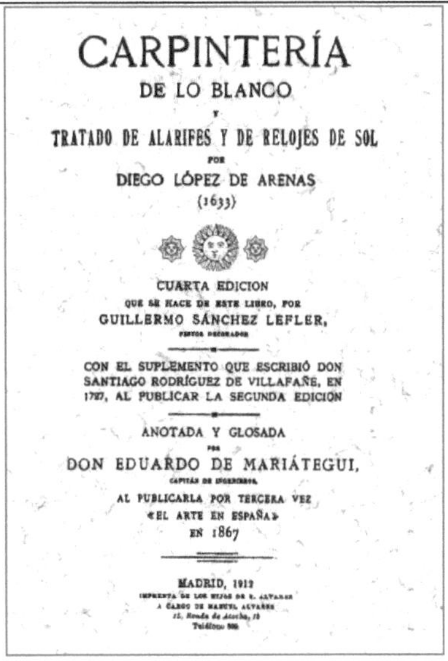

CARPINTERÍA
DE LO BLANCO
Y
TRATADO DE ALARIFES Y DE RELOJES DE SOL
POR
DIEGO LÓPEZ DE ARENAS
(1633)

CUARTA EDICION
QUE SE HACE DE ESTE LIBRO, POR
GUILLERMO SÁNCHEZ LEFLER,

CON EL SUPLEMENTO QUE ESCRIBIÓ DON SANTIAGO RODRÍGUEZ DE VILLAFAÑE, EN 1727, AL PUBLICAR LA SEGUNDA EDICION

ANOTADA Y GLOSADA
POR
DON EDUARDO DE MARIÁTEGUI,

AL PUBLICARLA POR TERCERA VEZ
«EL ARTE EN ESPAÑA»
EN 1867

MADRID, 1912

Figure 2.12 Al–Kashi (1380–1429) and de Arenas (–ca. 1650), the pioneers of documenting Eastern and Western *muqarnas* and the first pages of their books.

2.3.2 The Origin of *Muqarnas*

Regardless of the vast extent of research tracing the origins of *muqarnas*, there is still no certainty about the structure's chronological and geographical pattern and the issue remains open for discussion among scholars since the 19th century. What is however clear, is that by the 12th century stalactite becomes the one ornamental form that cements Islamic architectural expressions regardless of geography and building tradition from Spain to India. Even in many places it has been flavoured with the regional arts and traditional decorations or as Grabar puts it *"muqarnas is an all Islamic innovation that adorns not only mosques but all forms of prominent Islamic structures"* (Grabar, 1983).

Based on available findings, the origins of *muqarnas* are often sought in the 10th century in Iran (Behrens-Abuseif, 1993; Dold-Samplonius, 1992; Yaghan, 2001b). While some scholars believe that almost simultaneously but independently, it evolved in central North Africa, under the Fatimids (Dold-Samplonius, 1992; Sarre et al., 1911). Al–Asad as well expresses confusion on the geographical origins of the form, attributing it to any of the three typically presumed birthplaces, namely Turan of the 10th century, Baghdad or North Africa in the 11th century (Al-Asad, 1995). However, Diez credits no specific ethnicity for the invention, but rather believes that it emerged simultaneously at different localities and out of a common worldview (Diez, 1993). Likewise, Grabar considers it a simultaneous innovation that suddenly emerges out of the building traditions of North–Eastern Iran and North Africa (Grabar, 1992).

Necipoglu believes that *muqarnas* is the product of the same line of experiments that resulted in the discovery of knotwork, i.e. *gereh*. She believes that stalactites are not accidental innovations that may have simultaneously emerged in various building traditions out of a mysterious spiritual desire that would reflect an exclusive world–view of the Muslims. On the contrary, she adds that according to the scholars, stalactites were the product of deliberate experiments at the construction sites in Baghdad that disseminated through working drawings that came out of this specific city, but she does not provide any specific date or sample in the city as a proof (Necipoglu, 1995). Based on other investigations, explained in detail below, *muqarnas* is speculated to have emerged out of Greater Iran, Iraq or Syria; but rather than 10th century, the ornament is believed to be originally conceived in the late 8th and early 9th century.

Pope and Ackermann attribute the origins of the geometrical style to the Eastern lands, rather than Morocco and Egypt. They even go as far as claiming that the style originated in Greater Iran, at the advent of the Seljuk period, in the 11th century (Pope & Ackerman, 2005) and Pope refers to the geometrical framework promoted by the Seljuks, as a rather Persian innovation (Pope, 1965). In addition, the findings of Herzfeld and Sarre on the Jame' of Ibn–Tulun in Cairo, indicates the origins of the decorations to be from 9th century Baghdad (Sarre et al., 1911). Herzfeld assigns to Baghdad the same degree of stylistic primacy in the Muslim world, as enjoyed by Rome in Europe. He believes that the seat of the Caliphate and the centre of Seljuk political legitimacy would have been the most likely birthplace for two and three dimensional geometrical designs that would then disseminate into the surrounding lands (Herzfeld, 1942). Allen, too, posits that *muqarnas* is an innovation of the Abbasid era, developed in Iraq or Iran, and then introduced to the rest of the Muslim world (Allen, 1988).

Noting the similarities of the few elemental drawings unearthed at Takht–i Sulaiman attributed to a stalactite design from the 10th century, Doris Behrens–Abuseif speculates that contrary to popular association with the Fatimids, if the stalactites of Fustat are assumed to have belonged to the Abbasid era, it would then make sense to attribute the origins of the invention to Baghdad, where both works could trace their origins to (Behrens-Abuseif, 1993). On the hypothesis that posits *muqarnas* to be conceived in North Africa, Jonathan M. Bloom, in his paper, claims that the flat *muqarnas* vault and portal were unknown in Egypt prior to their introduction there by the Memluks from Syria. He points out to the hospital and *madrasa* of Nur Al–Din, 1168 C.E., shown in Figure 2.13, as two of the oldest structures in Damascus that incorporate the Iraqi style of conical stalactite (Bloom, 1988).

Figure 2.13   Nur Al–Din Hospital, Syria (Source: Dome.mit.edu).

Ibn Jubayr, the traveller from Andalusia, speaks of the introduction of *muqarnas* and its geometrical style into his area under the unified rule of the Berbers over Iberia and Morocco, between the 11th and 13th centuries (Ibn-Jabayr, 1907). Necipoglu admits that the spread of geometrical decorations and *muqarnas* from the East into Morocco was not incidental. She adds that it took place under the direct patronage of the Almoravids between 1056 and 1147 C.E. (Necipoglu, 1995). However, Necipoglu criticizes this search for the origins of the *muqarnas* in Iraq by mentioning that if the dome at Imam Dur, Samarra, (1075–1090), shown in Figure 2.4, is the oldest sample, it must be pointed out here that at least in Iran, *muqarnas* in the proper understanding of the phenomenon refers to the concave side of the built work, and does not apply properly to the resulting convexities at the back, even if such inevitable forms are deliberately incorporated in the ornamentation of the exterior (Necipoglu, 1995).

Diez, Aslanapa and Mulayim attribute the dissemination of two and three dimensional geometrical designs to the Turkic nomads who brought these artistic tastes from Turan, i.e. Eastern Iran and Central Asia, and introduced them into Iran and Anatolia. Claims as such, charged with racial–political agendas cannot be entertained on scientific levels, since they fail to present convincing evidence in their own support (Aslanapa, 1971; Diez, 1917; Mulayim, 1982). Even Aslanapa, in another part of his book, mentions that the two and three dimensional decorations, containing stalactites and geometrical bands, around the tomb entrances and over the cenotaphs of the Anatolian Seljuks in the 13th century, are a result of their cultural interactions with Iraq and Persia (Aslanapa, 1971).

2.3.2.1 Squinch

Muqarnas decorations were first used over the squinches supporting domes. These squinches that appear in pyramidal contours appear at the four top corners of the quarters, at the transition levels between walls and ceilings. Figure 2.14 shows one of the oldest existing samples, at the palace of Ardeshir, Fars, with details. The building dates back to 224 C.E. (Byron, 2007).

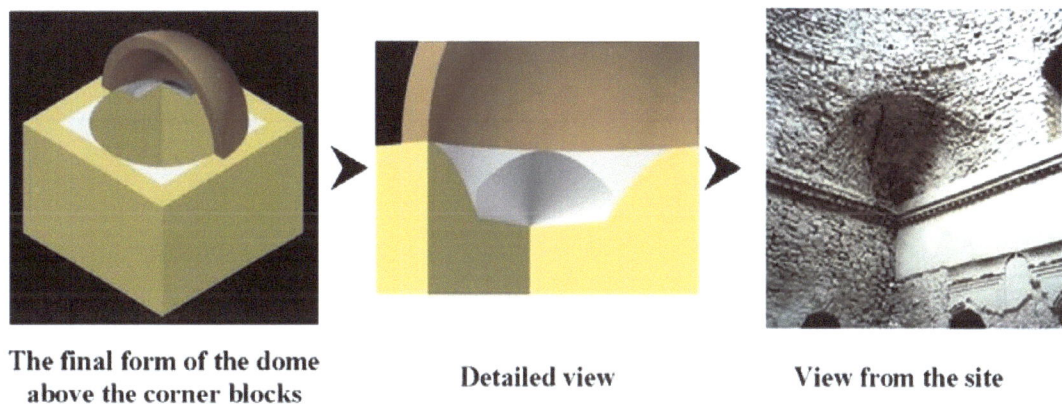

The final form of the dome above the corner blocks   **Detailed view**   **View from the site**

Figure 2.14   The details of one of the initial examples of squinch, the palace of Ardeshir, Fars, Iran (Elkhateeb, 2012).

Different methods have been applied in the architectural heritage of Iran for pier–dome transition. Godard categorized these methods into six main groups as illustrated in Figure 2.15. In the Sassanid period, domes were exclusively constructed over a rectangular base, while in the Islamic era, builders started experimenting with forms such as pentagons, e.g. Baba Rokn Al–Din, Isfahan, octagons, e.g. Gonbad Ali, Abarqu, decagons, e.g. Bastam, dodecagons, e.g. Shrine of Houd, and even rectangles, e.g. Shrine of Haftado–Do Tan (Godard, 1965).

It may have been that Muslim builders decided to combine squinches into complex forms from there, hence developing the *muqarnas*. Progress in design enabled these builders to develop a variety of *muqarnas* patterns suitable for vaults, cornices, minarets, capitals, and portals.

There is, however, a reliable theory by Michel Écochard introducing squinch as the functional origin of *muqarnas* in Iran, even though he does not present any evidence to demonstrate its gradual evolution. Squinches were used widely by the 10th century in Iran to provide smooth transition from the square edges of walls to the circular bases of the overhead domes. Écochard argues that these early tripartite corners are the primary structures leading to the development of *muqarnas* (Écochard, 1977).

Likewise, Stierlin attributes the origins of stalactite vaults to the squinch. He tries to explain the development by discussing that that the forms of *muqarnas* elements are the result of the shrinkage of the squinch units leading to the development of the *muqarnas* in the 13th and 14th centuries. He supports his proposed dates by referring to Mozaffarieh school, built within the Jame' Mosque of Isfahan, which displays this development of forms and has its structure dating back to the year 1366 (Stierlin, 1976).

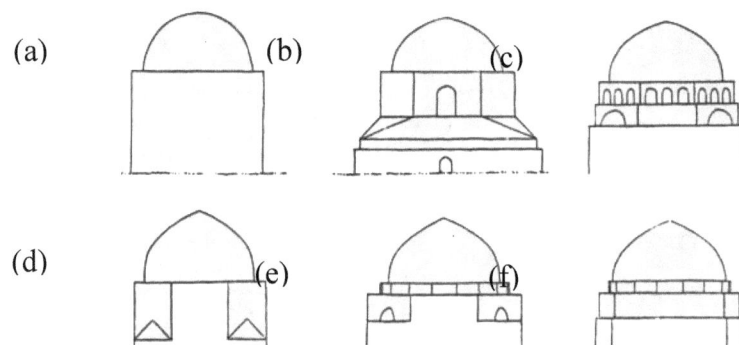

Figure 2.15 (a) Ismail Samanid Mausoleum, Bukhara, (b) Arsalan Jazeb, Sangbast, (c) Davazdah Imam Mosque, Isfahan, (d) Golpaygan Jame' Mosque, (e) Isfahan Jame' Mosque, and (d) Orumieh Jame' Mosque.

In another effort, Harb introduces pendentives, used in Roman architecture, as the connecting link between *muqarnas* and squinch (Harb, 1978). Furthermore, Hautecoeur mentions Almoravids as the developers of *muqarnas* in Morocco and Spain and focuses on the Al–Karaouine Mosque at Fes, Morocco, expanded in 1134, and in the Tlemcen Mosque (1136) in Algeria, as examples of domes over squinches, in which as he states, "the ceilings were replaced by a combination of overlapping elements similar to squinches and to other portions of vaults" (Hautecoeur, 1931).

2.3.2.2 *Patkaneh*

The composition of a semi–sphere and a cube, hence a dome resting on a square, has always been the most imposing form in the architecture of Iran. This is the beginning of the creation of the pendentive, squinch and quasi–squinch forms, as Tavassoli indicates, which in turn gives way to the creation of *patkaneh* (Tavassoli, 2004). Introducing *patkaneh* as the missing link between squinch and *muqarnas* is one of the key objectives that this manuscript pursues. Therefore, in this sub–section the author tries to bring about the available direct mentions or indirect implications of this structural form and its definition in the Persian literature.

The structural form of *patkaneh* owes a great deal to its component materials. Grütter speaks of the importance of materials, adding that construction with the right materials is a principle that is ignored today, while, before the advent of modernism and the freedom of possibilities it brought with construction techniques, architectural forms were largely dictated by building materials (Grütter, 2004).

Memarian describes *patkaneh* as a set of arctuated porches that are arranged in symmetrical and projecting order on top of each other. He explains that the phenomenon may serve as cover for awnings or as transition zones for vaults. He classifies *patkaneh* as both structural and decorative. In his explanation structural *patkaneh* is where each curved surface rests on two structural load–bearing ribs, a member that the decorative type lacks, necessitating a tensile support that connects the veneer to the vault behind it (Memarian, 2012). Figure 2.16 shows a sample of *patkaneh*, a mihrab at the Jame' Mosque of Isfahan, constructed entirely in bricks.

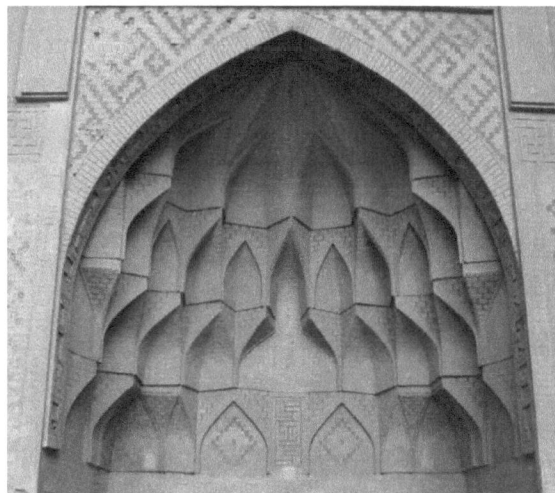

Figure 2.16   Brick *patkaneh*, Isfahan Jame' Mosque (Source: Author).

Pirnia considers *patkaneh* as a technique in transition and provides a definition based on form and structure, stating that *patkaneh* refers to multiple tiers of niches, one over another, that project outwards to complete a transition zone. He believes that *Patkaneh* is etymologically derived from the Persian word كانه, [*kaneh*], which means niche over niche. On the distinction between *patkaneh* and *muqarnas*, Pirnia posits: "In spite of their superficial resemblance, *muqarnas* is a dead load that hangs from the true ceiling behind, while *patkaneh* rests on its own strength (Pirnia, 1991).

In addition, in constructing *patkaneh*, the first tier is made to stand on its own while the second tier is built on top of this, with its load in turn resting on the lower tier, while the construction of *muqarnas* begins from the top tier and proceeds downwards" (Pirnia, 1991). It must be noted here that Pirnia's definition is limited to the *patkaneh* structures that used in squinch areas and does not extend to the models that cover the entire span of vaults and domes.

Among the famous and most comprehensive Persian language encyclopaedias, such as Dehkhoda and Mo'in, no explanation was found for the term *patkaneh*. Yet the technical encyclopaedia of traditional architecture by Fallahfar, explains the term with reference to Pirnia's proposed definition of the architectural form, i.e. a transition technique that is composed of tiers of niches with increasing projection, constructed one over the other (Fallahfar, 2010).

In an attempt to provide a definition of *kaarbandi*, i.e. a types of Iranian traditional rib vaulting, Bozorgmehri speaks of a complete sample of such phenomenon under the sanctuary dome at the Jame' Mosque of Nain, 10th century C.E., and considers the work as a load–bearing *kaarbandi*. She also speaks of the transition over Ismail Samanid Mausoleum in Bukhara, 9th century C.E., and Nezam Al–Molk dome at the Jame' mosque of Isfahan, 11th century C.E., as *kaarbandi*, and believes that *patkaneh* is the predecessor of *kaarbandi*. Although the author believes the abovementioned examples of *patkaneh* are structurally independent phenomena, they are the predecessors of other non–structural and merely decorative forms that eventually lead to the formation of *muqarnas* (Bozorgmehri, 1996).

Furthermore in his book, Besenval describes a traditional method of vaulting in the East as "arctuation with projection" (Besenval, 1984), which in the Persian translation of his book has been replaced with the term *patkaneh* (Besenval, 2000) and these are the only available sources in which the term *patkaneh* has been directly used. The multifaceted behaviour of *patkaneh*, however, borders on more structural forms like ribbed vaulting on one end of its spectrum, and on purely decorative *muqarnas* on the other end (Memarian, 2012).

Considering the structure and form of *patkaneh*, an overall view towards the phenomenon is necessary; one that would consider structural as well as form issues. The absence of such a view in most existing researches has resulted in the misrepresentation of the phenomenon. For example, Andre Godard indirectly mentions it, pointing at the samples in the Jame' Mosque of Isfahan and Barsian, claiming that these forms are purely decorative because of the fact that in the latter sample, although the *niches* have been completely destroyed, the dome appears hanging, resting solely on the bridging arches of the squinches. He therefore does not consider the form to be structural, dissecting it into two components, hence the structural arches, and the dead–load niches (Godard, 1990).

Furthermore, Al–Asad classifies *muqarnas* as a form of ribbed vault, in which *tasehs*, i.e. the niche–like elements of *muqarnas*, are arranged in tiers, with one tier carrying the next projecting one, with the end result appearing in the form of a corrugation that is often also called honeycomb or stalactite vault. But the fact is that Al–Asad's definition of "each tier carrying the next projected one" actually describes *patkaneh* in a more sufficient manner than *muqarnas* (Al–Asad, 1995). In addition, referring back to the proposed definition of *muqarnas* and its various types as described by Al–Kashi hence the simple, clay–plastered, arched and *Shirazi*, one can conclude that the first two types are in principle *patkaneh*, and the arched *muqarnas* is indeed identical to decorative *patkaneh*. The author believes that what Al–Kashi classified as *Shirazi muqarnas*, which is purely decorative is the only form that qualifies as *muqarnas*.

2.3.3 Mathematical and Computational Studies

One of the most important aspects of *muqarnas* is its geometry. Although many scholars including Jones and Goury in 19th Century, and Notkin and Écochard in 20th Century have analysed the subject, yet it remains mysterious for most architectural historians. This is probably because its analysis should be undertaken mathematically and geometrically. Of course, it is possible to analyze *muqarnas* on different levels in terms of mathematical depth, since not only scholars like Al–Kashi have analysed it, but that also artisans who design and construct it possess introductory understanding of Euclidean geometry (Écochard, 1977; Al-Kashi, 1954; Notkin, 1995; Goury & Jones, 1842). A renewed trend is observed over the past two decades, in the investigations focusing on computational visualization of *muqarnas* and its constituent elements, based on its plan. In addition, efforts have been devoted to interpreting and decoding the meanings of available two–dimensional patterns that do not follow the conventional architectural plan rules.

Takahashi was one of the pioneers of visualising *muqarnas*. He tried to measure and record about 1000 *muqarnas* samples from all over the world, and tried to create their two–dimensional pattern plans accurately (Takahashi, 1973). Two decades later, Mohammad Al–Asad and his research team in Harvard University, and Mohammad Ali Yaghan, a PhD student of the University of Tsukuba, Japan who started a new trend of visualizing *muqarnas* in form of three–dimensional patterns (Al-Asad, 1995; Yaghan & Hideki, 1995)

Al–Asad's team focused only on producing the *muqarnas* patterns of Topkapi Scroll computationally and tried to visualize their three–dimensional interpretations, using *Computervision CADDS 4X* (Al-Asad, 1995). However, Yaghan continued the path after his graduation and developed the skill to motivate many other scholars to model *muqarnas* and study its elements through computational visualisation. Table 2.1 shows the chronological trend, based on the publication dates.

2.4 Characteristics of *Muqarnas*

Muqarnas has always been special, due to its three–dimensional nature which could therefore fill a volume wherever it was used, providing any desirable depth or combination of elements. With its depth and curve being completely controllable, it could be used either architecturally in relationship with the vaults, or simply as a veneer. The details about the design, construction techniques and materials, as well as the structural role of the ornamental phenomenon from the perspectives of other scholars are gathered and analysed in the following sub–sections.

Table 2.1: Research accomplished about visualization of *muqarnas*.

Name of University	Country	Date	References
Harvard University	United States	1995	(Al-Asad, 1995)
University of Tsukuba	Japan	1995	(Yaghan & Hideki, 1995)
King Saud University	Saudi Arabia	1998	(Yaghan, 1998)
King Saud University	Saudi Arabia	2000	(Yaghan, 2000)
King Saud University	Saudi Arabia	2001	(Yaghan, 2001a; Yaghan, 2001b)
Heidelberg University	Germany	2002	(Dold-Samplonius et al., 2002)
King Saud University	Saudi Arabia	2003	(Yaghan, 2003a)
King Abdulaziz University	Saudi Arabia	2003	(Yaghan, 2003b)
University of Petra	Jordan	2005	(Yaghan, 2005)
Heidelberg University	Germany	2005	(Harmsen, 2006)
Heidelberg University	Germany	2007	(Harmsen et al., 2007)
Tarbiat Modares University	Iran	2009	(Safaeipour, 2009)
University of Catania	Italy	2010	(Garofalo, 2010)
Utrecht University	Netherlands	2010	(Hoeven & Veen, 2010)
German Jordanian University	Jordan	2010	(Yaghan, 2010)
Calgary University	Canada	2011	(Hamekasi et al., 2011)
American University Beirut	Lebanon	2011	(Hamekasi et al., 2011)
Iran University of Science &Technology	Iran	2012	(Memarian, 2012)
King Abdulaziz University	Saudi Arabia	2012	(Elkhateeb, 2012)

2.4.1 Design Characteristics

Numerous books have been published on the geometry of Iranian architecture, most of which focus on the appearance and the geometry of patterns (Abolghasemi & Omranipour, 2005). Fereshteh–Nejad refers to the medallion or *shamseh* found at the apex of the half–domes and portals around which the *muqarnas* is arranged, resembling the sun, as *shamseh* which transliterates as solar pattern having stars around it (Fereshteh-Nejad, 2010). Bryson also uses in his signature, terms to explain the aesthetics of *muqarnas* patterns that he believes are in complete harmony with the extensive "gazes" demanded by one-point perspective. Instead, the designs demand constant motion that would delay the perception of the underlying order of the designs, resulting in a sense of awe; that which is a prerequisite to the analytical function of logic and reason (Bryson, 1983).

This is while the structural calculations of Iranian buildings are comprehendible through their geometry, as well (Abolghasemi & Omranipour, 2005). Structural form is directly influenced by the architectural ideals for aesthetics. It may be incorporated into the actual façade or interior, or it may become part of the architectural design. That is why the structure in Islamic architecture can play a major role in the design by itself. In this light, keeping a building upright is not the only role that structures play, but that they will also have to convey the necessary aesthetic messages the designer intends (Golabchi, 2007). In the Islamic religious atmosphere, the relentless struggle of the Muslim with figurative ornaments, associated with idolatry, would encourage the builder to concentrate on abstract representations. This important factor in itself would find exaltation as a means to encourage imagination. Under such conditions, the design would turn into an endless repetition of units, turning columns and arches into *muqarnas* decorations (Hillenbrand, 2004).

2.4.1.1 Two–Dimensional Pattern Plans

Since the constituent elements do not overlap in *muqarnas*, one can project the three–dimensional structure to a plane surface, making a pattern plan. A plan defines the procedures and design methods adopted by the builder, especially when the building tradition under discussion lacks theoretical frameworks for construction. Yet under Middle Eastern traditions, plans appeared in very minimal and simple forms, with the builders zealously guarding them as trade secrets. Here it was rather the verbal traditions that transferred the knowledge from one generation to the next (Necipoglu, 1995).

With the advent of the Renaissance, scaled plans and measured drawings became the norm in European architecture, widening the gap with the traditions in the East that relied greatly on abstract geometrical motifs. Therefore, the remaining *muqarnas* patterns from the Middle East, that do not obey conventional drawing rules, were first suggested by Yaghan to be termed as *two–dimensional pattern plans* for *muqarnas*, abbreviated as 2DPP, in many recent publications (Yaghan, 2001b; Yaghan, 2003b; Yaghan, 2005; Hamekasi et al., 2011; Yaghan, 2000; Halaweh, 2010).

The remarkable point about these 2DPPs of *muqarnas* is the way they were designed and the fact that no one knows exactly how these two–dimensional pattern plans were transferred to three–dimensional structures, except those who were builders, the masons and craftsmen. During the middle ages, the methods of transformation of the 2DPPs into three–dimensional models were kept as a secret by craftsmen who had the responsibility to guard these methods and have them learnt by the next generations of the guild (Notkin, 1995)

Necipoglu further points to the dimensional ratios of the designs, elaborating that in order for the patterns to match the actual dimensions on site, the easiest solution was to adhere as much as possible to rectangular frames the dimensions and scales of which were not so much important as the fixed ratios of length to width (Necipoglu, 1995). It is worth mentioning that as a practical builder, and member of the guild, who has learned the secrets behind the traditional *muqarnas* patterns from his ancestors, the author can interpret the available old patterns and has a keen eye, capable of catching plans fast and accurately, and converting them into appropriate architectural plans, that has been done in following chapters.

2.4.1.2 Scrolls

Based on available resources, no architectural drawing survives in the Muslim World, predating the Mongol era, yet occasionally mention is made of patterns that were drawn on the ground (Necipoglu, 1995). It is worth reiterating that the oldest found example of a *muqarnas* pattern is the 13th century plaster tablet found in Takht–i Sulaiman, Nishapur, West Azarbaijan, Iran, in 1968, shown in Figure 2.10. This stucco plate shows one quarter of a *muqarnas* pattern of a vault. Parts of the actual *muqarnas* were also unearthed from the ruins where the plate was found. The elements of the *muqarnas* seem to be pre–fabricated and then mounted on the wall (Ghazarian & Ousterhout, 2001; Harb, 1978; Dold-Samplonius & Harmsen, 2005; Harmsen, 2006; Hoeven & Veen, 2010).

The Topkapi Scroll (16th century) and Mirza Akbar Scrolls (19th Century) are other samples in which two–dimensional pattern plans of *muqarnas* can be found (Necipoglu, 1995). The Topkapi Scroll contains a collection of 114 *muqarnas* patterns as documented in 1986. The valuable scroll is kept at the Topkapi Palace Museum, in Istanbul. It is the oldest available scroll of its kind as yet discovered intact (Hoeven & Veen, 2010).

Necipoglu is of the opinion that the plans of the Topkapi Scroll, as shown in Figure 2.17, are composed of radial orders adorned by intertwined stars and polygons in angular arrangements.

She further explains that the patterns are drawn on the basis of the radial lines projecting from the corner of the design. In the drawings, black, red and dotted lines are used to facilitate the reading of the plan by the builders. Further, as illustrated in Figure 2.18, elements such as *toranj*, *shaparak* and *parak* are coloured to distinguish them from other filling components, in addition to a clear distinction of the tiers. A more detailed explanation of *muqarnas* elements is available in Chapter four of this manuscript.

The distinction of the elemental contours and tiers with different colours in the plans resemble very closely the designs of the scrolls of Tashkent, and Mirza Akbar and the drawings of El–Bouri. The other common factor among all of these scrolls and drawings is that they represent direct or inverted ceiling plans with all of the tiers present (Necipoglu, 1995). Figure 2.19 shows one of the *muqarnas* patterns of the scrolls of Mirza Akbar, which is now kept in the Victoria & Albert Museum, London, United Kingdom.

Christie describes the drawings found in the Mirza Akbar scrolls, as shown in Figure 2.20, as a set of stars and polygons intertwined by concentric circles and dissected by radiant lines, turning them into a complex system of radial symmetries (Christie, 2010). This method in the latter's opinion enabled Muslim designers to redefine the pre–existing geometrical forms with their own inventive rules. The scholar considers this discovery as the greatest achievement of Muslim ornamentation, with yet limited effects on Western decorative traditions. From this Christie surmises that as the designers overcame the complexities of the designs, each pattern would degenerate into a repetitive module and passed around until the likes of scrolls in Mirza Akbar's possession would become indispensable tools at every significant work site.

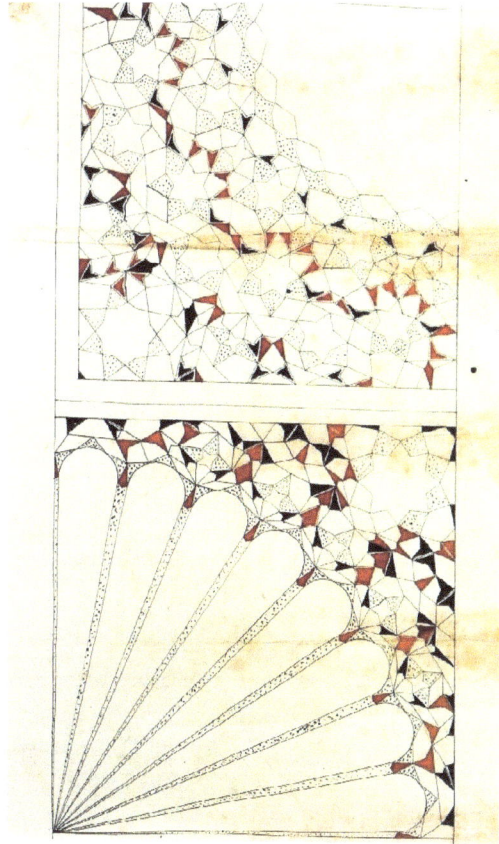

Figure 2.17   Topkapi scroll, radial design (Necipoglu, 1995).

Figure 2.18   Topkapi scroll, coloured elements (Necipoglu, 1995).

Figure 2.19   A *muqarnas* pattern in Mirza Akbar scroll (Source: Patterninislamicart.com).

Figure 2.20   Mirza Akbar Scroll, coloured elements (Source: Patterninislamicart.com).

Referring to the similarities between 2DPPs of *muqarnas* and Arabesque knotwork or interlace patterns, known as *gerehs*, Behrens–Abuseif writes:

> "It is not clear whether the development of the mukarnas in Persia started from the planar into the three–dimensional, or vice versa, or that it developed in parallel directions. Since the excavations in Takht–i Sulaiman, the ornamental rather than the structural origin [for *muqarnas*] is being favoured."

(Behrens-Abuseif, 1993)

In addition, Bulatov believes that the direct interchange of patterns between the two–dimensional *gereh* and the three–dimensional *muqarnas* has enriched the Islamic architecture of

Iran and Central Asia, with a congruity that unites the symmetrical order of the plans, walls and vaulted ceilings (Bulatov, 1988).

2.4.1.3 Al–Kashi's Method of Drawing *Muqarnas*

The proposed method of drawing a simple *muqarnas* by Al–Kashi is very significant, since all other extant drawings pertaining to *muqarnas* comprise of plans only, but this drawing comes with an explanation on how builders decide the curve of the arch. Figure 2.21 illustrates the pertinent instructions as they appear in the manuscript.

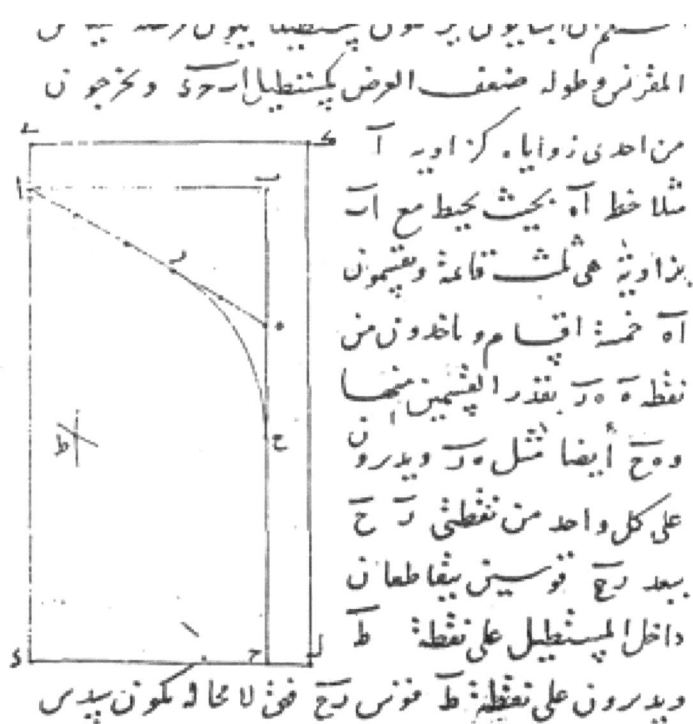

Figure 2.21   Al–Kashi's hand–drawing (Dold-Samplonius, 1992).

With reference to Figure 2.22, builders draw a rectangle (ABDG) to the width of the elemental profile and a height twice the width. From point A they draw a –30° angle. Then they divide AH into five, and mark the second division from H, hence R. using a compass they determine E on the vertical line BG, where HE=HR. using points R and E as the centres and RE as radius, they draw two curves to intersect at point T. now placing the needle of the compass on T, and with the same radius, they draw a curve to connect AR to GE.

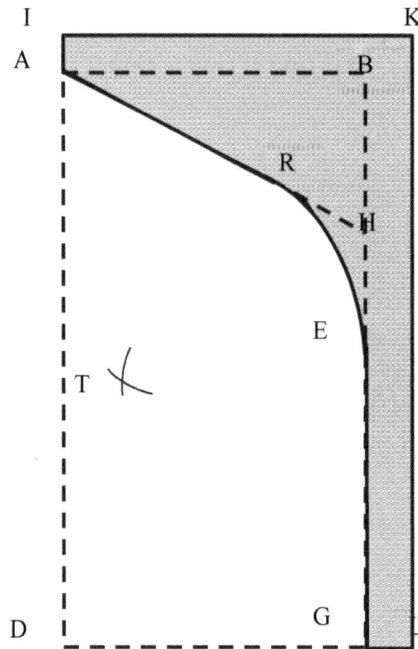

Figure 2.22   The rectangle of *muqarnas*.

Next they extend DA and DG to the desired lengths at points I and L, constructing IKLGA. This profile will be used for a mould to reproduce as many similar shapes as needed by the plan. They will then install each of these plaster profiles on the sides of an element at the workplace. It must be added that the height to width ratio is flexible, since the depth of the *muqarnas* elements is not constant at all points. However the height of the elements will be equal since the distances between the tiers are uniform (Jazbi & Al-Kashi, 1987).

2.4.2   Construction Characteristics

In fact, the spatial organization of *muqarnas* in a building and its pertaining construction technique depends on the location of the decorations in the building. This location is determined by either the intersection of two non–coplanar surfaces (i.e. cornices, corbels and capitals) or at the transition zone between the rectangular quarter and the spherical vault (i.e. pendentives, squinches, portals, and ceilings) (Garofalo, 2010).

Wheeler Thackeston describes the methods applied by masons and craftsmen in constructing *muqarnas*, by using inverted plans. He illustrates how builders apply a vertical rectangle with a width equal to the projection of the *muqarnas* element, and a height twice the width, to construct the decoration. After explaining the method by which the vertical profile of the *taseh* is determined, he moves on to analyze the technique by which each *muqarnas* tier is laid out on a plaster board and adjusted to the pertinent vault (Thackeston, 1989). Komaroff also believes that Al–Kashi wrote his treatises for a scientific audience rather than for artisans. She points to the fact that the scholar reiterates the otherwise alternative terms used by craftsmen for the mathematical phenomena he discusses, which indirectly hints to the fact that he is discussing this with and for a group of scholars (Komaroff, 1991).

Necipoglu speaks of a practical training by which builders would decipher the most complex plans of multi–layered stalactites for construction. In this craft, the masons would develop three dimensional views of the relevant elements, eliminating the need for elevation drawings, by which they would project the working plans onto the actual surfaces, as well relying on their practical expertise (Necipoglu, 1995). Pugachenkova posits that in the construction of complex vaults and arches, the old technique of drawing the plan on the floor in real scale, and without the need for elevations is still in practice. The elements of the *muqarnas* are constructed on the ground first and then, hung and attached to the structure by means of ribs (Pugachenkova & Rempel, 1965).

Sha'rbaf, who is a very famous expert traditional architect in Iran, also mentions that in constructing a *muqarnas* suspended layer technique is used over the vault (Sha'rbaf, 1996). For a better understanding of the building process of *muqarnas*, as illustrated on Figure 2.23, he always equips his *muqarnas* plans with line projections onto the elevations of the pertinent vaults (Sha'rbaf, 2006). Al–Asad posits that the most important factor in the execution of the *muqarnas* plan is the equal division of the height into the number of component tiers. Each tier represents a set of elements with uniform height and contains a number of *shaparaks*. On the elevation, each set of *shaparak* elements is accompanied by a tier devoid of *shaparak*. Unlike other tetragonal elements, the *shaparaks* get narrower as they recede, and this lack of uniformity disables the formation of a rectangular row (Al-Asad, 1995).

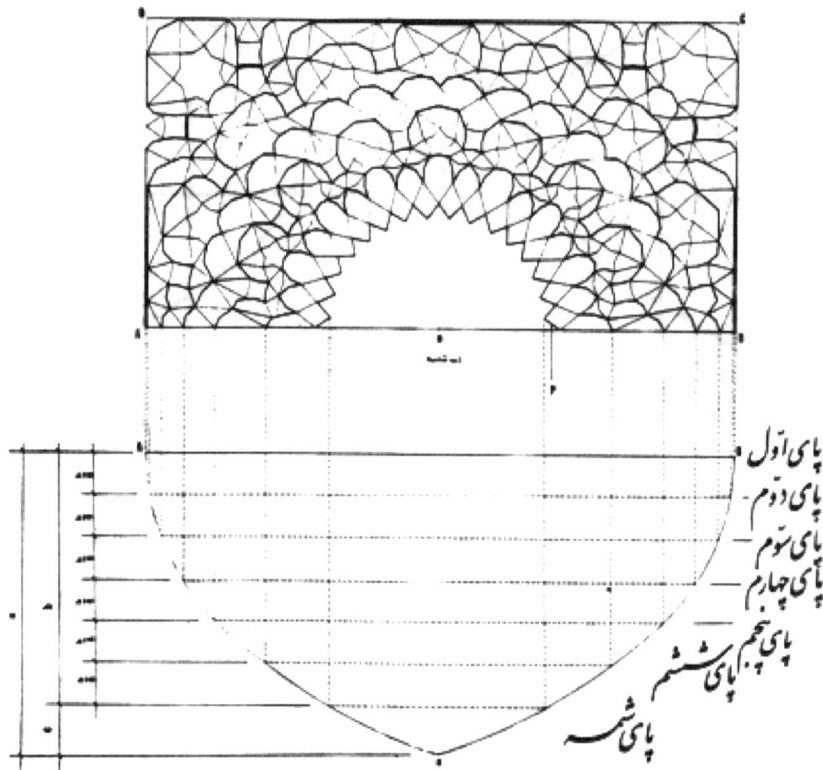

Figure 2.23   Sha'rbaf drawing, including projection lines (Sha'rbaf, 1996).

Clarke has been one of the earliest eyewitnesses to record the traditional method of constructing *muqarnas* in 1893. According to Clarke, the construction of a *muqarnas* vault commences with the execution of the plan on actual scale on a flat bed of plaster poured onto the ground that is levelled with a layer of ash. Builders would then imprint the lines carving them into the plaster, lubricating the surface with lard, out of which they produce a plaster template roughly half an inch thick. They repeat this procedure for all of the tiers in the plan, after which they install them into place on the designated site. It must be noted that because Eastern *muqarnas* contains straight edges, manual construction is not very difficult (Clarke, 1893). Forty-three years later, Myron Smith speaks of a trip to Isfahan in 1936, wherein he observes the construction of a *muqarnas* vault using basic tools like straight edges, strings and needle compasses. At the site, as Smith explains,

"The builders had laid the real scale plan carved on a plaster bed using a pick."

(Smith, 1947)

This statement means that the techniques had not changed since Clarke's observation in the 19th century. In addition, Snelling speaks of similar experience in 1995 in Turkey:

"I watched a local builder decorating the interior of a cafe and he made 30mm slabs of shaped Plaster of Paris "stucco" on a glass sheet bounded by wooden battens and with fresh plaster he could both stuck slabs together and then onto walls."

(Snelling, 1995)

2.4.2.1 Construction Materials

Based on the studies conducted in different historical periods, the employed material of the *muqarnas* varies by geography. Where stone is abundant and the technology to harness it available, *muqarnas* is built in stone. This is the case with Syria, Egypt, and Anatolia, while in North Africa, plaster and wood are mostly used. Table 2.2 shows the range of materials used in different parts of the Islamic world for constructing *muqarnas* (Yaghan, 2001b). In the modern era innovative construction materials may also be employed, such as fibreglass and its pertaining advanced technologies (Garofalo, 2010).

Table 2.2: *Muqarnas* construction material in Islamic world (Yaghan, 2001b).

	Europe	Egypt	Syria	Anatolia	Iraq	Iran	India
Brick		X			X	X	
Plaster and Stucco	X					X	
Ceramic and Terracotta						X	
Faience						X	
Stone		X	X	X			X
Wood	X	X					
Mosaic Tiling						X	

In Iran and Iraq, on the other hand, *muqarnas* is mostly made in bricks, with the common brick size of 20 x 20 x 5 cm^3 and at times covered in plaster or tiles (Stierlin, 1976). The latter veneers would help the mason change the orthogonal geometry of the brick and hence, provide a higher degree of freedom to the composition. By employing these veneers, it is not necessary for angles to be strictly 90° or 45° or 135°. Moreover, it makes it possible to solve the difficulties on the junctions, and to add more specific details (Garofalo, 2010).

It is not only *muqarnas*, but nearly all Islamic structures in Iran have been built in brick, in spite of the fact that major cities, such as Isfahan were surrounded by mountains, and had access to very fine quarries. Nevertheless, the application of sundried and baked bricks date back to two millennia in the history of the region. Persepolis is the only large scale Iranian construction in ashlars. Other than this it is difficult to come by samples that would speak of a continuing tradition in stone masonry (Stierlin, 1976).

2.4.2.2 Structural Role of *Muqarnas*

Structural Forms are those encapsulating spaces and linking them together for the purpose of providing shelters, (Mainstone, 1998). Speaking of the structural form of *muqarnas*, André Godard posits that unless under exceptional conditions, *muqarnas* vaults are non–lead–bearing and have little to do with the load–bearing vaults behind them that carry their weight by means of tensile elements. Brick stalactites are mounted on–site only after the completion of structural works and simultaneous with the application of wall finishes; and plaster stalactites in their many forms are merely prefabricated veneers that do not even come in contact with the structural surfaces beneath (Godard, 1990).

Furthermore, Henry Martin, states that *muqarnas* stalactites are definitely Muslim ornaments that intend to abstractly regenerate the stalactite formations found in nature. By his description, *muqarnas* comprises of seven geometrical components that are non–lead–bearing and rest over such load–bearing members as columns, in the form of capitals, pendentives, and arches. In Syria these ornaments are carved in stone, in Iran and Turkistan, in baked earth, and on numerous occasions prefabricated in plaster (Martin, 1964). Safaeipour also mentions the structural role of architectural forms, such as squinch and *patkaneh*, but he does not consider *muqarnas* a load–bearing structure (Safaeipour, 2009).

2.4.2.3 Constituent Elements of *Muqarnas*

Different approaches were taken to describe the constituent elements *muqarnas*. Many scholars rely on Al–Kashi's definition of *muqarnas* elements (Dold-Samplonius, 1996; Dold-Samplonius et al., 2002; Hamekasi et al., 2011; Harmsen et al., 2007; Hoeven & Veen, 2010; Krömker, 2007), while some other professionals define the constituent elements of their own (Yaghan, 2001b; Necipoglu, 1995; Sha'rbaf, 1996; Lorzadeh, 1981). Al–Kashi's explains *muqarnas* to be comprised of tiers that are in turn divided into cells, in round stair–like arrangements on the vault, where each cell connects to the next with a predetermined angle. In addition, there are intermediate elements connect the roofs of adjacent cells to each other (Al-Kashi, 1427).

The two–dimensional plans of Al–Kashi's described *muqarnas* cells are basically simple geometrical forms like square, half–square (right–angled triangle), rhombus, half–rhombus (isosceles triangle or equilateral triangle), almond (deltoid), jug (one quarter of an octagon), large biped (complement to a jug), small biped (complement to an almond) and sometimes rectangles (Dold-Samplonius, 1992). Since the Ilkhanid period, the above mentioned elements are used for constructing *muqarnas* without any change. Although more advanced forms of *muqarnas* with four– to seven–pointed star elements were erected during Ilkhanid period as well. Figure 2.24 shows the constituent elements of *muqarnas* as introduced by Al–Kashi.

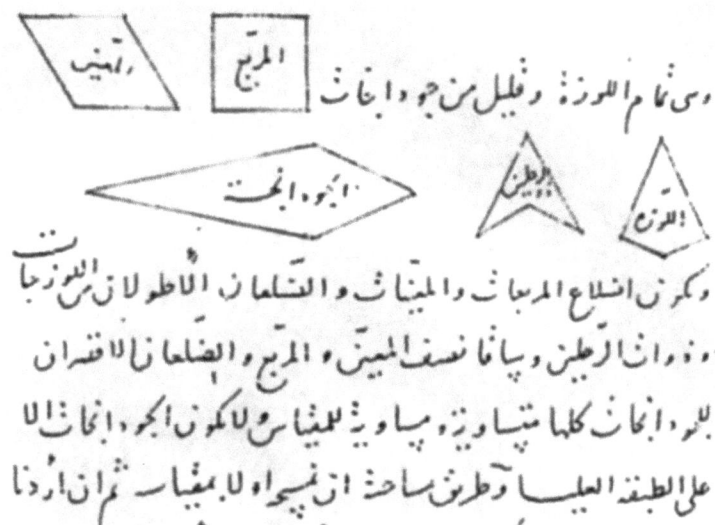

Figure 2.24   Constituent elements of *muqarnas* as introduced by Al–Kashi (Dold-Samplonius, 1992).

The basic *muqarnas* elements of Al–Kashi were extensively studied by different scholars and it was expanded to nine main elements, i.e. the basic elements shown in Figure 2.25, as well as the *stars*. In addition, as shown in Figure 2.26, Dold–Samplonius looks at the description of Al–Kashi's curved *muqarnas* components from a mathematical point of view and defines the geometrical definition of Al–Kashi's *muqarnas* elements.

Necipoglu has her own method of naming the components of *muqarnas*. She refers mostly to the two–dimensional pattern plans of *muqarnas*, found in Topkapi Scroll. She introduces stars, as well as six other main elements, namely *biped, arrow–shaped double biped, rhombus, rhomboid/kite–shaped rhomboid, irregular hexagon* and *elongated hexagon* (Necipoglu, 1995).

Among other non–Iranian scholars, Yaghan divides the constituent elements of *muqarnas* into four main groups. He looks at the issue from a more mathematical point for view and therefore, rather than trying to name the elements, he categorises *muqarnas* elements based on their form to *Point to Edge* (PTE), *Edge to Edge* (ETE), and *Edge to Point* (ETP) elements, as well as *roof–patches*. However, he introduces some basic elements like *mini squinches* or *niches, brackets* and *mini pendentives*, as the main constituent elements of *muqarnas* (Yaghan, 2001b). Figure 2.27 shows the above–mentioned categories and Figure 2.28 shows the three basic *muqarnas* components that he introduces.

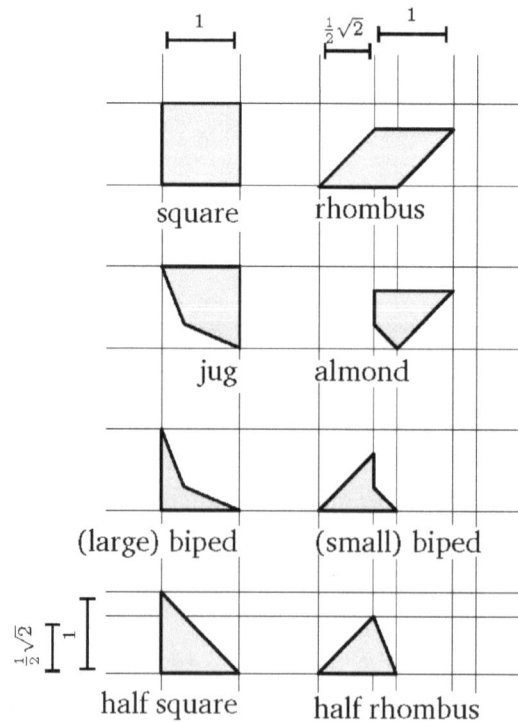

Figure 2.25   Basic *muqarnas* elements of Al–Kashi (Harmsen, 2006).

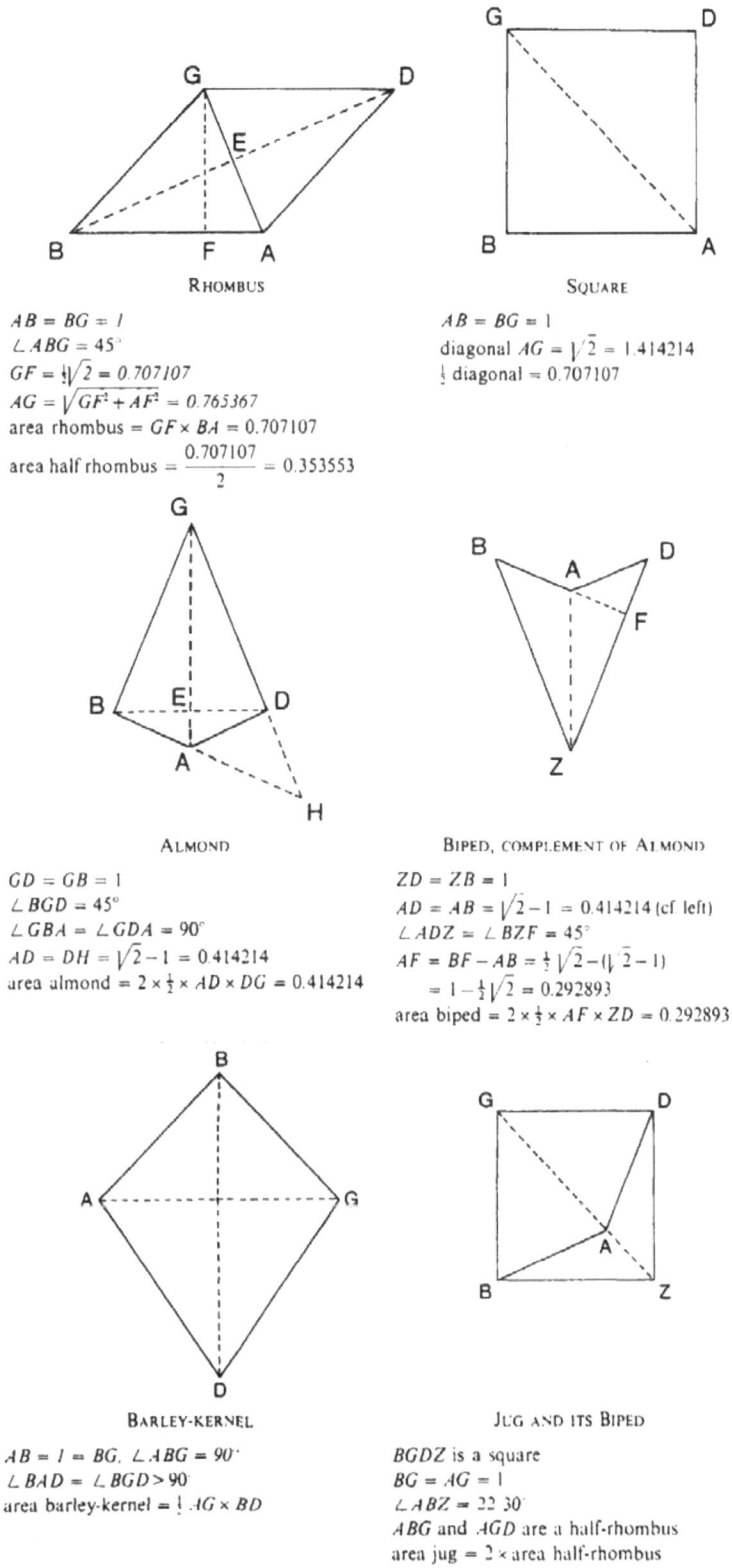

RHOMBUS

$AB = BG = 1$
$\angle ABG = 45°$
$GF = \frac{1}{2}\sqrt{2} = 0.707107$
$AG = \sqrt{GF^2 + AF^2} = 0.765367$
area rhombus $= GF \times BA = 0.707107$
area half rhombus $= \dfrac{0.707107}{2} = 0.353553$

SQUARE

$AB = BG = 1$
diagonal $AG = \sqrt{2} = 1.414214$
$\frac{1}{2}$ diagonal $= 0.707107$

ALMOND

$GD = GB = 1$
$\angle BGD = 45°$
$\angle GBA = \angle GDA = 90°$
$AD = DH = \sqrt{2} - 1 = 0.414214$
area almond $= 2 \times \frac{1}{2} \times AD \times DG = 0.414214$

BIPED, COMPLEMENT OF ALMOND

$ZD = ZB = 1$
$AD = AB = \sqrt{2} - 1 = 0.414214$ (cf. left)
$\angle ADZ = \angle BZF = 45°$
$AF = BF - AB = \frac{1}{2}\sqrt{2} - (\sqrt{2} - 1)$
$\quad = 1 - \frac{1}{2}\sqrt{2} = 0.292893$
area biped $= 2 \times \frac{1}{2} \times AF \times ZD = 0.292893$

BARLEY-KERNEL

$AB = 1 = BG, \angle ABG = 90°$
$\angle BAD = \angle BGD > 90°$
area barley-kernel $= \frac{1}{2} AG \times BD$

JUG AND ITS BIPED

$BGDZ$ is a square
$BG = AG = 1$
$\angle ABZ = 22°30'$
ABG and AGD are a half-rhombus
area jug $= 2 \times$ area half-rhombus

Figure 2.26 Geometrical definition of Al–Kashi's curved *muqarnas* components (Dold-Samplonius, 1992)

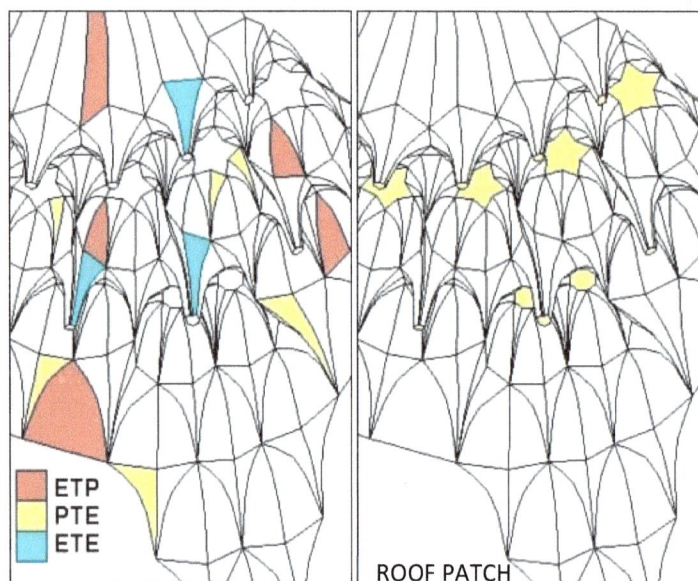

Figure 2.27   Yaghan's method of categorizing *muqarnas* components (Source: *Muqarnas*.org).

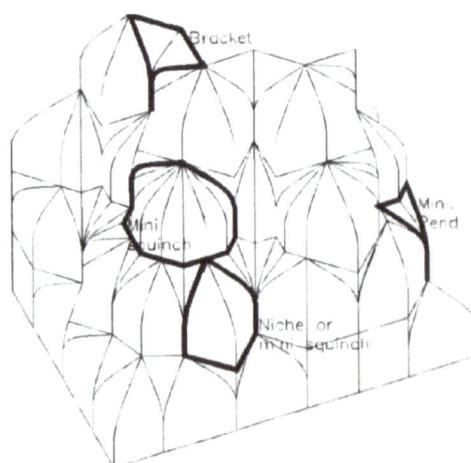

Figure 2. 28   Yaghan's basic constituent elements of *muqarnas* (Yaghan, 2001b).

Vincenza Garofalo simply compares the constituent elements *muqarnas* to portions of vaults ranging from a few centimetres in Moroccan *muqarnas*, to meters in Jame' Mosque of Isfahan, in dimension. She adds that material is one of the important factors that influence the size of the *muqarnas*. Big *muqarnas* can be made by stone, while the small one can be made by wood. In addition to material, historical period can also affect the dimensional aspects (Garofalo, 2010).

Among Iranian maestros and scholars, Lorzadeh introduces a set of constituent elements for *muqarnas*, comprising of eleven elements that are explained below and shown in Figure 2.29:

(1) *Shaparak*: A three–sided element, responsible for connecting the tiers and units together.

(2) *Takht*: Regular horizontal star–shaped element which many be four–, five–, six– or more pointed.

(3) *Irregular Takht*: A horizontal element which has an irregular shape.

(4) *Shamseh*: The medallion on the apex of vaults or iwans, responsible for finishing the ornament.

(5) *Toranj*: The element that comes immediately after *shamseh*, as well as under regular *takhts*.

(6) *Taseh*: A famous element that always appears between *shaparaks*, and sometimes between two *toranjes* or two *T*'s.

(7) *Tee*: A narrow element that appears between *Tasehs*.

(8) *Madani*: An element that fills the space between two *takhts* from two different tiers.

(9) *Double–Madani*: This element is a *madani* that has two legs instead of one.

(10) *Lozi*: Equivalent to *Darvazeh*, meaning gate. A curved rhomboid responsible for creating hanging components; and finally,

(11) *Susan*: a small *taseh*, that looks like an isosceles triangle (Lorzadeh, 1981).

Sha'rbaf, on the other hand, adds two more elements to the collection, namely *pabarik*, and *tanoureh* (Sha'rbaf, 1996), while Pirnia adds *ahou–pa* too and posits *ahou–pa* as the hanging portion of the *muqarnas* (Memarian, 2008). In this collection, what that is call *tanoureh* is identical to Lorzadeh's *madani* element and *pabarik* is in fact an elongated *toranj*.

Although the constituent elements of *muqarnas* are mostly similar in Iran and in other countries, different names had been given to them by different scholars and masons. Even within the borders of an identical country that is Iran, different names had been given to an identical element during different periods or regions. In order to minimize the problem, the author tried to first minimize the number of elements by differentiating elemental units from basic elements, and then to gather all available variable names for each element and present them in form of tables. Hence, as a part of this research's objectives, Chapter 4 provides the above information and a clear definition for each constituent element of *muqarnas*. This step was necessary as the evolution of *muqarnas* can only be observed through the evolution of its constituent elements, in terms of variety and complexity.

Four-Pointed *Takht*   Six-Pointed *Takht*   *Tee*

Eight-Pointed *Takht*   *Shamseh*   Irregular *Takht*

Five-Pointed *Takht*   *Taseh*   *Toranj*

Rhombus *Takht*   Two-pointed *Madani*   *Madani*

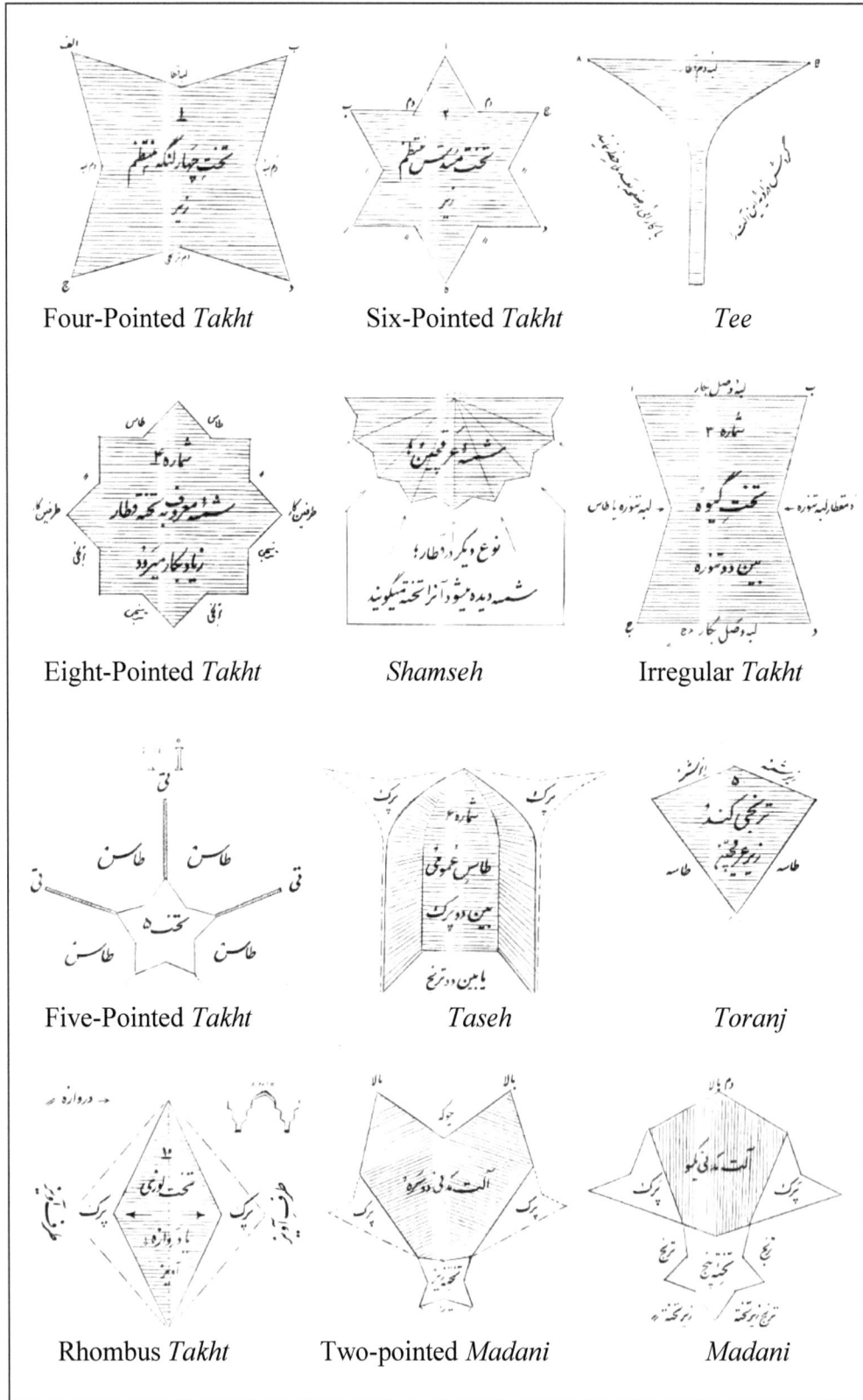

Figure 2.29   Constituent elements of *muqarnas*, from Lorzadeh's point of view (Source: adapted by Author from (Lorzadeh, 1981)).

2.5 Conclusion

In spite of the vast amount of researches focusing on the subject of *muqarnas* over the past few decades, only a portion of the existing gap in the knowledge is filled. To obtain a comprehensive chronology of *muqarnas*, its origin, evolution and maturation process, its design and construction methods, as well as its meaning, reliance on the available information is still not enough and further investigation on the subject is demanded.

To overcome these shortcomings, not only do we need to survey, record and document the extant samples of ancient works, but we also need to pay attention to what may be found in poems and literary works of the past, which has less attracted the attention of the architectural historians of the Muslim world. Only through these kinds of interdisciplinary research is it possible for a scientific investigation to come out of pure speculations and get closer to the truth. Figure 2.30 shows the increasing desire between the scholars to discover the beauties behind the most complex decoration of traditional architecture, which is now a signature decorative element of Islamic architecture.

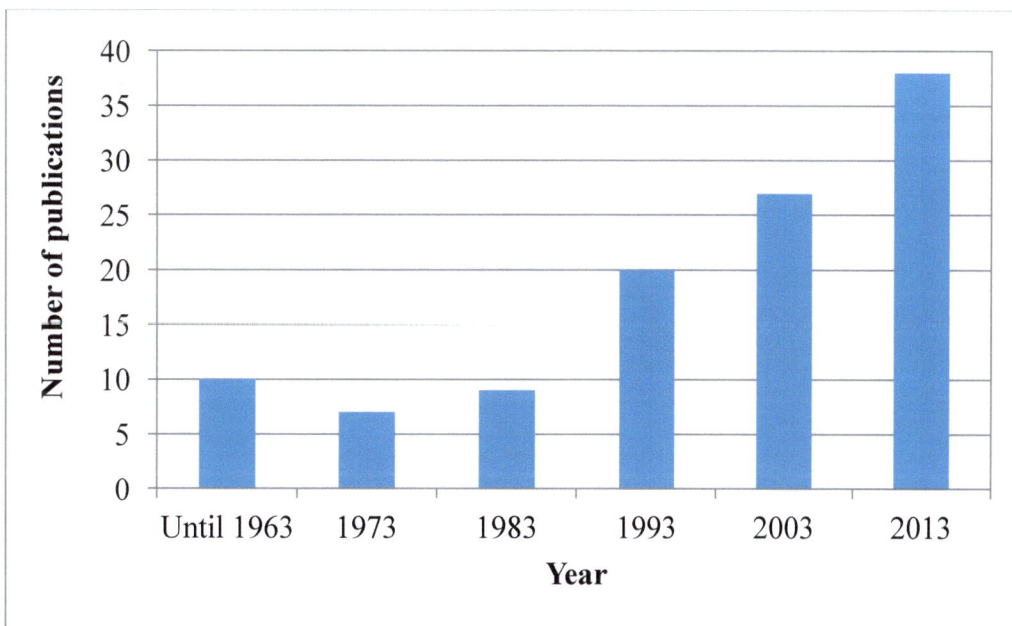

Figure 2.30   Number of publications about *muqarnas*, in each decade (Source: Author).

With the rich literature in hand, obtained from academic endeavours in many world-known universities as divided by topic in Table 2.3, it is noteworthy that still many vital aspects of the knowledge on *muqarnas* are unspoken and require systematic scientific analysis. With this in mind, this manuscript strives to take the necessary steps toward achieving a comprehensive understanding of *muqarnas*, through the scientific methods described in Chapter 3. Hence, in the subsequent chapters following Chapter 3, the process of the gradual creation of *muqarnas* and its maturation from squinch to its contemporary state is demonstrated, which leads to introduction of the missing links between the two known structures. For that, proposing a detailed definition of the structures and their constituent elements is vital, which will in turn provide a valuable insight and knowledge necessary for differentiating *muqarnas* from other pseudo–*muqarnas* architectural forms.

Table 2.3:　*Muqarnas* related topics studied in different universities worldwide.

Name of University	Country	Topic of Research					
		History	Typology	Math	Plan	3D	Elements
Tama Art University	Japan		X		X		
MIT University	USA	X				X	
Harvard University	USA	X				X	
University of Tsukuba	Japan		X		X		
King Saud University	Saudi Arabia				X	X	
Heidelberg University	Germany			X	X	X	
King Abdulaziz University	Saudi Arabia				X	X	
University of Petra	Jordan				X	X	
Tarbiat Modares University	Iran	X	X				
University of Catania	Italy			X	X	X	
Utrecht University	Netherlands			X	X	X	
German Jordanian University	Jordan				X	X	
Calgary University	Canada			X	X	X	
American University Beirut	Lebanon			X	X	X	
Iran University of Science &Technology	Iran		X		X	X	X

CHAPTER 3

Investigation Method

3.1   Introduction

The purpose of this chapter is to explain the methodology of this research, including explaining details of data collection and the pertaining analysis techniques, which are *qualitative* in nature. The research design is qualitative as the type of data with which the research is dealing is non–numerical. In research about the evolution of *muqarnas*, visual data is the main data available for collection, although it is later processed by several stages of coding and analysing.

Despite the variety in technical, conceptual and philosophical approaches of architectural research methodology, they are more about people and objects as separate entities causing an important gap between architectural research and practice (Franz, 1994). Therefore, Grounded Theory, a well–established qualitative research method is taken as the guiding research design and adapted to fit the target of this manuscript which is a research into art history and seeking the evolution of a well–known complex architectural ornament, i.e. *muqarnas*, from a simple architectural transition technique, i.e. squinch.

As explained in Chapter 2, there are two major theories stating that the origin of *muqarnas* should be found in Iran and in squinches. Michel Écochard and Henri Stierlin are the two famous French architects and archaeologists who introduced squinch as the functional origin of *muqarnas* in Iran (Écochard, 1977; Stierlin, 1976). Écochard does not provide any evidence to show how the evolution has taken place and Stierlin posits that the elements of *muqarnas* of Mozaffarieh School, Isfahan are in fact several small squinches. Figure 3.1 shows a *muqarnas* sample in Mozaffarieh School. Starting from the abovementioned theories, and combining them with the author's in depth experience and knowledge about traditional and contemporary techniques of designing and constructing *muqarnas*, as well as gathering a reliable database of 100 *muqarnas* samples in Iran, the research is designed to prove how the gradual change from squinch to *muqarnas* has taken place.

Figure 3.1   A *muqarnas* sample in Mozaffarieh School, Isfahan (Based on (Kazempourfard, 2006)).

3.2   Grounded Theory Research Approach

Research about architecture is in fact a research in seeking new ideas and knowledge for building, which can be categorized into several groups, namely history of art and architecture, building technology, environmental behaviour study, and computational technology (Mahgoub, 2008). Compared to many professional fields of research, architectural research relates to a variety of different methodological choices situated within a full range of research approaches (Groat & Wang, 2013). Depending on the research topic and the nature of research problem, different research methods may be applied, ranging from experimental studies to theoretical approaches.

Studying a phenomenon like *muqarnas*, as an element of traditional architecture guides the research towards *art and history* discipline in the research spectrum. Hence, different qualitative research types were examined to find the most suitable research technique. As the historical aspects of the decorative elements plays an important role in the progress of the research, similar to quantitative studies, this investigation requires as much evidence as possible to provide enough data for seeking the progress of the phenomenon (Groat & Wang, 2013).

A qualitative research may be conducted in five major approaches, namely Ethnography, Grounded theory, Case studies, Phenomenological research and Narrative research (Creswell, 2013). In the grounded theory technique, the researcher is expected to induce a general theory to describe a phenomenon, a process, an action, or interaction, which involves different stages of data processing through an inductive approach (Creswell, 2013; Patton, 2001; Corbin & Strauss, 1990). The bottom–up theory development technique has two main characteristics, i.e. continuous comparison of the obtained data, as well as explaining the properties of different groups and clarifying their similarities and differences (Creswell, 2013). The primary data collection technique is either by interview or by observation, depending on the topic and the data analysis involves three main stages, namely open coding, axial coding and selective coding.

Open coding means that the collected data is reviewed in detail and probable coding concepts are extracted from it. Then, during the axial coding step, the processed data is organized based on a single variable and reviewed in order to provide suitable leading concepts for the next step, with the target of generating a reliable theory that should be general. Finally, in the selective coding stage, the data is studied based on the obtained concepts and the samples are compared with each other to seek their similarities and differences. The process of the grounded theory research technique is only finalized when saturation takes place and no new concept appears in the comparisons. In the narration process, the achieved ideas and concepts are abstracted and as there are no new concepts, the theory is validated and expressed (Patton, 2001). Figure 3.2 shows a diagram of grounded theory process and Table 3.1 provides a quick reference about the details of each step.

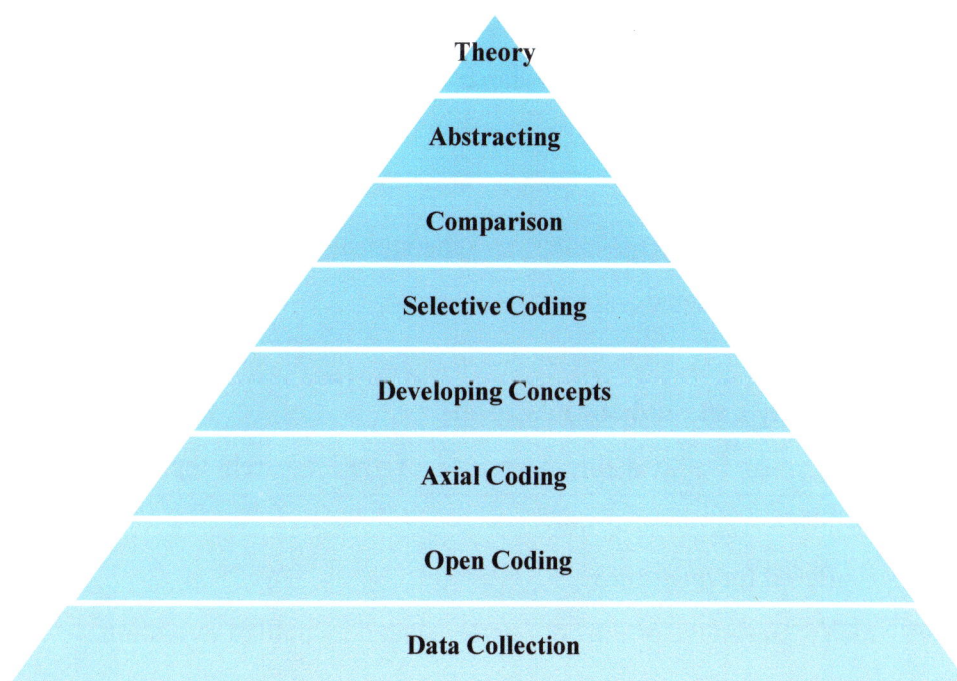

Figure 3.2   Grounded theory process (Source: adapted by author from (Creswell, 2013; Patton, 2001; Corbin & Strauss, 1990)).

Table 3.1:   Grounded theory, quick description (Source: University of Missouri).

Grounded theory		
	Purpose - theory development	
		• Used in discovering what problems exist in a social scene &how persons handle them • Involves formulation, testing, & redevelopment of propositions until a theory is developed
	Method - steps occur simultaneously; a constant comparative process	
		• Data collection - interview, observation, record review, or combination
	Analysis	
		• Concept formation • Concept development - reduction; selective sampling of literature; selective sampling of subjects; emergence of core concepts • Concept modification & integration
	Outcomes - theory supported by examples from data	

3.3   Research Design

The design process of a qualitative type of research is normally associated with four main stages, although this type of research has an active nature and the stages of research are actively involved with each other (Groat & Wang, 2013). Hence, after preliminary study, the current work was planned to be carried out in four main stages, as schematically explained in Figure 3.3 and then the achieved concepts are abstracted and narrated clearly.

3.3.1   Preliminary Study and Study of Feasibility

In the preliminary studies, the available literature of the research topic was evaluated and categorized into six different groups, namely:

(1) Definitions offered for *muqarnas*,

(2) Studies which focus on the origins of *muqarnas* and its path of evolution,

(3) Studies considering *muqarnas* in specific geographical regions,

(4) Publications which consider *muqarnas* from its historical point of view,

(5) Papers about the geometry, form and structure of *muqarnas*; and,

(6) Publications available on the topic of *muqarnas* construction techniques.

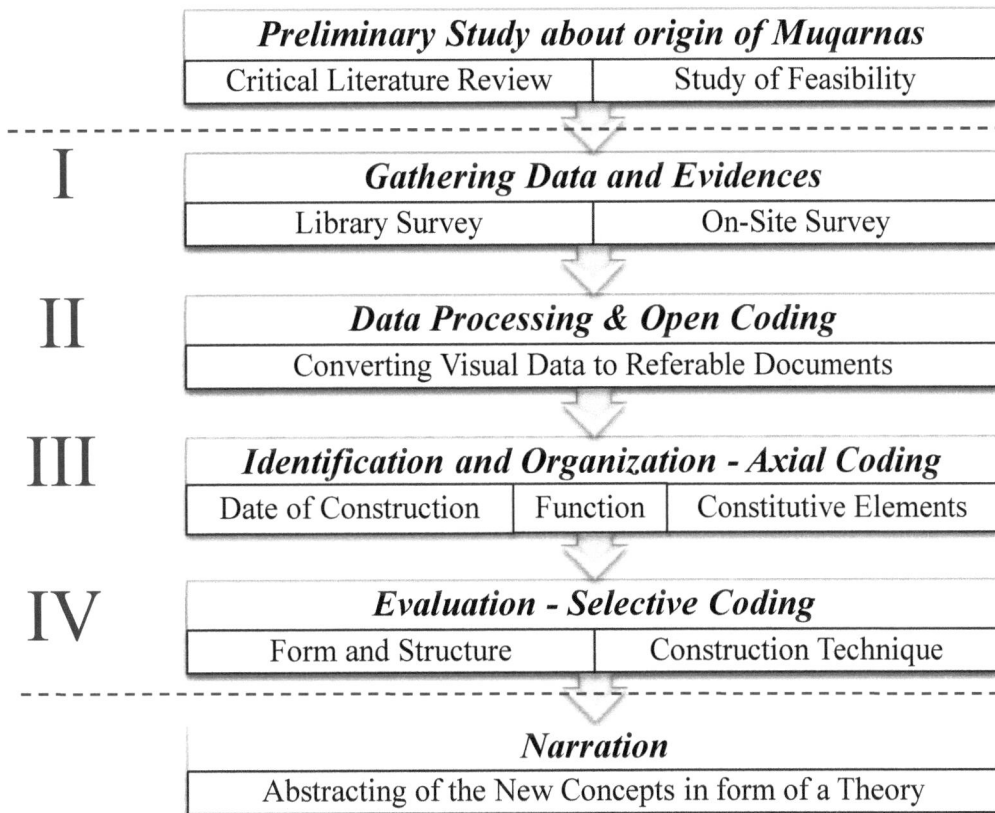

Figure 3.3   Qualitative research stages (Source: Author).

According to the gathered literature, a general image of the topic was obtained, based on which the feasibility of the research was analysed . As the historical aspects of the decorative element play an important role in the progress of the research, similar to quantitative studies, this investigation requires as much evidence as possible to provide enough data for seeking the progress of the phenomenon through passage of time (Groat & Wang, 2013).

3.3.2   Gathering of Data and Evidence

The data required to achieve the research goal was gathered in this stage by two main approaches. As the diagram of Figure 3.4 demonstrates, the first approach was on–site data gathering and recording available *muqarnas* samples based on the scope of this study in Iran, i.e. photography, measurements and recording of the dimensions of the work as well as its building and decorative material.

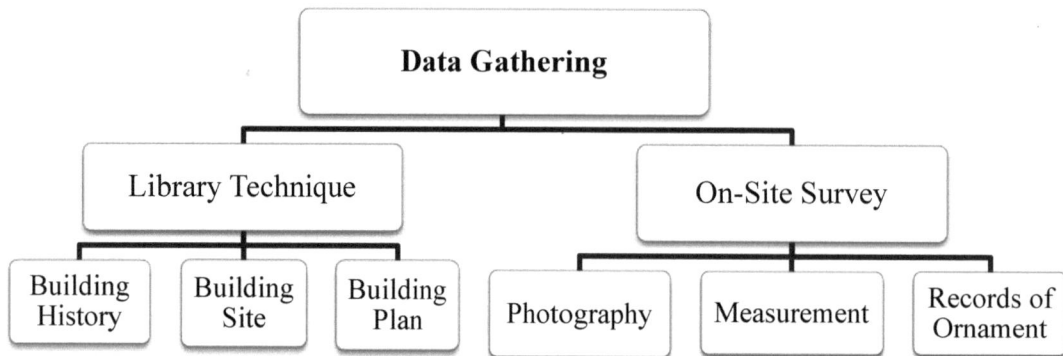

Figure 3.4   Data gathering flowchart (Source: Author).

Before continuing to the next stage, the architectural and historical details of the buildings, details about their construction techniques, and other related information about the recorded *muqarnas* samples were collected from available resources. These details also include the building date and details of its renovations, building location on city map, as well as the location of the studied *muqarnas* on the plan of the building. As almost all of the visual data gathered for this research is recorded from on–site measurements and is not obtained from external sources, it is categorized as an *interactive* analytical research (Strand & Weiss, 2005).

3.3.3   Data Processing and Open Coding

The gathered visual data are required to be modified in a way that they could be used for the purpose of qualitative analysis and inductive comparison by coding and developing new concepts (Emisson & Smith, 2000). Hence, as illustrated in Figure 3.5, data processing is carried out with the purpose of finding suitable variables and concepts for open coding of the gathered data. The two–dimensional plan of each *muqarnas* was created using the appropriate photo from under the *muqarnas* and then, with reference to the pertaining measurements the three–dimensional pattern of the *muqarnas* was developed. The front and rear views of the structure were then saved as a picture file and compared with the identical photo of the same view. In the next phase, the *muqarnas* components and constituent elements were taken out of the three–dimensional pattern and their details such as their function in the building and the number of tiers of each ornament were recorded for further investigations.

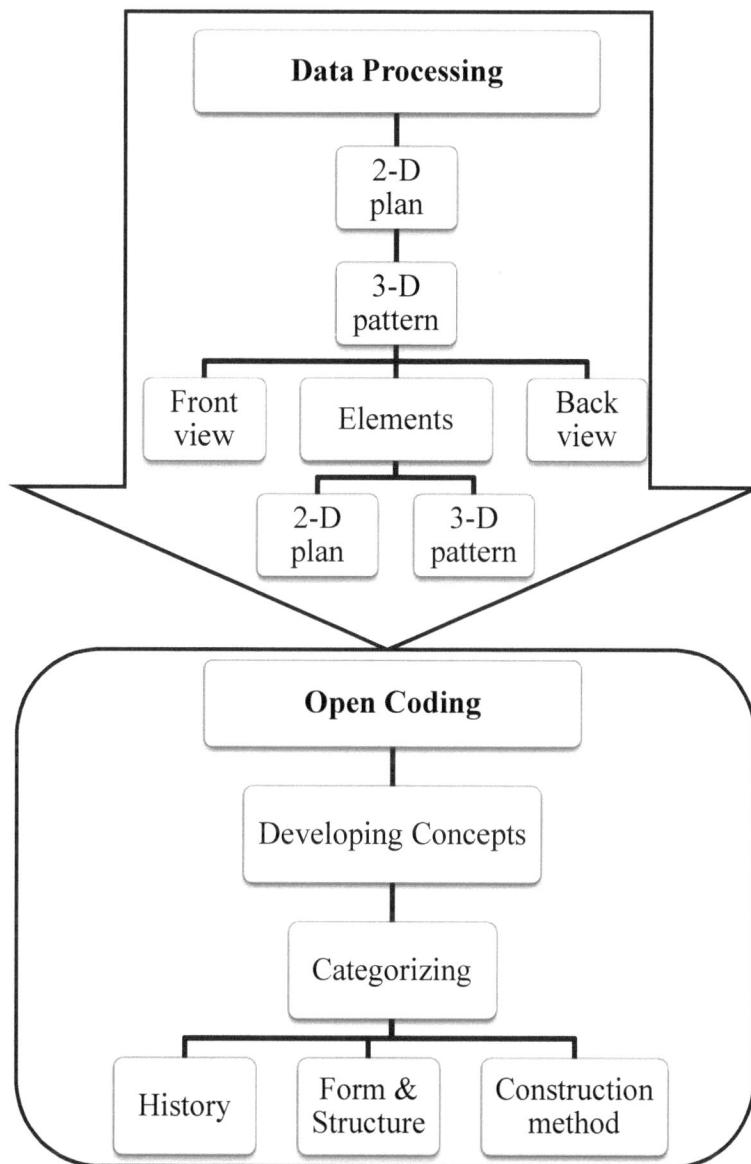

Figure 3.5   Data processing flowchart (Source: Author).

The software packages used for the purpose of data processing were mainly 3DMax, Adobe Photoshop and AutoCAD. Based on the historical data of the decorative structures and its constituent elements, three main concepts were defined to conduct the inductive study of the samples, namely the date of construction, form and structure, and construction method. The selected samples will finally go through comparison investigations which are based on these three main categories. The mentioned categories will then be divided into smaller groups to form the final selective coding. This part of investigation will be conducted by matrix analysis approach which will be explained later in this chapter.

3.3.4 Axial Coding

With attention to the goal of the current research, i.e. seeking the origin of *muqarnas* among its primitive samples and finding its path of progress and evolution from three aspects, namely form and structure, construction techniques and chronology and history, an inductive analysis was opted. Mainly by means of the qualitative grounded theory approach, the research is expected to achieve its goals through several stages of coding, resulting in definition of new concepts which are general and which fit to all existing phenomena.

Although initial measurements and recordings of data were accomplished from 2005 to 2007, further data collection was in progress actively until 2011, based on the requirements of the research. During the mentioned years, a set of buildings and their *muqarnas* works were chosen as valuable samples for the purpose of this investigation by the author and Dr. Memarian, the co–supervisor of the study. The chosen *muqarnas* structures, scattered in more than 25 different cities and villages in Iran, were photographed, and measured accurately and the architectural details and historical contexts of them were recorded for further reference. Finally, all the collected measurements and photos were converted to two–dimensional plans and three–dimensional figures and a comprehensive database was created. Table 3.2 provides the list of the 100 recorded samples, their dates of construction and cities of origin. The list is arranged chronologically and the samples highlighted in yellow are the 20 selected samples.

3.3.4.1 Justification of the Selected Samples

The next important step in axial coding is to achieve another set of criteria to narrow up the samples towards finding and reviewing the most remarkable samples, based on research objectives. For that, the following selection criteria was developed based on which 20 more noteworthy samples were opted from 100 measured and recorded ones in database. The selection was conducted based on the following concepts:

Table 3.2: List of samples, arranged axially based on their date of construction.

#	Name	Date	Page	#	Name	Date	Page
1	Firuzabad Palace	224	256	51	Mozaffarieh-Mihrab	1366	294
2	Sarvestan Palace	420	125	52	Darb Imam-Portal	1453	296
3	Samanid Mausoleum	914	129	53	Darb Imam-Iwan	1453	296
4	Jurjir Mosque	939	132	54	Jame Yazd-Portal	1470	290
5	Na'in Mosq–Mihrab	960	137	55	Jame Isfahan-Minaret	1480	268
6	Na'in Mosq–Iwan	960	139	56	Jame Isfahan-W portal	1501	269
7	Gonbad Qabus	1006	258	57	Jame Isfahan-E portal	1501	270
8	Davazdah Imam	1036	144	58	Jame Isfahan-Majlesi	1501	269
9	Na'in Mosq.–Portal	1062	260	59	Jame Isfahan-Hakim	1501	270
10	Varzaneh Mosque	1100	262	60	Qazvin Jame-Iwan	1501-99	298
11	Barsian–Squinch	1114	264	61	Shazdeh Hosein Portal	1568-99	300
12	Barsian–Mihrab	1114	148	62	Ali Qapu, Iwan	1590-91	302
13	Jame Isfahan-Mihrab	1118-57	267	63	George Church	1611	303
14	Jame Isfahan-Nave	1118-57	160	64	Sheikh Lotf Allah	1617	305
15	Jame Isfahan-Darvish	1118-57	267	65	Khan School-Portal	1636	307
16	Jame Isfahan-Saheb	1118-57	268	66	Agha Noor Mosque	1639	308
17	Jame Isfahan-Ostad	1118-57	162	67	Hakim Mosq-Portal1	1656-62	310
18	Bastam Minaret	1120	152	68	Hakim Mosq-Portal2	1656-62	312
19	Sin Jame' Mosque	1134	155	69	Hakim Mosq-Mihrab1	1656-62	311
20	Bayazid Mausoleum	1299	170	70	Hakim Mosq-Mihrab2	1656-62	311
21	Pir Bakran Mausoleum	1303-12	177	71	Hakim Mosq-Mihrab3	1656-62	312
22	Abd Al-Samad –Iwan1	1304-25	273	72	Hasht Behesht-Vault	1670	314
23	Abd Al-Samad –Iwan2	1304-25	272	73	Hasht Behesht-Pavilion	1670	314
24	Abd Al-Samad-Minar	1304-25	272	74	Molla Abdollah-Portal	1678	316
25	Jame' of Natanz-Dome	1304-25	186	75	Nimavar School-Portal	1691	318
26	Jame' of Natanz-Iwan	1304-25	190	76	Nimavar School-Iwan1	1691	319
27	Abd Al-Samad-Portal	1304-25	273	77	Nimavar School-Iwan2	1691	319
28	Soltanieh-Iwan	1312	182	78	Char Bagh School	1706-14	321
29	Jame Ashtarjan-Dome	1315	275	79	Ali Gholi Agha	1710	323
30	Jame Ashtarjan-Portal	1316	196	80	Bagh Nazar Garden-1	1749	325
31	Jame Varamin-Dome	1322	277	81	Bagh Nazar Garden-2	1749	326
32	Jame Varamin-Portal	1322	200	82	Bagh Nazar Garden-3	1749	326
33	Imamieh School-Iwan1	1325	279	83	Bagh Nazar Garden-4	1749	327
34	Imamieh School-Iwan2	1325	280	84	Bagh Nazar Garden-5	1749	327
35	Imamieh School-Iwan3	1325	280	85	Nabi Mosque-Portal	1787	328
36	Imamieh School-Iwan	1325	205	86	Nabi Mosque-Mihrab1	1787	329
37	Sheikh Safi	1334	282	87	Nabi Mosque-Mihrab2	1787	329
38	Jame Eziran	1335	284	88	Jame Shiraz-Portal1	1793	331
39	Jame Kerman-Portal	1349	209	89	Jame Shiraz-Portal2	1793	332
40	Jame Qom- Portal 1	1350	286	90	Sepahsalar Mosque-1	1804	334
41	Jame Qom- Portal 2	1350	286	91	Sepahsalar Mosque-2	1804	334
42	Jame Qom-Portal3	1350	287	92	Imam Mosque	1810-25	336
43	Jame Qom-Iwan	1350	287	93	Seyed Mosque-Portal1	1815	338
44	Jame Qom-Dome1	1350	288	94	Seyed Mosque-Portal2	1815	339
45	Jame Qom-Dome2	1350	288	95	Seyed Mosq.-Mihrab	1815	339
46	Jame Yazd-Dome	1365	291	96	Molla Hassan Shrin	1821	341
47	Jame Yazd-Mihrab	1365	291	97	Agha Bozorg-Iwan	1832	343
48	Mozaffarieh-Portal	1366	293	98	Agha Bozorg-Minaret	1832	343
49	Mozaffarieh-Iwan1	1366	293	99	Nasir Al-Mulk-Portal	1876	345
50	Mozaffarieh-Iwan2	1366	294	100	Imam Mosque	1931	347

Selected samples

(a) As the chronological path of evolution of *muqarnas* structure was one of the key factors considered in this research, several samples were chosen from each historical period in *muqarnas* evolution timeline and from the main stages of the structure's development. It is worth considering that many of the studied buildings were renovated or even reconstructed several years after their date of construction. Hence, accurate investigation about the date of the specific part of the building under study was vital. Therefore, efforts have been made to find reliable dates for each building and the specific date of construction of that particular decoration. In traditional architecture of Iran, the date of erecting each specific part of the building and the mason's name used to be written in some form on an inscription band or among the patterns of the decorative veneer. Furthermore, dates provided and published by archaeologists and those involved in restoration of the buildings are another reliable source of seeking each sample's date of construction.

(b) The next criteria for choosing these samples were the importance of not missing any of the varieties of forms and structures of *muqarnas* during the studied eras.

(c) There have been various method for building *muqarnas* throughout its history in Iran, the construction technique of the *muqarnas* was also an important criteria in choosing the final *muqarnas* samples.

(d) Another very important criterion for choosing the samples was the first time that a constituent element of *muqarnas* was invented and used in a structure; and finally,

(e) The function of the structure in the building was also a factor that was considered for choosing the appropriate examples from the collected database.

3.3.5 Selective Coding

After identifying and organizing sources, based on fact–gathering, note–taking, modelling and observation, in terms of measurement and photographical records, the obtained results were evaluated and a set of selection criteria was defined. Evaluation of the data was accomplished through analysis and assessment of the organized sources, preparing descriptions and verifying them by inductive approaches (Groat & Wang, 2013).

Table 3.3 provides the names of the selected structures and introduces briefly the reasons behind their significance. Each column is related to a criterion as explained above and the short explanations after them are the evidence of the claim. Furthermore, in order to demonstrate the chronological dispersion of the samples within the scope of this study, Figure 3.6 is designed to illustrate the evolution time–line of *muqarnas*.

Although twenty samples were only selected for more detailed investigations and more in–depth studies, the existence of all of one hundred recorded *muqarnas* works was a necessity to provide a comprehensive database for choosing the best samples from them.

The complete list of selected *muqarnas* samples is introduced in Table 3.4, where the names of the buildings and cities to which they belong have been also mentioned. The date of construction and the position of the *muqarnas* in the building are the other details provided in this table. Furthermore, Figure 3.7 shows the cities of the selected *muqarnas* samples on the map of Iran.

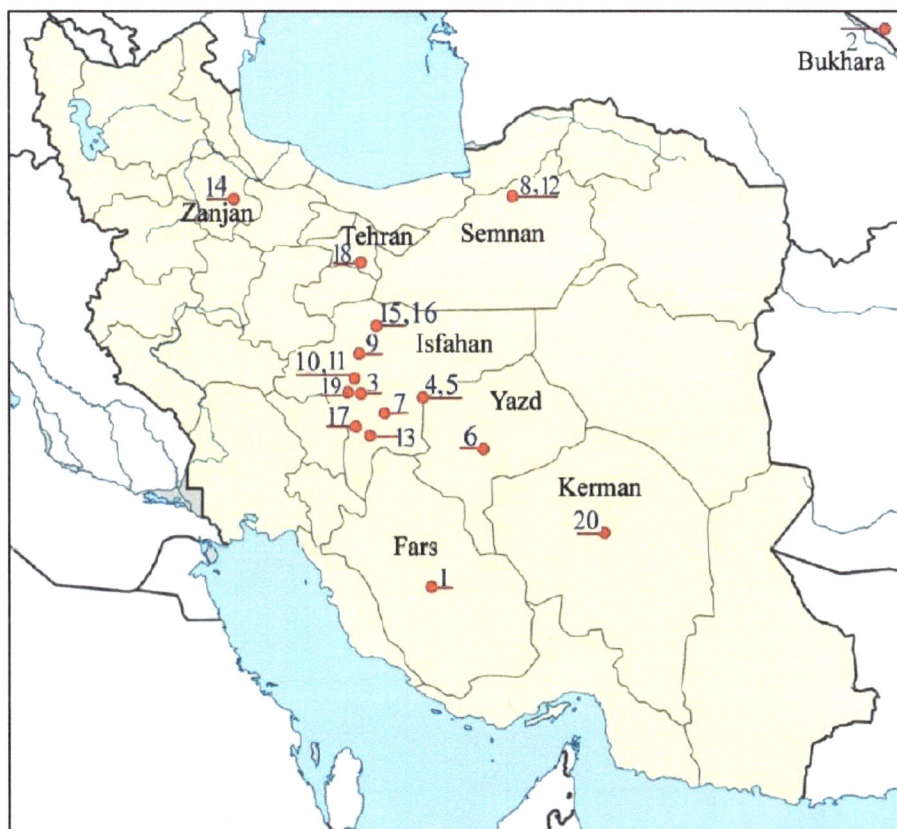

Figure 3.6   Location of the selected twenty *muqarnas* samples on map of Iran (Source: Author)

Table3.3: The significant samples selected based on axial coding criteria.

#	Sample Name	Chronolo	Structure	Construct	Constitue	Function	Details
1	Sarvestan Palace						The only pre-Islamic sample; The only amalgamated construction method
2	Samanid Mausoleum						The first Islamic sample; The first use of rib in construction
3	Jurjir Mosque						The first example of squinch being used on portal instead of dome chamber.
4	Nain Mosque, Mihrab						Creation of plumb *Shaparak* element; The first example of whole vault veneer
5	Nain Mosque, Iwan						Creation of *Shamseh*, the medallion; First example of whole iwan vault
6	Davazdah Imam						The first ever double-tier squinch and transition technique
7	Barsian Mosque						First Seljuk sample; The first three-tier sample; Creation of half-*Shamseh* for the first time; First use in mihrab
8	Bastam Minaret						The first pole table plan; The only sample build by suspended units; The first function in minaret
9	Sin Mosque						Innovative combination of squinch with radial ornament; First mixed plan type; First whole dome
10	Isfahan Jame', Nave						The first time function as a skylight vault inside *shamseh*
11	Isfahan Jame', Iwan						Creation of basic elements *Tee*, 3D-*Shaparak* & *Espar*; Creation of elemental unit *Tanureh*
12	Bayazid Mausoleum						The first seven-tier sample; First sample built by suspended layers; Creation of elemental unit multiple-*shaparak*
13	Pir Bakran Shrine						First Ilkhanid sample; Fragments of similar ornament, showing construction method
14	Soltanieh Dome						First nine-tier sample with 110 *taselts*, in a world famous building, having the largest brick dome
15	Jame of Natanz, Dome						Sample's plan is different from dome base; First 10-tier sample; Creation of elemental unit double-*Parak*
16	Jame of Natanz, Iwan						Uncommon elements used for the first and last time in this sample; Inception of *takht* element
17	Ashtarjan Mosque						Creation of *takht* element for the first time: 4-pointed star; First *muqarnas*
18	Jame' of Varamin						Creation of *takht* element: rhombus; Uncommon aspect ratio: one by four
19	Imamieh School						First pole table plan type in rectangle base
20	Jame' of Kerman						Most complete *muqarnas* with all constituent elements; Creation of *takht* element: 8pointed star; Last Ilkhanid sample

83

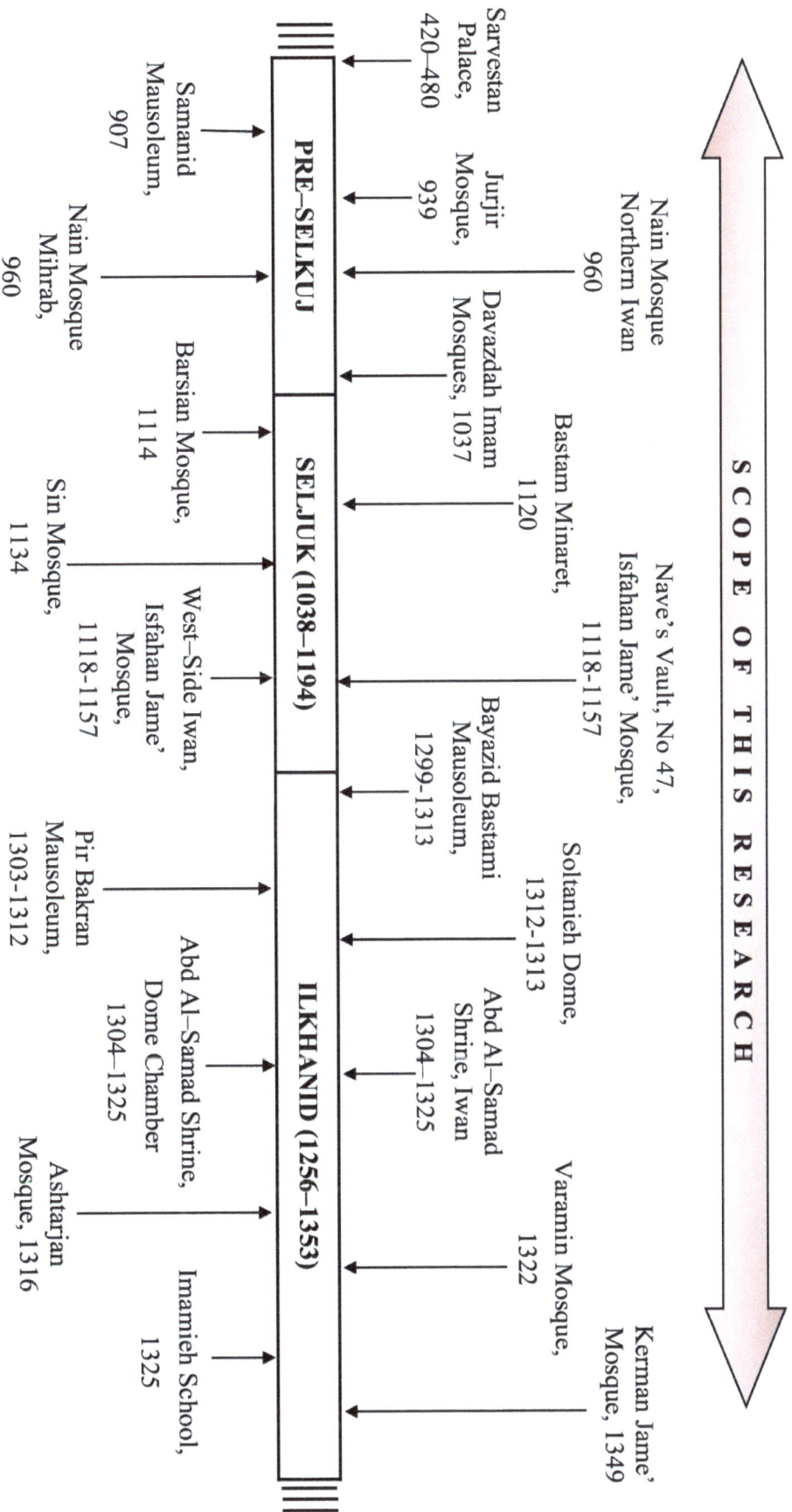

Figure 3.7 *Muqarnas evolution time–line, the scope of this research (Source: Author).*

Table 3.4: The list of the selected *muqarnas* samples and their details.

No	Name of Building	Date	City, State	Location of *Muqarnas*
1	Sarvestan Palace	420–480	Sarvestan, Fars	Corner
2	Samanid Mausoleum	907	Bukhara, Uzbekistan	Corner
3	Jurjir Mosque	939	Isfahan, Isfahan	Entrance Portal
4	Nain Mosque, Mihrab	960	Nain, Isfahan	Mihrab
5	Nain Mosque, Iwan	960	Nain, Isfahan	Iwan
6	Davazdah Imam Mosq.	1037	Yazd, Yazd	Corner
7	Barsian Mosque	1114	Barsian, Isfahan	Mihrab
8	Bastam Minaret	1120	Bastam, Semnan	Minaret
9	Sin Mosque	1134	Sin, Isfahan	Dome Chamber
10	Isfahan Jame', Nave	1118-57	Isfahan, Isfahan	Nave, Skylight
11	Isfahan Jame', Iwan	1118-57	Isfahan, Isfahan	Iwan
12	Bayazid Mausoleum	1299-1313	Bastam, Semnan	Entrance
13	Pir Bakran Mausoleum	1303-1312	Lanjin, Isfahan	Iwan
14	Soltanieh Dome	1313	Soltanieh, Zanjan	Iwan
15	Jame' of Natanz, Dome	1304–1325	Natanz, Isfahan	Dome Chamber
16	Jame' of Natanz, Iwan	1304–1325	Natanz, Isfahan	Iwan
17	Ashtarjan Mosque	1316	Ashtarjan, Isfahan	Entrance
18	Varamin Jame' Mosque	1322	Varamin, Tehran	Entrance
19	Imamieh School	1325	Isfahan, Isfahan	Iwan Vault
20	Kerman Jame' Mosque	1349	Kerman, Kerman	Mihrab

3.3.5.1 Methods of Analyzing Visual Data

A comprehensive and worthy database of 100 *muqarnas* samples was created by collecting, measuring and recording available samples in Iran. Besides, the two–and three–dimensional patterns of the recorded samples were also produced and the historical contexts of them were all recorded to be analysed and evaluated systematically. The process of analysis includes identification, categorizing and grouping the characteristics of recorded samples, in which the constituent elements of each structure played an important role.

One of the main focuses of this research is recognizing the forms and details of the physical characteristics of the constituent elements of *muqarnas*, in a way that a reliable and general set of components could be defined. Therefore, in the process of visual analysis of data, based on the abovementioned investigations, a chapter on defining the constituent elements of *muqarnas* was written, as a prerequisite of stepping forward to the final coding and analysis of the studied samples. The final set of general constituent elements of *muqarnas* includes *shamseh*, *toranj*, *taseh*, *parak*, *shaparak*, *takht*, *Tee*, and *espar*, which are going to be further explained in Chapter 4.

The characteristics of different types of elements were studied after being recognized and the extensible *muqarnas* elements were identified. Based on the visual and physical properties of these elements, a set of variables were defined. The arrangement of these elements with respect to each other and some limits regarding the quality of these arrangements were also interpreted as a part of this investigation.

3.3.5.2 Validation Process

As mentioned before, the objectives of this research should be accomplished in a sequential order. Therefore, after achieving the first objective, i.e. collection of 100 *muqarnas* and pseudo-*muqarnas* samples, in Iran, open coding was accomplished with the purpose of finding an extensible set of constituent elements that is capable of covering all elements of all *muqarnas* structures of all different historical periods in Iran.

Hence, before proceeding to the next objective, it was necessary to validate the result of the second objective, by examining the generality of the introduced constituent elements. Therefore, referring back to the database of 100 samples, one of the most complex *muqarnas* structures was chosen to control if the introduced set could cover all the components of it. Stepping forward toward the third objective, which was about characterizing the specifications of *muqarnas* and pseudo-*muqarnas* structures, was only possible after proof of validity of the extensiveness of the introduced set of constituent elements. The last objective, which is to illustrate the gradual evolution of *muqarnas* from squinch, is in fact a systematic interpretation and abstracting of the results of the second and third objectives.

3.3.5.3 Matrix Analysis Approach

In order to achieve a systematic inductive analysis, matrix analysis approach was applied in this research. For instance, in order to study the constituent elements of *muqarnas* samples, as a representative of the complexity of the studied *muqarnas*, a matrix that represents the variety of different types of elements and distributions of each element in each tier of a *muqarnas* work was produced.

In addition, for the qualitative study of the evolution of the *muqarnas* samples, several characteristic matrices were developed based on the main concept categories achieved in open coding stage. The categories are then divided into several groups and selective coding is conducted using the obtained matrices. Some of the finer variables used for the purpose of matrix analysis are as follows:

(a) The location of *muqarnas* in the building, or in other word, its application, i.e. mihrab, dome chamber, portal, vault, etc.,

(b) The material of the decorative covering of *muqarnas* or the finishing material and comparing it with the building's construction material,

(c) The aim of the construction of *muqarnas*, e.g. height–, size– and form–transition,

(d) The structure's number of tiers,

(e) The structure's plan type; as well as,

(f) Applied construction technique, which implies either the ornament has load–bearing role or it is merely decorative.

Using these matrices will help significantly in achieving an in–depth understanding from the application of different element types, in different *muqarnas* samples of different buildings. The obtained insight will subsequently lead towards a more practical categorization of *muqarnas* and in analyzing it more efficiently.

3.4 Research Ethics

The raw materials of the current research, i.e. the gathered and processed visual data, are of interactive type in nature. Using these gathered information requires considering the following etiquettes:

(1) Under the complete supervision of author, part of the gathered data, as well as part of the processing phase was accomplished by 74 architecture students of the author and Dr. Memarian in Iran University of Science and Technology, and Shahid Rajaee University of Tehran.

(2) Defining the variables and categorizing *muqarnas* works into different groups is a result of the author's insight gained from a knowledge that he obtained during his life from his

family and his masters, like maestro Sha'rbaf and Dr. Memarian, who are all professionally involved in traditional architecture of Iran.

(3) There is hardly any scientific reference available studying *patkaneh*, as the topic is newly introduced to the body of knowledge. Therefore, in order to find proper literature on *patkaneh*, the pertaining section in the manuscript of the author's colleague, Mr. Hadi Safaeipour, was a great help. Furthermore, Dr Memarian's book on Iranian traditional architecture was another main source for the topic, although as described in Chapter 2, some indirect mentions are made by different scholars about structures that are in fact *patkaneh*.

3.5 Conclusion

The main goal of this research is identifying and analyzing the characteristics of *muqarnas* samples in Iran, during Seljuk and Ilkhanid historical periods, with the purpose of seeking the structure's path of evolution.

This research looks for the practical origins of *muqarnas* and tries to find the structure's path of development by looking both at the details of the structures, i.e. their constituent elements, as well as their constructional characteristics. These characteristics are obtained by analyzing three main variables, namely, form and structure, construction technique and chronological order and the research design shows that the gathered data is obtained from two main methods, i.e. on–site measurement and library studies.

One of the main merits of the current research, compared to other available studies, is the comprehensive database created for the purpose of this study. This database contains 100 *muqarnas* samples from all over Iran, from which 20 samples were selected after three steps of open coding, axial coding and selective coding of the gathered data. The gradual evolution of *muqarnas* which is expected to be originated from squinches is illustrated as the outcome of this manuscript. Furthermore, a detail analysis of the constituent elements of *muqarnas*, leading to contributing and introducing an extensible set of elements, is another important achievement of this study, which is presented as a separate chapter.

CHAPTER 4

Constituent Elements of Muqarnas

4.1 Introduction

Geometrical analysis of *muqarnas* is an inevitable step in the process of understanding the characteristics of the structure. For that, one must segregate various components of the structures that are the tiers, and the constituent elements and identify the plan type. Here, a different perspective is needed on the subject, where focus is taken away from the whole and is attracted into the details. Each geometrical analysis of *muqarnas* somehow falls under the category of volumetric, arising from its three dimensional nature. At each scale from the elemental components on to the tiers and then the entire composition, *muqarnas* always deals with the three parameters of width, height, and depth; and a mere illustration of it in plan, or elevation, or both, proves inefficient in conveying it's three dimensional nature.

Hence, in this chapter, as a prerequisite chapter before being able to discuss the characteristics of the selected samples, after a detailed review of the shapes of the components of different *muqarnas* and pseudo-*muqarnas* structures, and referring to the available categorization of *muqarnas* components in literature, as described in Chapter 2, eight main elements were selected and introduced as the global constituent elements of *muqarnas*. These elements are global, meaning that there are no other elements, except them, and any elemental unit in any *muqarnas* structure can be defined as being either one of this global element or a combination of them.

Furthermore, with the purpose of categorizing the *muqarnas* samples, Takahashi used two specific terms, namely square lattice and pole table as plan types. However, no reliable definition was presented, except notation. Hence, in order to use the categorization variables of Takahashi, a clear definition of the mentioned plan types should also be introduced before relying on the categorizing technique throughout the manuscript. As another necessary step, explanations are also provided in this chapter about the most common method of constructing *muqarnas*, i.e. suspended layer technique.

4.2   Analysis of Different Types

As mentioned before, one of the practical methods of categorizing *muqarnas* is the one used by Shiro Takahashi (Takahashi, 1973). He introduces three main groups, i.e. square lattice, pole table and other, based on the method of drawing the plan. Although he uses the mentioned plan types to categorize all 1000 *muqarnas* plans that he collected, no definition is provided by him about the meaning of these plan types.

Hoeven *et al.* tries to interpret the meaning behind these names by defining the square lattice type as a *muqarnas* plan that consists of squares and 45° rhombuses, with a four-fold symmetry (Hoeven & Veen, 2010). In addition, referring to Takahashi's database of *muqarnas* plans, it is concluded that square lattice *muqarnas* structures were developed in the 11th century and they are equivalents of curved *muqarnas* types, introduced by Al-Kashi (Krömker, 2007; Post et al., 2009). The pole table *muqarnas* is considered to be developed in the 17th century (Krömker, 2007). It has a 4–, 7- to 11–pointed medallion on the apex and it is an independent decorative structure which should be attached to the building (Hoeven & Veen, 2010). Although the definitions are helpful, they are not clear enough to be used in predicting plan types of various *muqarnas* and pseudo-*muqarnas* structures. Hence, referring to 1000 *muqarnas* database, following definitions are presented for the abovementioned plan types.

4.2.1   Square Lattice

The two-dimensional plan of a square lattice plan type can be drawn mainly on a checker lattice of square grids, drawn by repeating equally-spaced parallel and perpendicular lines. The reference lines of plan are either the grid lines or the diagonal line of each square. This rule is general, though some exceptional angles other than 45° and 90° may also be observed. In addition, a square lattice plan may also be created in a lattice of co-centric rotating square grids, like the skylight vault of the nave in Isfahan Jame' mosque, which will be studied in the next chapter.

4.2.2   Pole Table

The reference lines in a pole table plan are beams of ray distributed from one or several origins on the plan. Existence of sunburst medallion or *shamseh* is inevitable in this plan type, having the main origin of beams on its centre point. The other origins, if any, are normally located on the corners of the sample plan. Although *shamseh* is observed in a many square lattice plan type too, any plan with a *shamseh* whose number of star point is not an integer multiple of four, is without doubt a pole table plan type. Examples include Khan School in Shiraz, which may be found in Appendix A. The first pole table plan type was observed in Bastam Minaret, built in 1120, i.e. in early 12th century. There are many examples in which a mixture of both plan types can be identified. These plan types are categorized as mixed type. Figure 4.1 provides an example for each plan type:

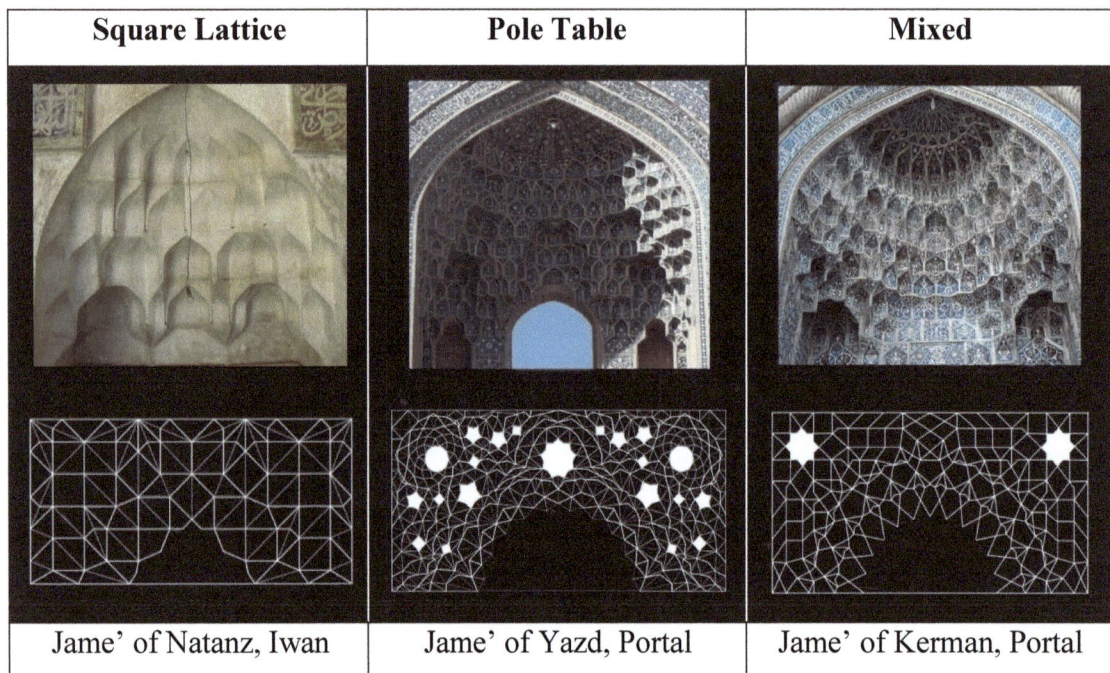

Square Lattice	Pole Table	Mixed
Jame' of Natanz, Iwan	Jame' of Yazd, Portal	Jame' of Kerman, Portal

Figure 4.1   Examples of different plan types (Based on (Kazempourfard, 2006)).

4.3   Constituent Elements

As the next necessary step before proceeding to inductive analysis of the characteristics of the selected samples, constituent elements of *muqarnas* should have been introduced in a way that they could be used to define all emerging components of different *muqarnas* samples, from different periods in Iran. Therefore, a quick look is taken to the methods of defining categorizing variables by other scholars and then referring to the database of 100 samples from different historical periods and different cities of Iran, the set of eight basic elements is defined as follows.

4.3.1   Categorizing Variables

In order to categorize the constituent elements of *muqarnas*, as explained in Chapter 2, various approaches had been taken by different scholars, mostly looking at the plan of the ornament. However, the three-dimensional arrangement of *muqarnas* is not something that could be comprehended by looking merely at its projection plan. Among the available categorizations, therefore, reference will be given to those of Yaghan and Al-Kashi, since they offer practical variables for this categorization. Yaghan describes that the three-dimensional units of *muqarnas*, which may appear as line or compartment in the plan, are either flat or with curvature. If the flat units are horizontal, he calls them "roof patches", otherwise, the units will have an edge and a top. Hence, he divides the rest of the units into the following three groups:

(a) Edge to Point (ETP) element are those starting from an edge-base and finishing with a point on top,

(b) Point to Edge (PTE) elements are starting from a point and finishing with an edge on top; and finally

(c) Edge to Edge (ETE) elements are surfaces starting and finishing with edges (Yaghan, 2001b).

This method of categorizing is very much helpful in understanding the tier by tier, bottom-up expansion of *muqarnas*. As explained in Chapter 2, *muqarnas* and pseudo-*muqarnas* structures are considered to have a bottom-up nature. In other words, as these structures used to be constructed from bottom to top, their tiers are always counted same way by masons, having *shamseh* on the last tier. Hence, as Yaghan's ETP, PTE and ETE categorizing variables meet the terms of the same bottom-up rule, they are considered as a practical method of categorizing element for this manuscript.

Al-Kashi, on the other hand, used the terms *basic* and *intermediate* for categorizing the constituent elements of *muqarnas* at his time (Al-Kashi, 1427), where *taseh* is a basic element for instance and *shaparak* an intermediate one. This method of categorizing is also referred in some cases in this manuscript. Aside from the above mentioned categories, as the evolution of *muqarnas* is under consideration of this study, the date of invention of each element is another key factor that should be taken into account. Based on mentioned variables, after detailed review of 100 samples gathered for the purpose of this research, a general and minimized set of constituent elements of *muqarnas* is introduced below.

4.3.2   Extensible Set of Constituent Elements

The available methods of describing the constituent element of *muqarnas* and pseudo-*muqarnas* samples by Iranian scholars, varies depending on the region and time. As mentioned before, Lorzadeh presents the most complete set with 11 members and Sha'rbaf add another elements to the collection, introducing a sum of 12 constituent elements for *muqarnas*, namely *Shaparak, Takht, Irregular Takht, Shamseh, Toranj, Taseh, Tee, Madani* or *Tanoureh, Double–Madani, Lozi* or *Darvazeh, pabarik*, and finally, *Susani* (Sha'rbaf, 1996; Lorzadeh, 1981). However, many of the proposed elements are in fact a combination of some others and in many cases, there is not a clear definition for the introduced constituent elements. Hence, the available collection of introduced constituent elements was abstracted to the following eight that are introduced as the general set of element. These elements are introduced here based on their date of creation and later, the famous element units are also explained and shown in detail.

4.3.2.1   *Taseh*

Taseh is the most extensively used basic element of all *muqarnas* and pseudo-*muqarnas* structures. The symmetric ETP component was first introduced as a stand-alone single squinch in ancient transition zones before Islam. Based on the number of segment, a *taseh* can have as low as one to maximum eight segments. A single-segment *taseh* may be two- or three-dimensional, whereas all other number of segment create three-dimensional surfaces. *Taseh*, which literary means scooper in Persian, can be imagined as a small single squinch or as two crossing concave vault-sections standing on an edge. Figure 4.2 shows the architectural details of *taseh* and figure 4.3 shows some two-and three-dimensional examples of the element.

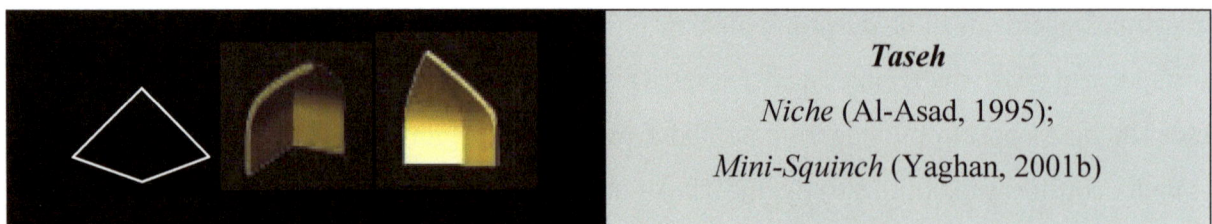

	Taseh
	Niche (Al-Asad, 1995);
	Mini-Squinch (Yaghan, 2001b)

Figure 4.2   Details of *muqarnas* elements; *Taseh* (Source: Author).

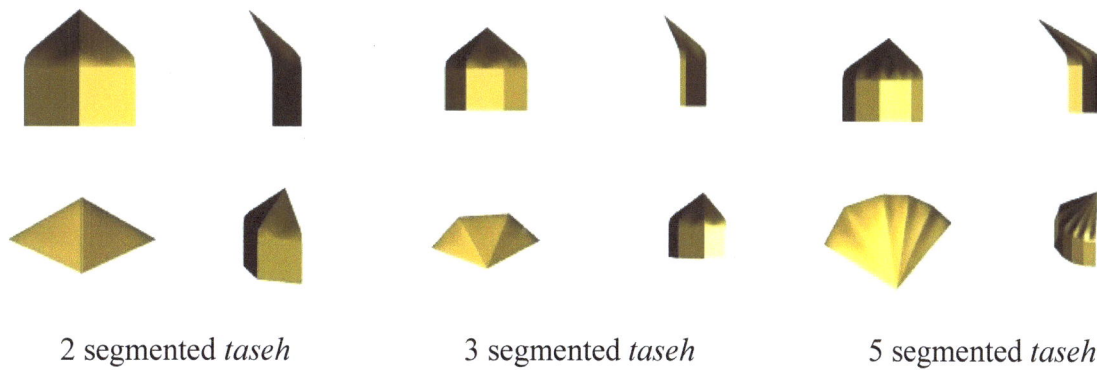

| 2 segmented *taseh* | 3 segmented *taseh* | 5 segmented *taseh* |

Figure 4.3   Examples of different *tasehs* (Source: Author).

4.3.2.2 *Shaparak*

The next most extensively used element and the second created component of *muqarnas*, is *shaparak*. Invented in the 10th century, this PTE element is a flexible intermediate element capable of covering various angles efficiently. Figure 4.4 illustrates the architectural details of *shaparak* and its variant names.

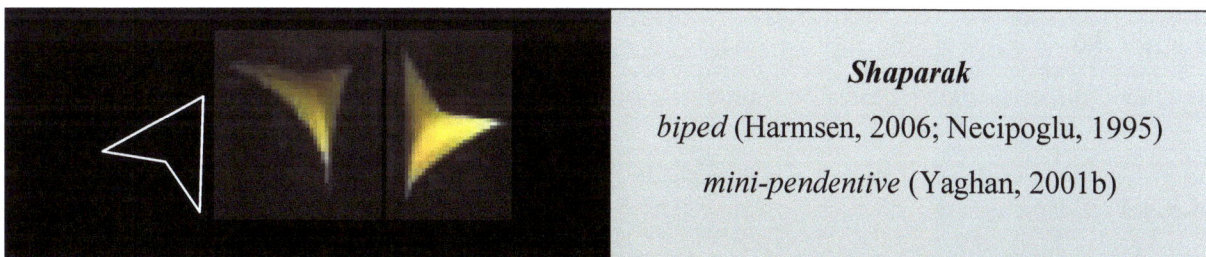

Figure 4.4   Details of *muqarnas* elements; *Shaparak* (Source: Author).

Shaparak literary means butterfly, where depending on its situation with respect to its neighbours, the wings of this butterfly may be wide open or almost closed. This element has four sides from which, as illustrates in Figure 4.5, the two top one are always equal, but the other two may have different lengths. Being made up of two adjacent triangles intersecting in one edge, *shaparak* can be observed in plumb or three-dimensional forms. Figure 4.6 shows the plan of two plumb and three various three-dimensional *shaparaks*.

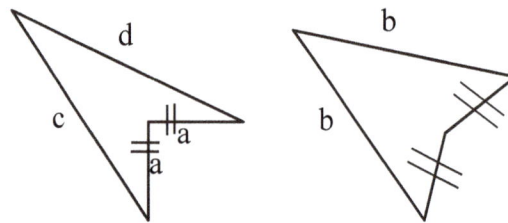

Figure 4.5   Different *shaparaks* (Source: Author).

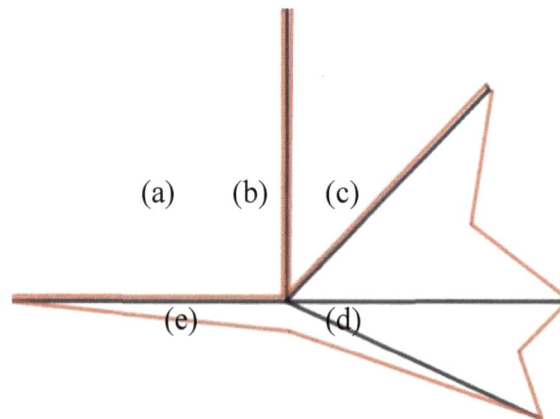

Figure 4.6   Plan of five variously angled *shaparaks*; (a) and (b) plumb *shaparaks*; (c), (d), and (e) three-dimensional *shaparaks* (Source: Author).

4.3.2.3 *Parak*

Created at the same time with *shaparak*, *parak* element is the next most extensively used component of *muqarnas*. Similar to *shaparak*, *parak* which is a PTE triangle was first created in plumb form as an intermediate element and then in 12th century, it appeared in three-dimensional form. Other than some exceptional examples, *paraks* are most of the times symmetric. Figure 4.7 shows the architectural details of *parak*.

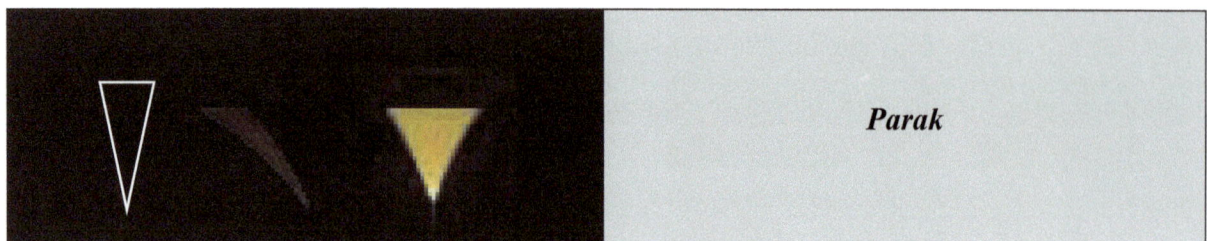

Figure 4.7   Details of *muqarnas* elements; *Parak* (Source: Author).

4.3.2.4 *Shamseh*

Shamseh is always seen on the apex of the ornament in the last tier. Literary meaning sun, *shamseh* is a several-pointed three-dimensional star with a radial symmetry, which may appear as a whole or a half-*shamseh*, depending on its function, as for covering a vault or an iwan. Although there is normally only one *shamseh* in each ornamental arrangement, there are examples in which both half- and whole-*shamsehs* are observed, like the sample of Atiq Mosque of Shiraz. Figure 4.8 shows the architectural details of *shamseh* and its variant names and Figure 4.9 shows the above mentioned sample, having two *shamsehs* in one example.

Figure 4.8   Details of *muqarnas* elements; *shamseh* (Source: Author).

The star of *shamseh*, which has normally more than 6 points, should not be mistaken by the *takht* star elements, like those marked with blue in Figure 4.9. Although both elements look similar in the two-dimensional plan, *takht*–stars are horizontal and they are never used at the edges, but *shamseh* and half–*shamseh* are three–dimensional and half–*shamseh* is always seen on borders of the work.

4.3.2.5 *Toranj*

Toranj is a tetrahedral element, with axial symmetry, which may be compressed or elongated based on its position in the ornament with respect to its neighbor elements. Figure 4.10 shows different views of a *toranj* and provides its variant names.

Figure 4.9   1: Half-*shamseh* and 2: Whole-*shamseh* on the portal of Atiq Mosque of Shiraz. The blue stars are *takht* elements (Based on (Kazempourfard, 2006)).

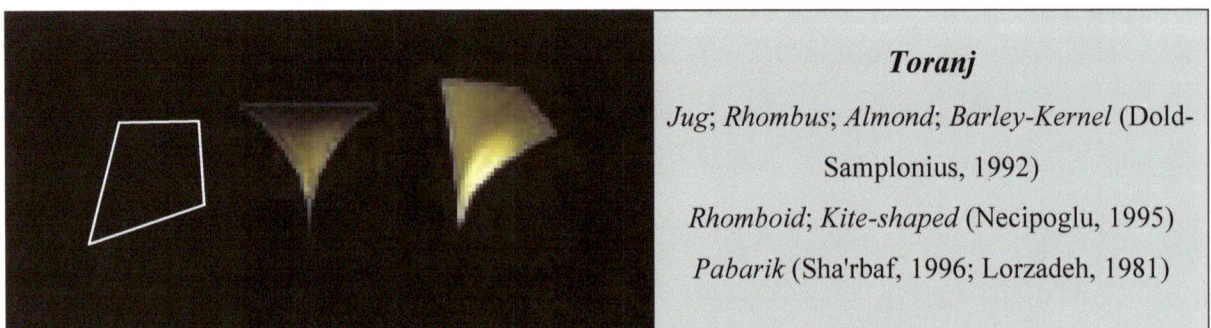

Figure 4.10   Details of *muqarnas* elements; *Toranj* (Source: Author).

This element is called *toranj*, which literary is the name of a fruit, i.e. bergamot orange, perhaps because of it non-circular shape which is elongated at one end. It is an intermediate element that is normally attached to *shamseh* or *takht* stars to fill the space between their pointed out wings. *Toranj* which was observed for the first time in Bastam Minaret in 12th century may be seen in different forms as shown in Figure 4.11. Some scholars tend to name these different shapes and different elements but in this text, all of these forms are considered to have rhombic shape and hence, they are all regarded as being *toranjes*.

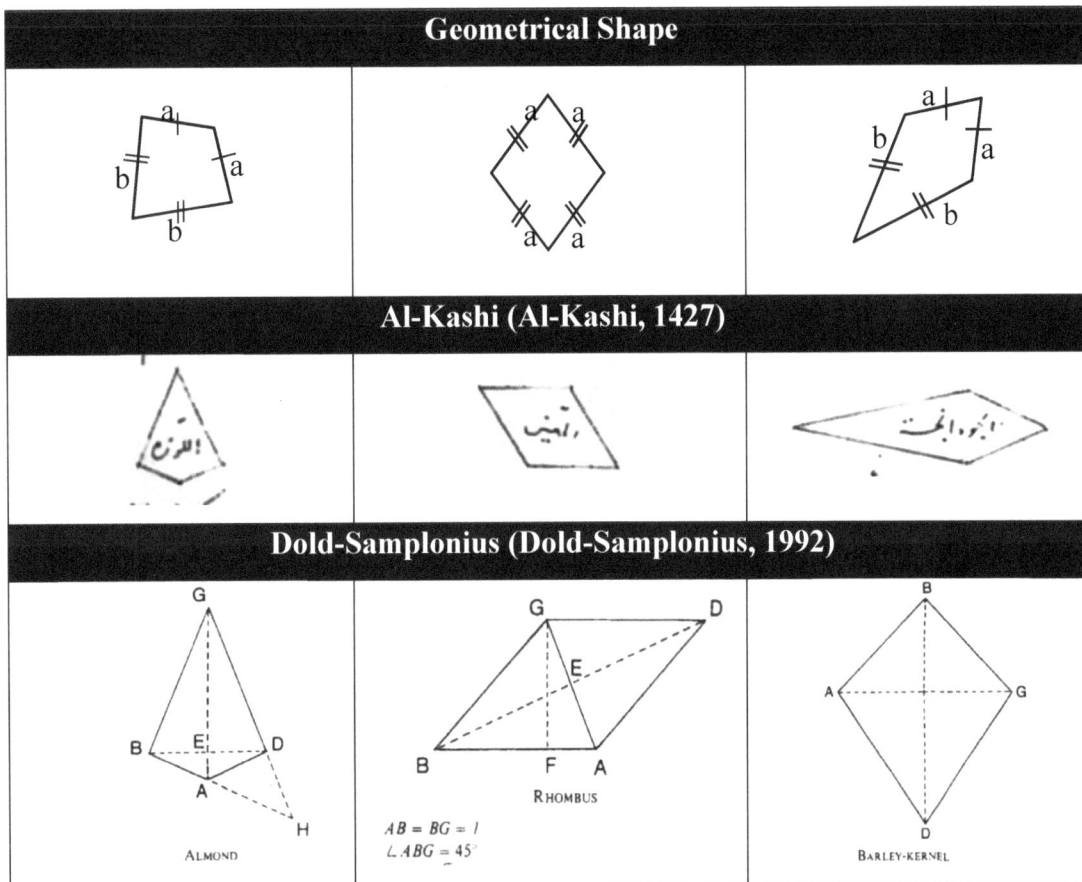

Figure 4.11   Different forms of *toranjes* (Source: Author).

4.3.2.6 *Espar*

Espar is the first ETE element created in 12th century, with the lower edge bigger than the top one. The edges are connected to each other by means of two curved lines. The word which literary means shield is the Persian equivalent of "entablature" as an architectural term. *Espar* is a plumb element that can be observed in the first few tiers of *muqarnas* on the walls and in the space between two *shaparaks*. In iwans, half-*espar* may also be observed. In many cases the top edge of the element is attached to the bottom edge of a *taseh*, creating an elemental unit named *susani* which will be illustrated later in this chapter. Figure 4.12 shows an *espar* element.

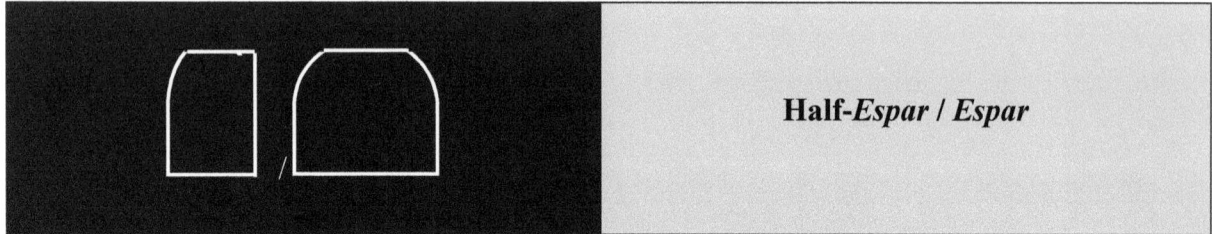

Figure 4.12 Details of *muqarnas* elements; *Espar* (Source: Author).

4.3.2.7 *Tee*

The next ETE plumb element is *tee*. The intermediate element is in fact a long rectangle which is normally attached to a *takht* element on its lower edge and has two flanking *paraks* as its lateral neighbours, making the whole combination similar to letter T. There are many samples in which the edge of *tee* and its flanking *paraks* made the lower edge of a *taseh*, creating an elemental unit named *tanureh* which will be shown in the next sub-section. Figure 4.13 shows a *tee* element.

Figure 4.13 Details of *muqarnas* elements; *Tee* (Source: Author).

4.3.2.8 *Takht*

Takht elements are the only horizontal elements which are specific to *muqarnas* and cannot be seen in any pseudo-*muqarnas* structures, before 14th century. They may have regular or irregular geometric shapes. Regular flats are star shaped flats, like three-, four-, five-, six-, seven- and eight-pointed stars and other shaped are defined as irregular flats. The existence of any flat element in an ornament shows that the ornament is undoubtedly a *muqarnas*. Figure 4.14 shows some regular and irregular *takhts* and provides the element's variant names.

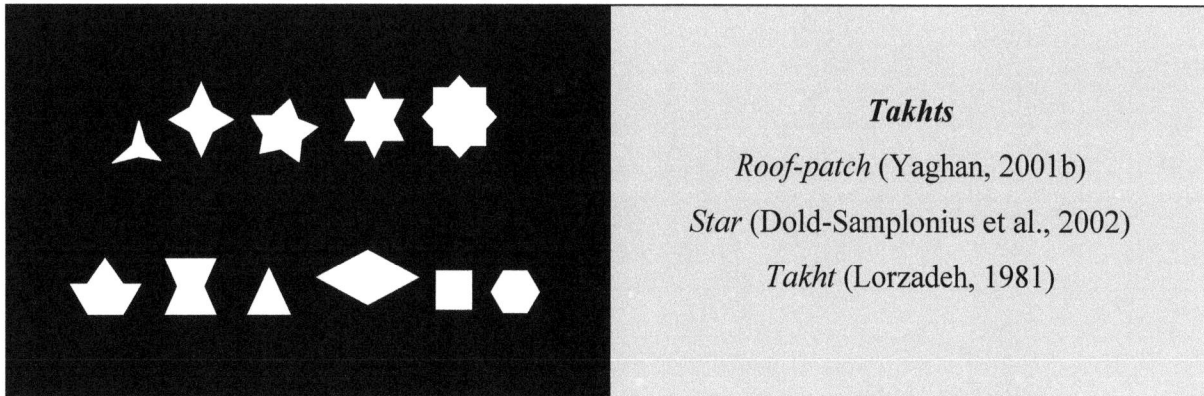

Figure 4.14   Details of *muqarnas* elements; *Takht* (Source: Author).

4.3.3   Famous Elemental Units

The basic elements introduced above are general and any element in any *muqarnas* or pseudo-*muqarnas* ornament fits to the definition of one of these elements. However, there are some famous elemental units, which are a specific combination of some of introduced elements, which will be illustrated as follows.

4.3.3.1   *Susani*

As explained before, if the top edge of an *espar* is attached to the bottom edge of a *taseh*, an elemental unit named *susani* will be created, as illustrated in Figure 4.15.

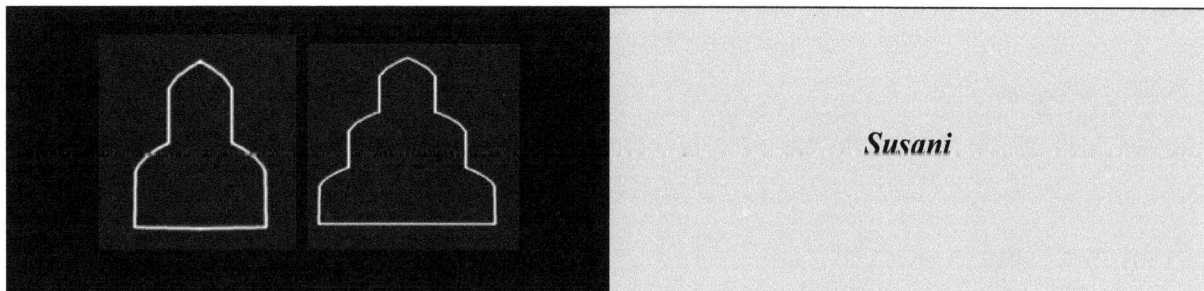

Figure 4.15   Details of *muqarnas* elemental unit; *Susani* (Source: Author).

4.3.3.2   *Tanoureh*

Tanoureh is another common element unit which consists of a *T* with flanking *paraks* carrying a *taseh* in their top edges. Figure 4.16 shows a *Tanoureh* elemental unit.

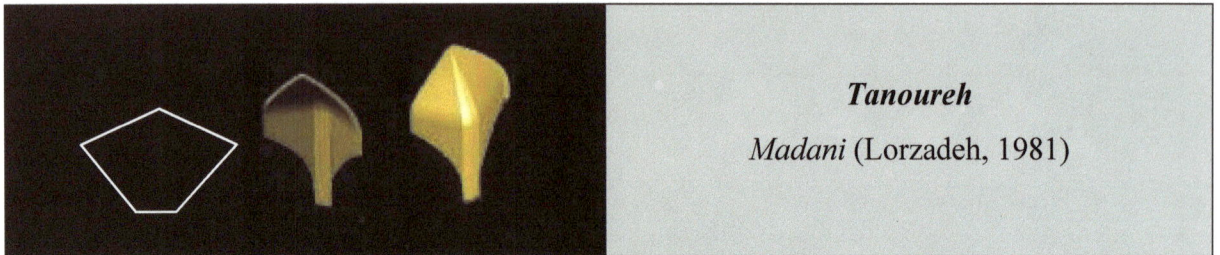

Figure 4.16   Details of *muqarnas* elemental unit; *Tanoureh* (Source: Author).

4.3.3.3 *Multiple Shaparaks*

In some cases, two to several *shaparaks* are attached together creating another elemental unit, as illustrated in Figure 4.17. In case of two adjacent *shaparaks*, the double *shaparak* combination is named as arrow shaped double-biped (Necipoglu, 1995).

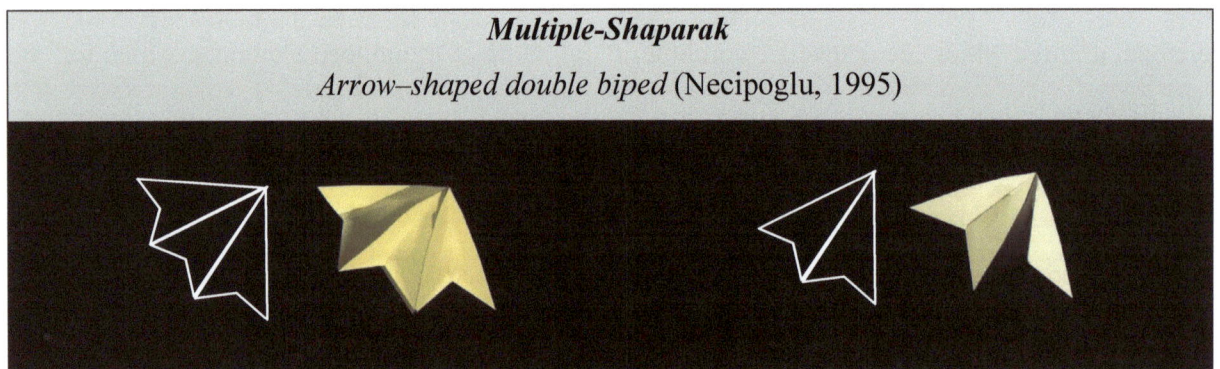

Figure 4.17   Details of *muqarnas* elements; *Multiple-shaparak* (Source: Author).

4.3.3.4 *Double Paraks*

Darvazeh, meaning gate, is the last elemental unit that will be introduced here. This unit is created when two *paraks* are attached together at their top edges, creating a shape that resembles a gate, as shown in Figure 4.18.

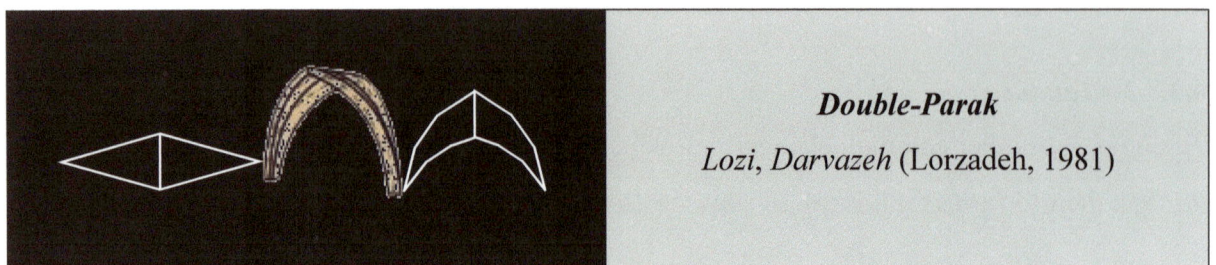

Figure 4.18   Details of *muqarnas* elements; *Double-parak* (Source: Author).

4.4 Validation

As explained before, before proceeding to the next objective, the results of this objective should be validated using samples from database (Patton, 2001). Hence, referring back to 100 sample database, one of the most complex *muqarnas* structures was selected for the purpose of controlling the generality of the proposed set of eight basic constituent elements.

The selected sample is the only recorded sample with two *shamsehs*. The *muqarnas* of the portal of Shiraz Jame' Mosque, or Atiq Mosque, shown in Figure 4.19, is one of the most complex known *muqarnas* ornaments in Iran. Built in 1793, the ornament is constructed about five centuries after the first recorded *muqarnas* with flat element, and 4 centuries after the scope of this research. Hence the selected sample was a good candidate to show either the abstracting of constituent elements of *muqarnas* was a successful one or not.

The sample, as explained in Table 4.1, has a total of 1594 different elements arranged in 18 tiers, namely one *shamseh*, one half-*shamseh* 524 *tasehs*, 514 *shaparaks*, 295 *paraks*, 96 *toranjes*, 69 flats, 46 *T* elements, 20 half-*taseh* and 28 *espars*. Hence, as one of the most complex recorded samples of the database, it was chosen to test the generality of the introduced constituent elements.

4.4.1 Elements on Studied *Muqarnas*

In this part, the introduced constituent elements and elemental units are shown as parts of the studied *muqarnas* of the portal of Shiraz Jame Mosque. An example is provided for each element and elemental unit introduced. For instance, as explained before, this unique portal has one *shamseh* and one *shamseh* at the same time, as shown in Figure 4.20. Furthermore, in Figure 4.21, *taseh*, *parak*; *T* and *shaparak* are also marked with orange, red yellow and white, respectively. On the right hand side two types of *shaparak* with two different angles can be distinguished. Three types of *taseh* are also marked with orange. As explained before, a combination of a *T*, two *parak* and a *taseh* can make an elemental unit named *tanoureh*. This unit which can be seen as separate elements on the left hand side is marked with black on the right part of Figure 4.21. *Espar* can only be observed in the first few tiers of each *muqarnas*. Figure 4.22 illustrates the *espars* of the studied *muqarnas* samples.

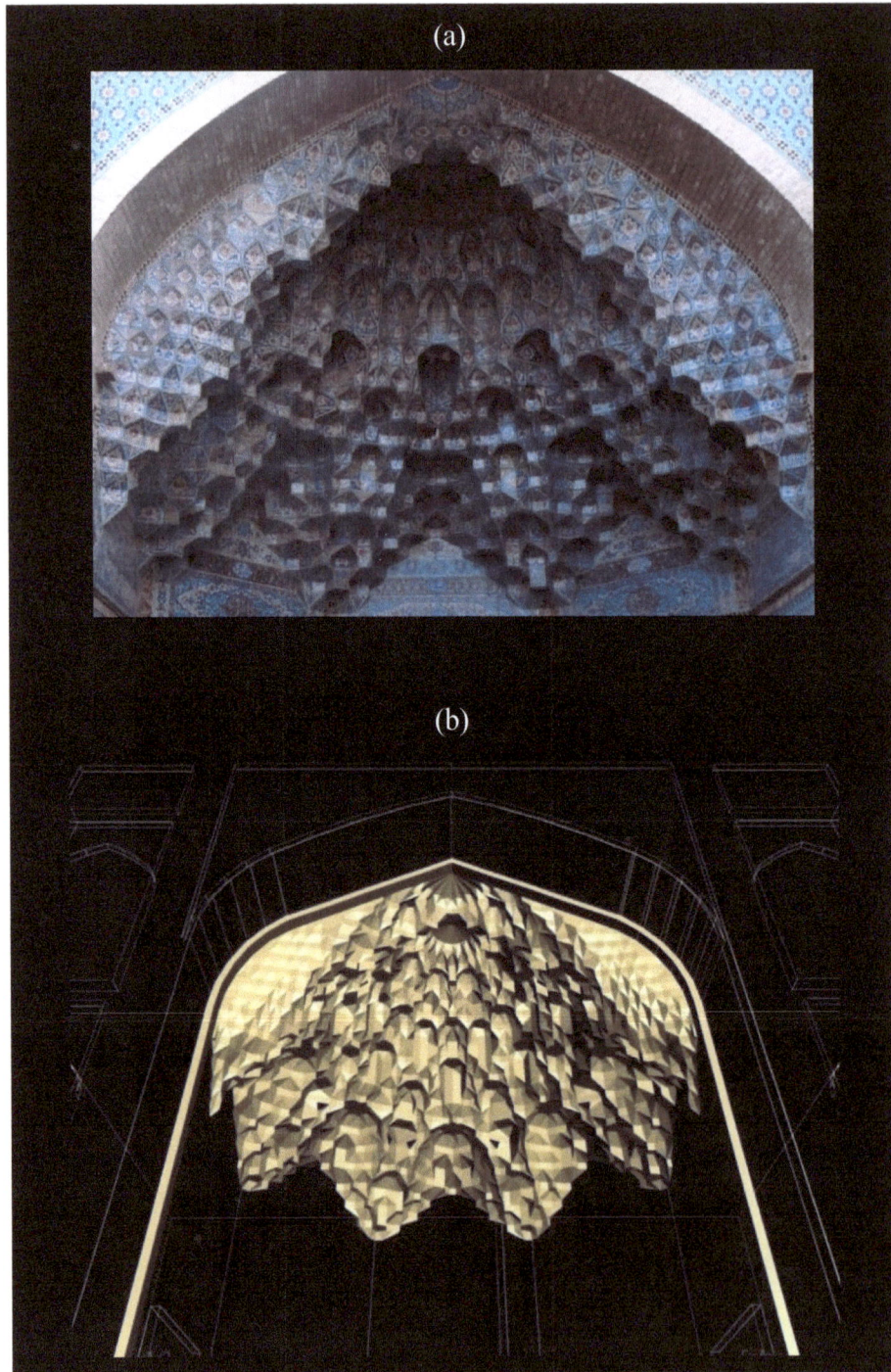

Figure 4.19 (a) Image, (b) Three-dimensional figure of Shiraz Jame' Mosque, Portal (Based on (Kazempourfard, 2006)).

Table 4.1: Constituent elements table, portal of Shiraz Jame' Mosque.

Element → / Tier ↓	Toranj	Tee	Taseh Type whole	Taseh Type half	Shaparak	Parak	Shamseh No	Shamseh Type half	Shamseh Type whole	Espar No	Espar Type half	Espar Type whole	Takht No	Takht Type
One			12		8	16				7	2	5	2	5pointed
Two		4	12	10	30	16				7	2	5	2	4pointed
Three		12	18		28	32				7	2	5	8	4pointed
Four			30		64	24				7	2	5	4 / 2	5pointed / 4pointed
Five		8	22	2	40	16								
Six			48		32	8							2	4pointed
Seven	30		52	2	32	32							4 / 5	4pointed / 8pointed
Eight		4	42		28	10								
Nine			30		30	8							2 / 4	4pointed / 6pointed
Ten			32		36	8								
Eleven			37	2	64	20							5	5pointed
Twelve		7	30		40	14								
Thirteen	20		27		20	20							2 / 4 / 5	4pointed / 5pointed / 7pointed
Fourteen	2	4	23		16	12							2	3pointed
Fifteen	2		28	2	34	9							2 / 5	3pointed / 4pointed
Sixteen	2	7	40	2	12	50							1 / 8	4pointed / 6pointed
Seventeen	24		25				1	√						
Eighteen	16		16				1		√					

Figure 4.20 Half-*shamseh* and *shamseh* on *muqarnas* of the portal of Shiraz Jame' Mosque.

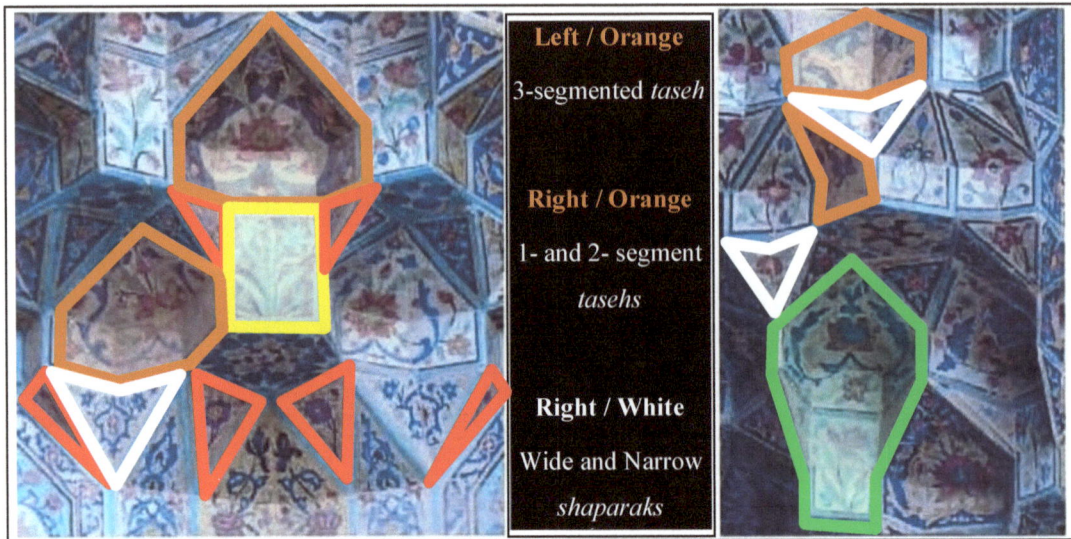

Figure 4.21   Orange: *Taseh*; Red: *Parak*; Yellow: T; White: *Shaparak*; and Green: *Tanoureh* (Source: Author).

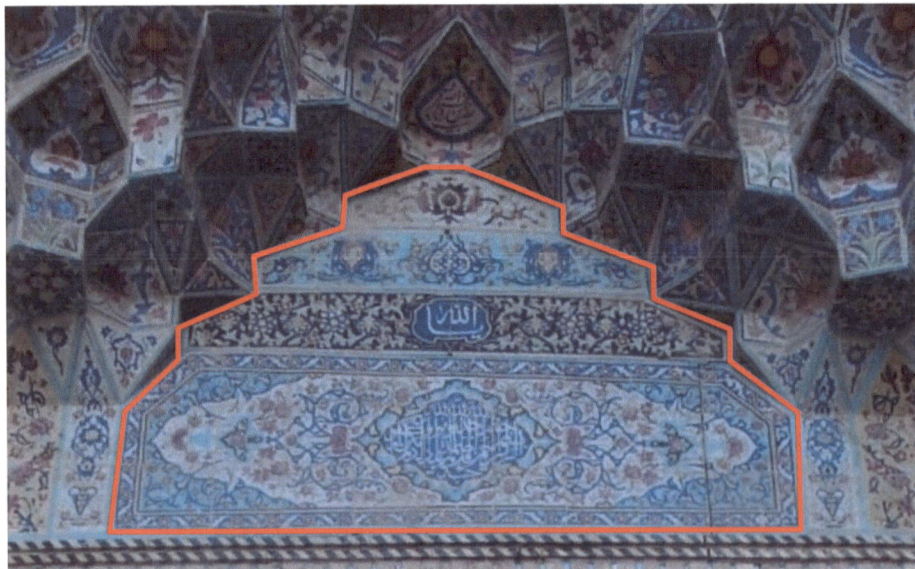

Figure 4.22   *Espar* elements in *muqarnas* of the portal of Shiraz Jame' Mosque (Source: Author).

There are a lot of various *takht* elements in the *muqarnas* of the portal of Shiraz Jame' Mosque. Figure 4.23 shows the plan of *muqarnas* and on right, the positions of the flat on plan are shown by circles. In addition, a four-pointed flat star and its neighbour *toranjes* are displayed in Figure 4.24, whereas Figure 4.25 shows different types of flat elements

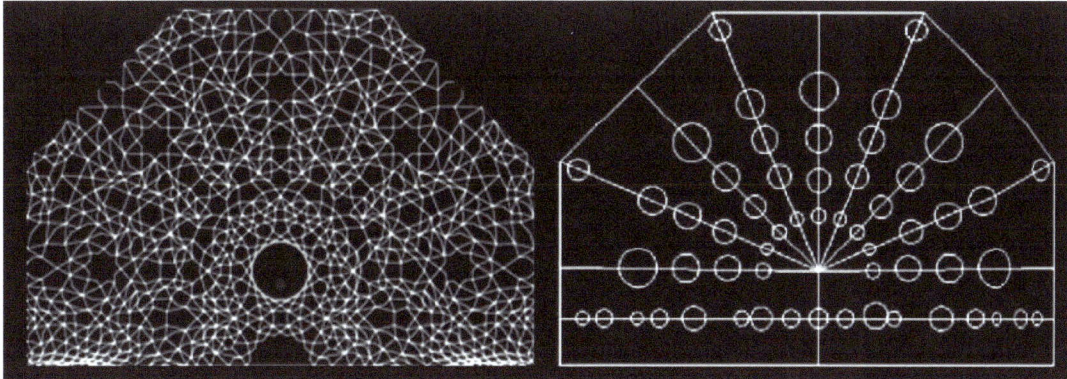

Figure 4.23 Flats on symmetry rays of *muqarnas* of Shiraz Jame' Mosque (Based on (Kazempourfard, 2006)).

Figure 4.24 *Toranjes* (red), and four-pointed star flat element (Source: Author).

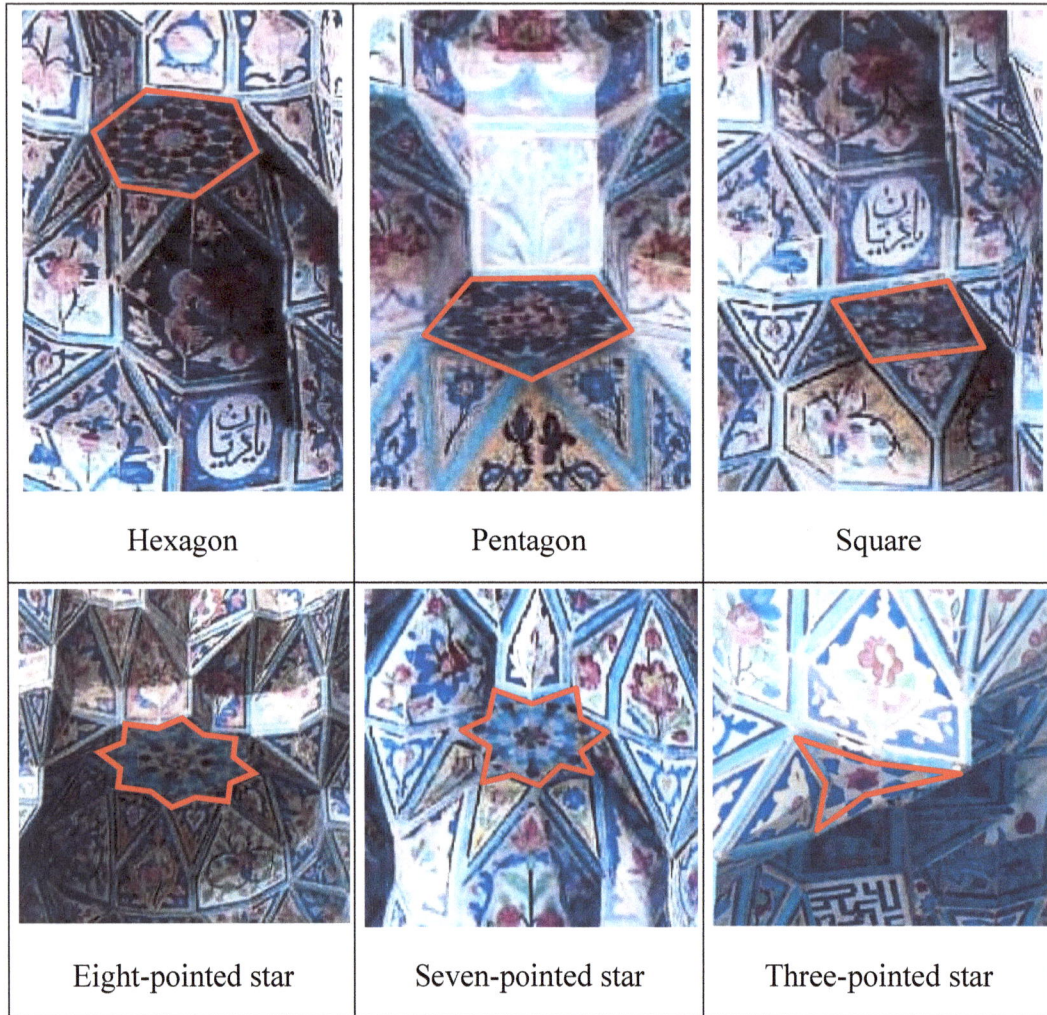

Hexagon	Pentagon	Square
Eight-pointed star	Seven-pointed star	Three-pointed star

Figure 4.25   Different types of flats (Source: Author).

4.4.2 Tier by Tier Construction Technique

The *muqarnas* of the portal of Shiraz Jame' Mosque is constructed in 18 tiers. As explained before, due to the construction technique of *muqarnas*, the tiers of the structure are always counted from bottom to top. Figure 4.26 illustrates the 18 tiers of the *muqarnas* sample.

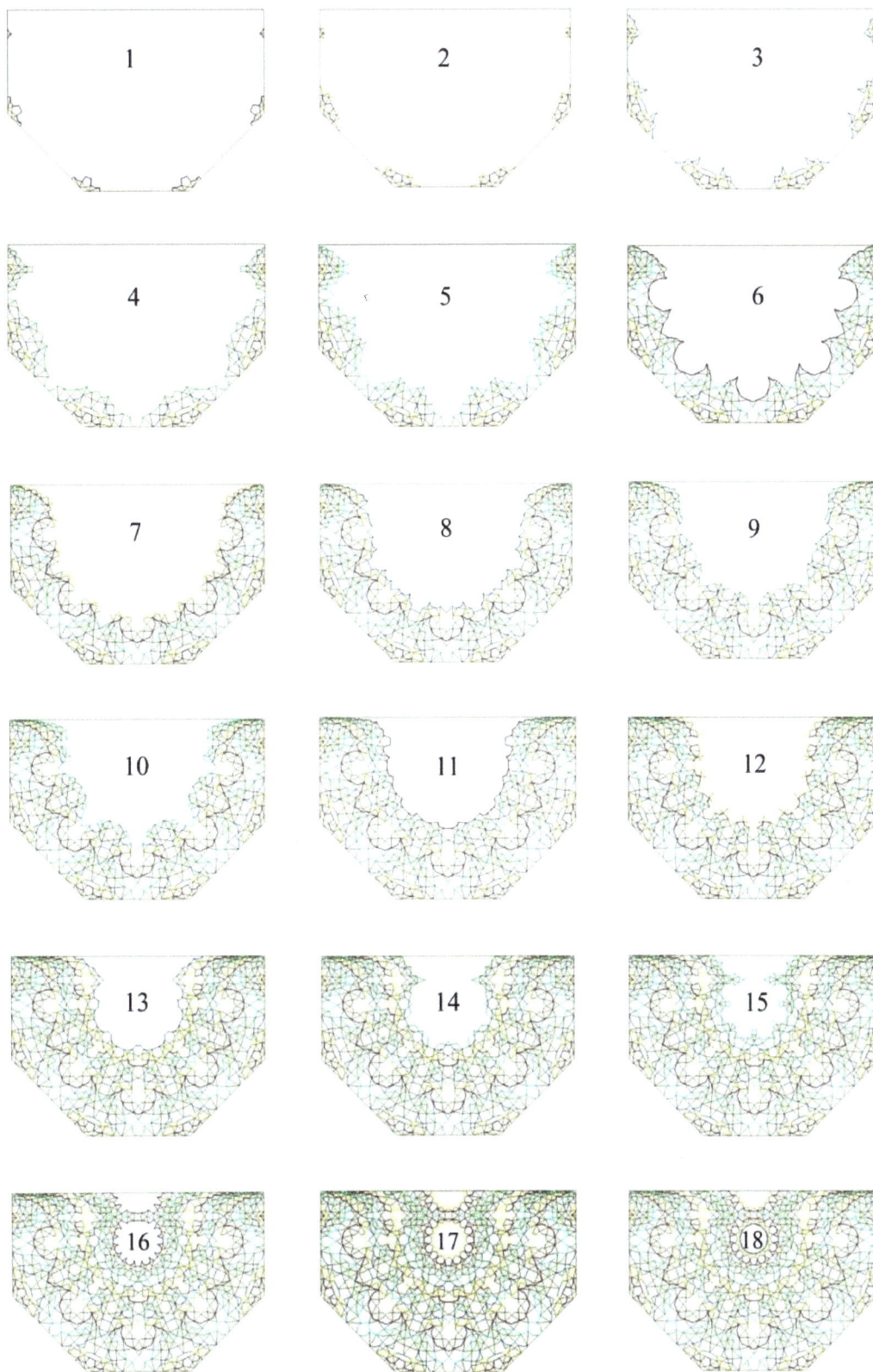

Figure 4.26 Eighteen tiers of *muqarnas* of the portal of Shiraz Jame' Mosque (Based on (Kazempourfard, 2006)).

In general, there are two major techniques for constructing *muqarnas* worldwide, i.e. suspended layer and suspended unit techniques (Yaghan, 2001b). As the ornament is purely decorative and adds dead load to the building, methods should be applied to transfer the load to the walls or ceiling of the main body of the structure. Figure 4.27 shows the fragments of a *muqarnas* in Isfahan in which the timbers and ropes that carry the load of the ornament to the ceilings and walls can be identified.

Form the two mentioned common techniques of building *muqarnas*, suspended unit is hardly used in Iran. Though there are three examples which had been constructed by suspended unit technique from which, Bastam Minaret will be studied in detail in Chapter 5 and the other two, which are unfortunately destroyed, are Takht–i Sulaiman and Dash Kasan Temple. Figure 4.28 shows the remnants of the *muqarnas* of Dash Kasan Temple. Suspended layer technique, on the other hand, is vastly used all over Iran, since Ilkhanid period. The method involves making some plaster tablets for each tier on the ground, based on the plan of the ornament and mounting them to the ornaments targeted location and attaching them there by means of rope and timber at appropriate heights.

The supportive ropes and timbers are then covered by plaster survive longer against corrosion. The ropes used in ancient time in Iran, was made of natural material, from palm's trunk and bough, which was very fatigue resistant and had a high tensile strength. Figure 4.29 shows this rope which is called *Sazu*.

Figure 4.27   A fragmentary *muqarnas* in iwan of Harun Velayat Mausoleum, Isfahan (Source: ne.jp)

Figure 4.28 Suspended units of the fragmentary *muqarnas* of Dash Kasan Temple (Source: chn.ir).

Figure 4.29 *Sazu*, the traditional rope used for hanging layers of *muqarnas* (Source: Author)

In suspended layers technique, the plaster tablets are responsible to transfer the dead load of the decorative veneer to the walls and ceiling of the building. The space between these tablets is then filled by appropriate constituent elements and only the edges will be seen from them. These edges play an important role in the final appearance of the decorative veneer, as they show the main line of each tier. Most of the times, the edges are decorated with a different material, mainly glazed tile, to boost the beauty of the decoration. However, in modern constructions, innovative material is also used for the purpose. Figure 4.30 shows how edges are exaggerated traditionally to show the beauty of the ornament, whereas Figure 4.31 shows how author used uncommon decorative material, i.e. bronze, to decorate the edges of a *muqarnas* in new Jame' Mosque of Varamin.

However nowadays, as shown in Figure 4.32, in most of *muqarnas* constructions in Iran, iron rebar is used to shape the layers and to install the layers and attach them to the ceiling. In addition, the steps of the process of using suspended layer construction technique are illustrated in Figure 4.33. In this figure, a small *muqarnas* is being prepared by author for a conceptual art biennale in Tehran, to show that the art is at the risk of extinction.

Figure 4.30   Traditional edge decoration by glazed tile (Based on (Kazempourfard, 2006)).

Figure 4.31   Author's innovative edge decoration by bronze (source: Author).

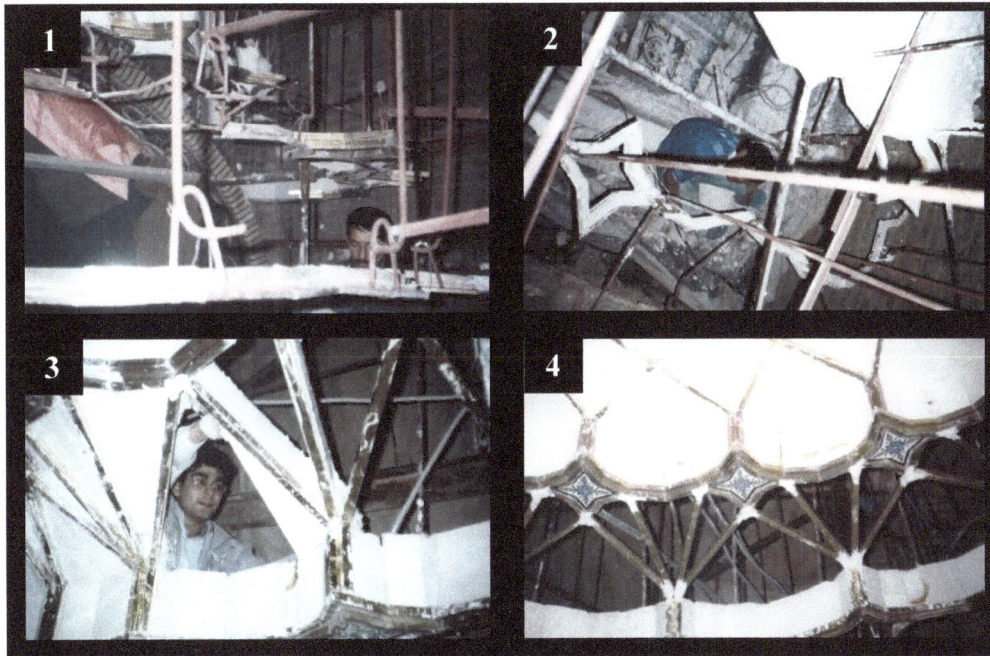

Figure 4.32   Using iron rebar to shape edges (Source: Author).

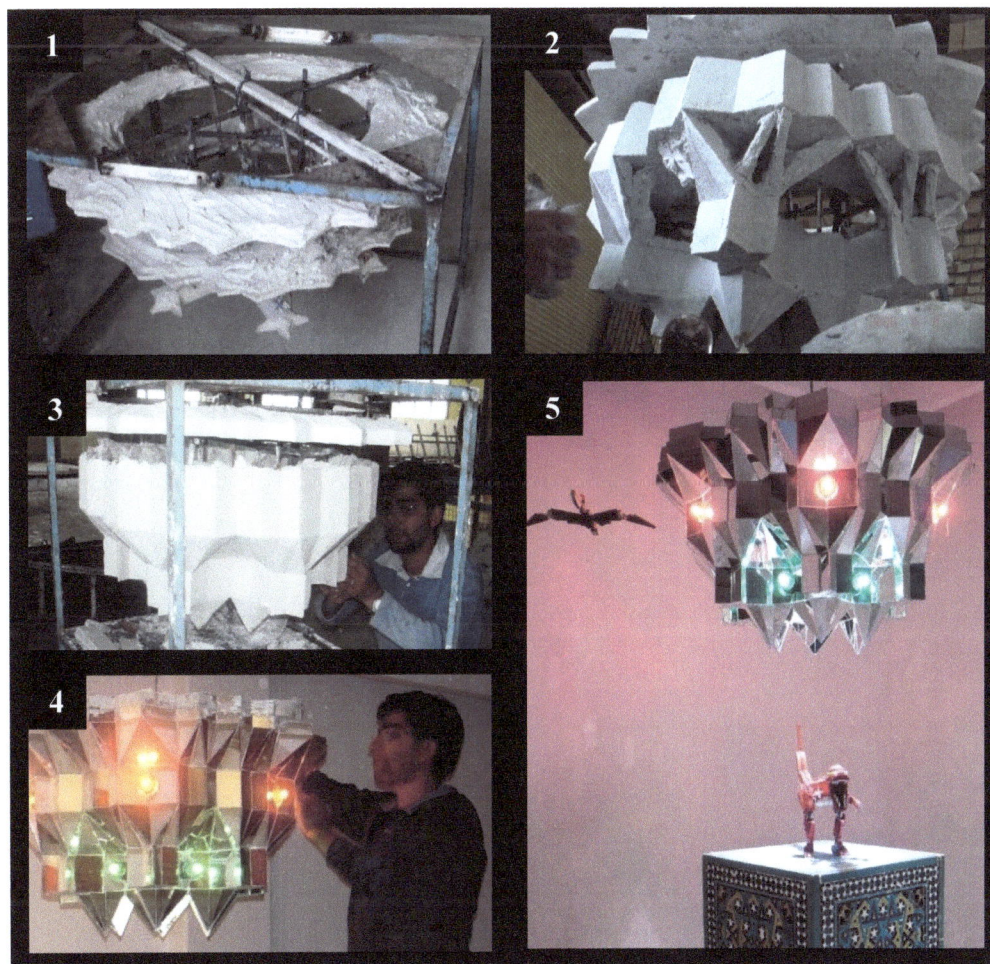

Figure 4.33   Suspended layer construction technique (Source: Author).

4.5 Conclusion

In the process of analysing the visual data, it was necessary to consider the constituent elements of the ornament as an important factor of development, hence, recognise their forms and the details of their characteristics. The characteristics of different types of constituent elements of *muqarnas* were studied after being recognised. Based on the visual and physical properties of these elements, a set of variables were considered for better understanding of the arrangement of these elements with respect to each other, namely being a basic or an intermediate element, as well as being an edge-to-point, point-to-edge or edge-to-edge element type.

Upon the investigation of the existing samples, it is discovered that the elements of *muqarnas* can be categorized as *shamseh*, *toranj*, *taseh*, *parak*, *shaparak*, *tee*, *espar*, and *takhts*; where *taseh* and *shaparak* are the most common elements. *Shamseh* is always located on the apex of *muqarnas* and *toranjes* are its next neighbours to fill the space between the outward points of *shamseh* or flat stars. The *takht* stars never appear towards the edges, and the corner areas are usually covered with flexible elements such as the *tasehs*, *shaparaks* or *paraks* because these geometrically flexible elements enable the adjustment of the design with the curve and tilt of the load–bearing arch. *Tee* is used to connect several tiers vertically whereas *espar* which is the other vertical element of *muqarnas* is only used in the first few tiers to connect the walls to the decorative veneer.

CHAPTER 5
Evolution of Muqarnas

5.1 Introduction

It has been about 2000 years since domes were constructed for the first time. As a prime example, one may refer to the building known as the palace of Ardeshir, introduced in Appendix A, as one of the remaining examples dating back to about 1800 ago. Domes were first built over cylindrical structures but later, builders became capable of constructing the semi–spherical dome over a square base, which issue was not resolved easily. Figure 5.1 shows a schematic dome over a cubic base. Many different methods have been tried before *squinches* were invented as the optimum solution by Iranian architects (Hautecoeur, 1931; Memarian, 2008; Écochard, 1977). Furthermore, in Roman architecture, in order to solve the issue of a smooth transition from walls to domes, pendentives were invented as the counterparts of squinches in Iran (Écochard, 1977).

There are still some ancient constructions in Iran, in which this method of transition from square to circle was used; namely, Zoroastrian fire temples at Niasar and Firuzabad, shown in Figure 5.2, and Sarvestan Palace, which will be studied later in this chapter.

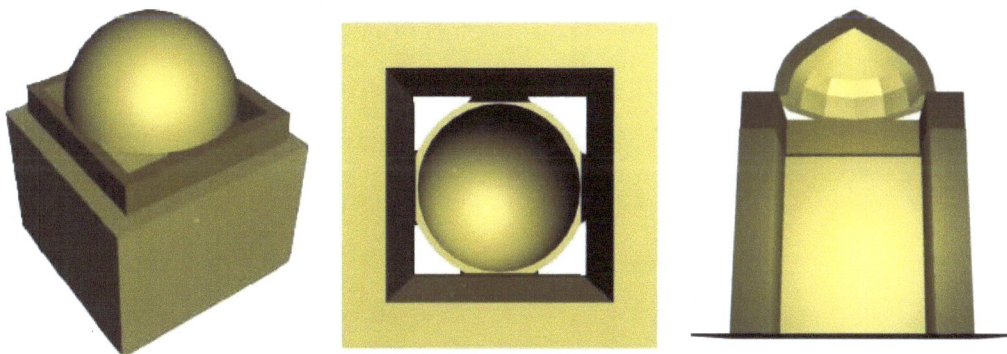

Figure 5.1   Domes over a cubic base (Source: Author).

Figure 5.2   Zoroastrian fire temples at (a) Firuzabad, Fars and (b) Niasar, Kashan (Source: cais–soas.com).

Iranian domes can be categorized into Sassanid and Islamic (Godard, 1990). In pre–Islamic architecture of Iran, there were three approaches for transition. To put it more specifically, transitional solutions were used to fill the upper angle between the walls and the dome. As illustrated in Figure 5.3, these methods included *Patkin*, *Filpush* and Squinch (Memarian, 2012), from the third method of which is of specific interest in this manuscript and will be explained in the next sub–section. More details about the *Filpush* in Ardeshir Palace are available in Appendix A.

Patkin consists of several layers of building material, arranged from small to large to cover the empty corner between the walls and the dome, whereas *Filpush*, literary meaning elephant ear, can be imagined as $1/8^{th}$ of a sphere, placed over the right angle of two intersecting walls, moving up to compensate the curvature of dome. The abovementioned techniques continued to be used in Islamic architecture of Iran, e.g. in Imamzadeh Ja'far Mausoleum, Damghan, in Molla Isma'il Mosque, Yazd, and in Qazvin Jame' Mosque, Qazvin. Figure 5.4 provides the sketches of these samples.

(a) Patkin (b) Filpush (c) Squinch

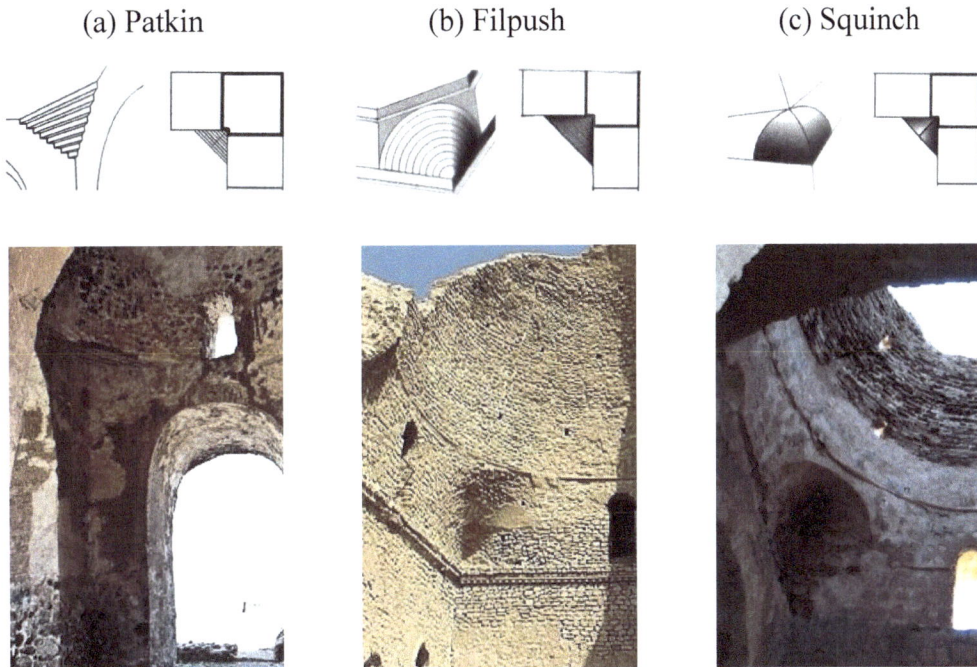

Figure 5.3   Transition technique on corners in ancient Iran (a) Patkin, Bazeh Hur Temple, (b) Filpush, Ardeshir Palace, and (c) Squinch, Sarvestan Palace (Memarian, 2012).

(a)   (b)   (c)

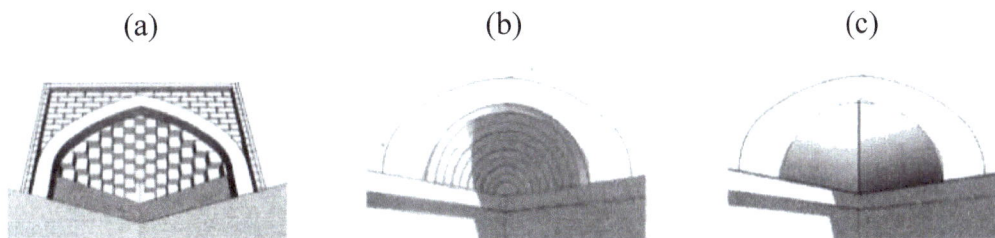

Figure 5.4   Sketching of more recent applications of ancient transition techniques in Iran; (a) *Patkin*, Imamzadeh Ja'far Mausoleum, Damghan, (b) *Filpush*, Molla Isma'il Mosque, Yazd, and (c) Squinch, Qazvin Jame' Mosque, Qazvin (Memarian, 2012).

5.2   Squinch

Squinch is a cornering technique which consists of two concave vault–sections intersecting each other at right angles, in a line. The intersection of these vault fragments creates the squinch. Figure 5.5 shows different views of the three-dimensional pattern of a squinch. Each vault–section can be imagined as a curved triangle that is in fact a semi–*Taseh*, as explained before in Chapter 2. In other words, a *taseh* is somehow derived from a squinch in terms of form.

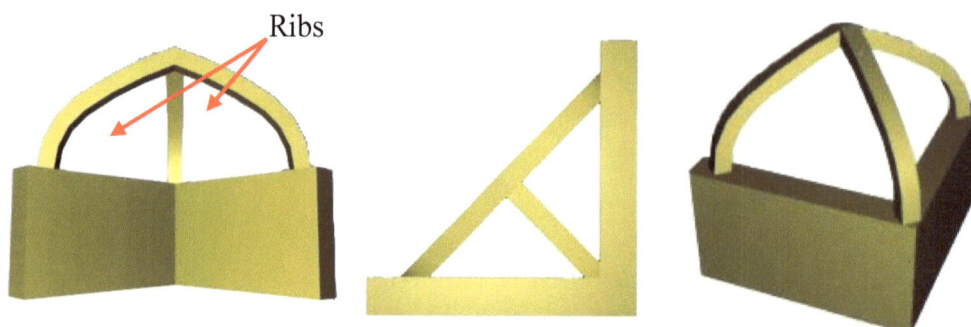

Figure 5.5   Different views of a squinch (Source: Author).

Transition techniques in Sassanid buildings differ from those of Islamic structures, in the sense that in pre–Islamic constructions, part of the load was directly carried on by the adobe layers and as illustrated in Figure 5.6, there were no distinct borders between the squinch and the dome. This is while in Islamic counterparts, the load from the dome is transmitted through vault's ribs, as auxiliary intermediates, as well as the rear support ribs to the walls (Figure 5.5). These ribs were made normally by an odd number of square bricks. The amount of adobes used to construct the ribs depended on the amount of tension the arch was expected to carry (Godard, 1990).

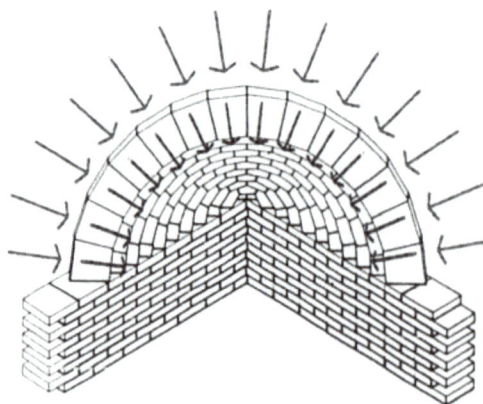

Figure 5.6   Load–bearing role of pre–Islamic corners (Jazbi & Al-Kashi, 1987).

5.2.1 Sarvestan Palace

Sarvestan palace, also known as Bahram Palace, shown in Figure 5.7, is a Sassanid Place near Shiraz. Built between 420 and 480 C.E. it is rated among the first buildings in which squinches were used (Homayooni, 1996; Bier, 1986). Besenval introduces Sarvestan as the first sample of squinch in the Middle East and believes that the squinch used in this building is the filiations of

Islamic buildings' transition techniques on corner zones (Besenval, 1984; Besenval, 2000). Figure 5.8 shows the plan of the ancient building and the position of the studied squinch on it.

Sarvestan Palace belongs to the Sassanid king, Bahram. Using the term "palace" may not be exactly correct as the building is believed to be rather a hunting lodge for the king or may have been even a sanctuary. The building is made of stone and mortar with fine decorations only parts of which have survived. As one of the pioneering buildings in which squinches were applied on the corners of dome chamber, this is the only pre–Islamic building studied in this manuscript. Table 5.1 provides some key information about the building of the studied structure and Table 5.2 shows its constituent element which is a single *taseh* in this case. Furthermore, Figure 5.9 demonstrates the isometric view and plan of the structure.

Figure 5.7   Sarvestan Palace (420–480 C.E.), Fars, Iran (Source: Wikipedia).

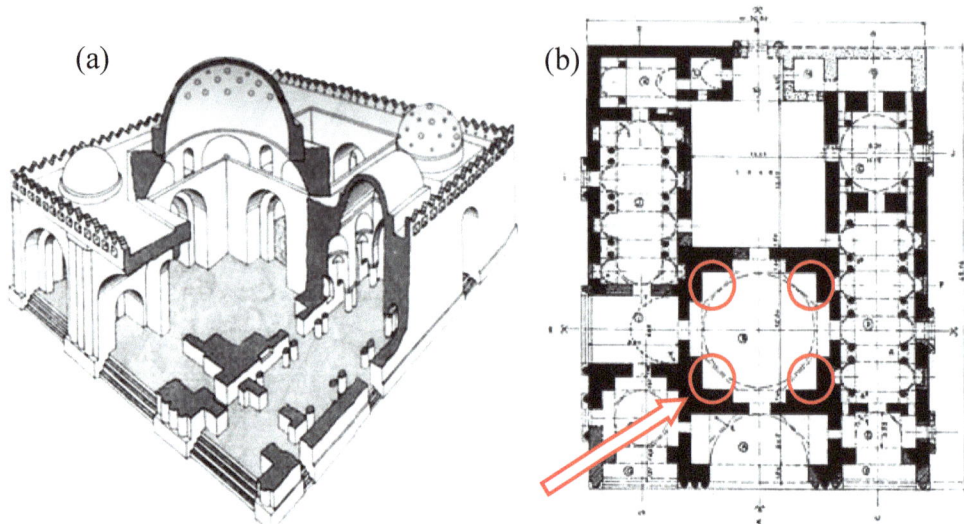

Figure 5.8 (a) Schematic appearance of Sarvestan Palace, (b) Building plan showing the position of studied squinches (Besenval, 2000).

Table 5.1: Constituent elements table, Sarvestan Palace.

Element	*Toranj*	*Tee*	*Taseh*						*Shaparak*			*Parak*			*Shamseh*			*Espar*			*Takht*	
			No	Whole		No	Half		No	Type		No	Type		No	Type		No	Type		No	Type
↓ Tier →				2D	3D		2D	3D		2D	3D		2D	3D		half	whole		half	whole		
One			1	√																		

The squinch in this construction has only one tier, to provide height and transition from walls to the dome. The material of the squinch is identical to that of the construction, in stone and mortar. Transition zone of Sarvestan Palace has structural role and is constructed amalgamated with the building. Furthermore, the studied squinch is categorized as of having a square lattice plan and it is one–sided, i.e. the back part is not seen. In fact, as mentioned before, there have been samples in which both sides of the decoration are visible, like the domes over Imam Dur Mausoleum, Samarra, Iraq and Nur Al–Din Hospital, Damascus, Syria, but nearly all of *muqarnas* and pseudo–*muqarnas* decorations found in Iran are one–sided.

Table 5.2: Sarvestan Palace building information.

Sarvestan Palace (29.19556°N, 53.23083°E)	
Variant Names	Bahram Palace, Bahram V Palace
Location	Sarvestan, Iran
Position	Corner
Date	420–480
Period	Sasanian
Century	5th
Building Type	Palatial
Building Usage	Palace
Copyright	Hamidreza Kazempour (3D and Plan)

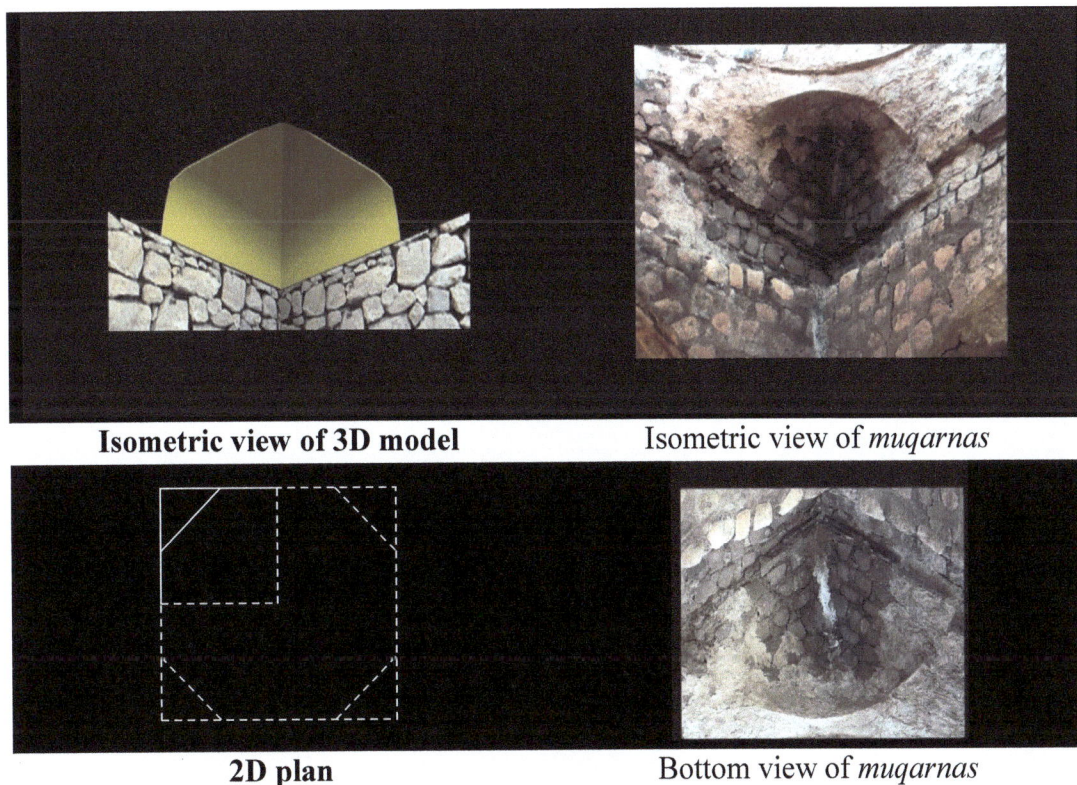

Isometric view of 3D model   Isometric view of *muqarnas*

2D plan   Bottom view of *muqarnas*

Figure 5.9 View of squinch in Sarvestan Palace (Source: Author).

5.2.2 Samanid Mausoleum

The Samanid Mausoleum in Bukhara was built between 914 to 943 C.E. Although the building is in fact the funerary spot of three persons, it is known as Ismail Samanid tomb (Grabar, 1966; Michell & Grube, 1995). The building has a brick structure with a simple form, i.e. a spherical dome over a cubic room. Each façade has an identical design, with a decorated arched opening, and joins the other with a cylindrical baluster. The exterior and interior are simple and elegant; patterned brickwork and some stucco decoration is considered an innovative method in that period (Hill, 1964).

Although the Islamic squinch in which the so called *ribs* were used became very common in the Seljuk period, the pre–Seljuk squinches in Ismail Samanid Mausoleum demonstrate one of the first and the most beautiful samples of the elegant combination of Sassanid tradition with Islamic rib–based vaulting technique. Figures 5.10 and 5.11 are the exterior views and the interior dome chamber of the Samanid Mausoleum, respectively, whereas the plan and the schematic elevation of the mausoleum, as well as the positions of its squinches on the plan are shown in Figure 5.12.

Figure 5.10   Samanid Mausoleum, Exterior (© Dr Gholam Hossein Memarian).

Figure 5.11   Samanid Mausoleum, Interior dome chamber (© Dr Gholam Hossein Memarian).

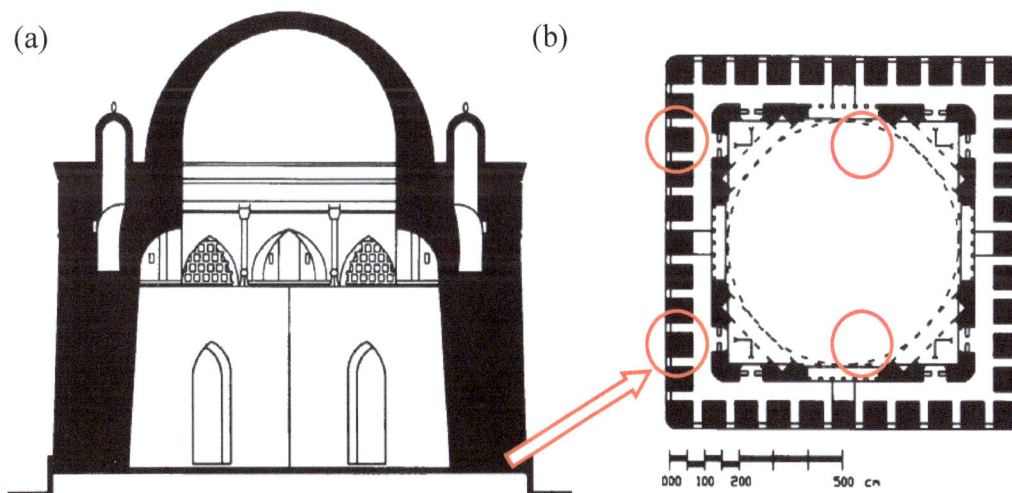

Figure 5.12   (a) Schematic appearance of Samanid Mausoleum, (b) Building plan showing the position of studied squinches (Pirnia, 2003).

Table 5.4 provides some key information on the building of the studied structure and Figure 5.13 demonstrates isometric view and plan of the structure. As illustrated in Figure 5.14, there are four ribs on the four corners of the dome chamber, each resting on the building walls by another supporting rib behind them. The supporting ribs are visible in this building and are used as a kind of decoration. The Space between these ribs is filled with two *Tasehs*, which can be considered as one of the first steps towards the creation of *muqarnas*. The details of the gradual development of the form will be more elaborated later in this chapter. Table 5.3 shows the details of the structure's constituent elements.

Table 5.3:   Constituent elements table, Samanid Mausoleum.

Element ↓ Tier →	Toranj	Tee	Taseh						Shaparak			Parak			Shamseh			Espar			Takht	
			No	Whole		No	Half		No	Type		No	Type		No	Type		No	Type		No	Type
				2D	3D		2D	3D		2D	3D		2D	3D		half	whole		half	whole		
One			2		√																	

The transition zone in Samanid Mausoleum provides size, height and form transition in a single tier from the polygonal base of the walls to the spherical dome. The material of the squinch is identical to that of the building itself, i.e. brick and patterned brick. The squinch has a structural role and benefits from the structural assistance of ribs. In addition, the studied squinch has a square lattice two dimensional pattern plan. Like many other Iranian corner and vaulting decorations, the squinch is considered as one–sided structure.

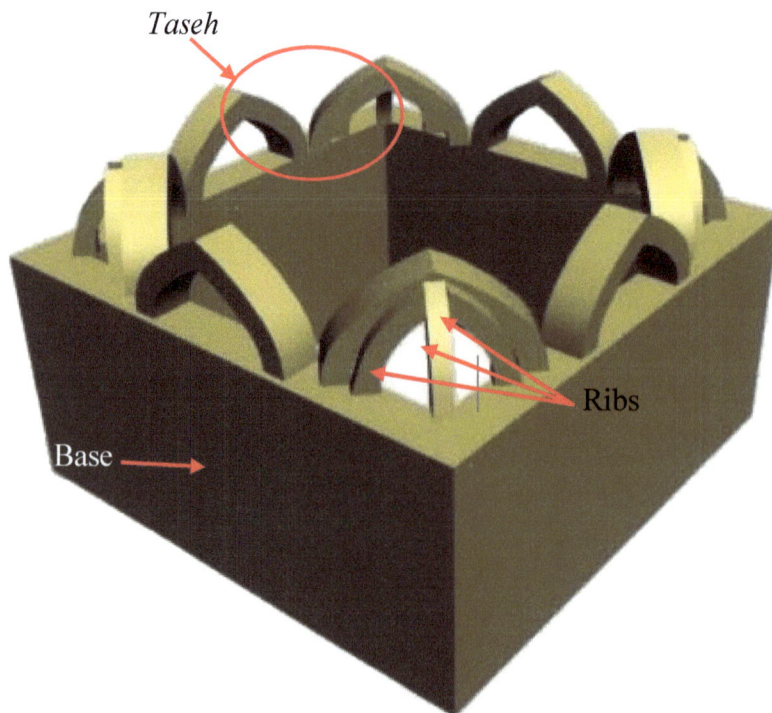

Figure 5.13   Three–dimensional view of the structure behind squinches of Samanid Mausoleum (Source: Author).

Table 5.4:   Samanid Mausoleum building information

Samanid Mausoleum (39.776944°N, 64.400556°E)	
Variant Names	Mausoleum of the Isma'il Samani, Tomb of Isma'il Samani, Abu Ibrahim Isma'il ibn Ahmad Samani Tomb, Mausoleum of Ismail the Samanid, Ismail Samani Ghabr, Ismail Samani Qabr, Qabr–i Isma'il Samani, Ghabr–i Isma'il Samani, Mausoleum of the Samanids, Tomb of Amir–I Adil, Tomb of Amir–e Adel, Tomb of Amir–i–Adil, Tomb of Amir–e–Adel, Amir–i–Mazi, Amir–e–Mazi
Location	Bukhara, Uzbekistan
Position	Corner
Date	914–943/301–331 A.H.
Period	Samanid
Century	10th
Building Type	Funerary
Building Usage	Mausoleum
Copyright	Author (3D and Plan); Images: Dr Memarian

Isometric view of 3D model   **Isometric view of *muqarnas***

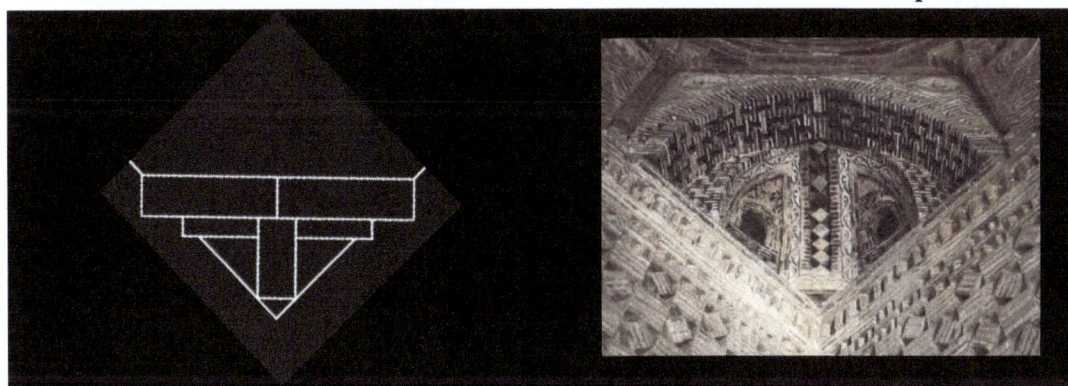

2D plan   Bottom view of *muqarnas*

Figure 5.14   View of squinch in Samanid Mausoleum (Source: Author).

5.2.3 Jurjir Mosque

The next example is observed in Buyid period in Isfahan. The entrance of Jurjir Mosque from which only parts of the exquisite portal is remaining is now located on the North–West side of Hakim Mosque in Isfahan. The incomplete Jurjir façade is a 10th century structure now a part of the 17th century Hakim Mosque which was discovered during restoration of the historical site in 1955 (Grabar, 1990; Memarian, 2012). The typical Buyid decoration of the façade is the only remnant of the original Jurjir Mosque, consisting of an entrance portal which has a tripartite semi–dome hood. Figure 5.15 shows the entrance of Jurjir Mosque.

Fragmented veneer

Figure 5.15   Jurjir Mosque portal (Source: Author).

The Squinch of Jurjir Mosque differs from that of Samanid Mausoleum structurally. The thick supporting back rib of Samanid Mausoleum is replaced here with two narrower load–bearing ribs, providing an open space in the middle. There are two *tasehs* flanking the portal decorations, creating a window above entryway and the bricks are covered with carved patterns as a decorative veneer. Referring back to Figure 5.15, one may discover that the decorative brick veneer is fragmented on top left side of the portal.

As mentioned before, this method of decoration was common in Buyid period. In addition to carved bricks forming patterns in relief, there was another method for creating repeated geometric patterns, mostly on stucco, as used on the outer layer of rib arches in this sample. This latter method was normally used over rounded surfaces by moulding (Grabar, 1990; Hillenbrand,

1987; Blair & Bloom, 1994). Table 5.5 provides some key information about the building of the studied structure and Figure 5.17 demonstrates isometric view and plan of the structure.

Table 5.5: Jurjir Mosque building information.

Masjid Jurjir (32.660833°N, 51.675556°E)	
Variant Names	Masjed–e Jurjir, Jurjir Mosque, Darvazah Jurjir, Jorjir Mosque, Jurjir Mosque, Jurjir portal, Jorjir Portal, Masjid–i Hakim, Jurjir Portal, Masjid–i Hakim, Hakim Mosque
Location	Isfahan, Iran
Position	Corner of Entrance
Date	939/328 AH
Period	Buyids
Century	10th
Building Type	Religious
Building Usage	Mosque
Copyright	Hamidreza Kazempour

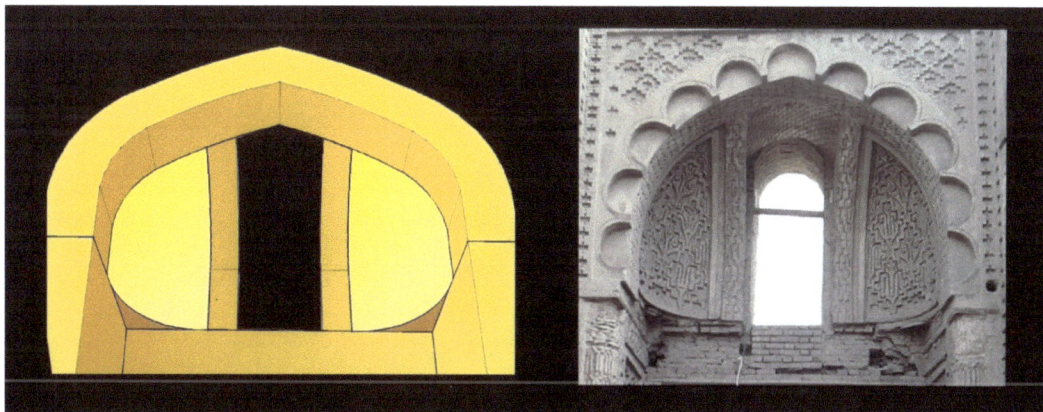

Isometric view of 3D model   Isometric view of *muqarnas*

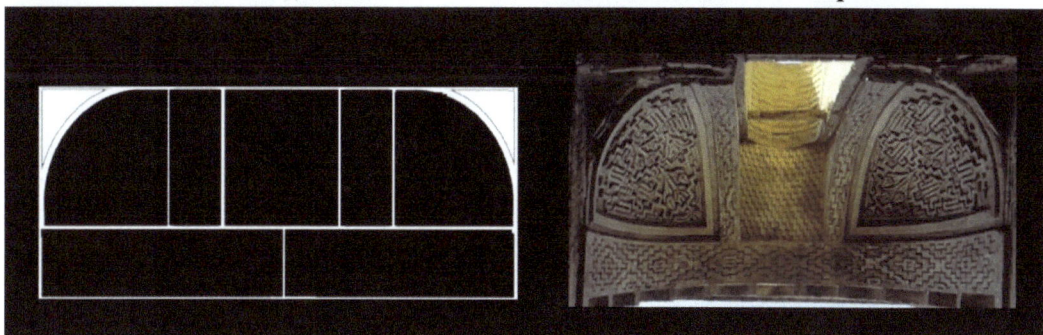

2D plan   Bottom view of *muqarnas*

Figure 5.16   View of squinch in Jurjir Mosque (Source: Author).

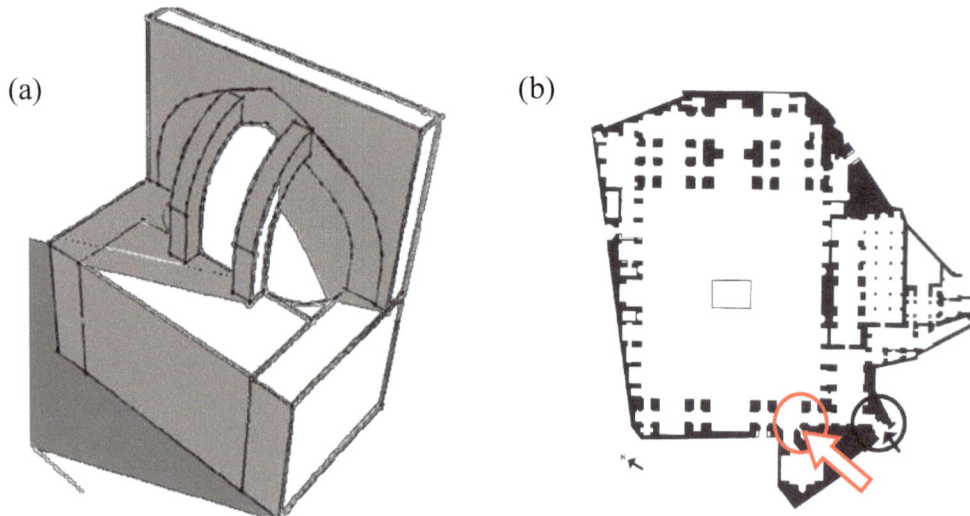

Figure 5.17 (a) Squinch in the entrance portal of Jurjir Mosque, (b) Building plan showing the position of the portal (Memarian, 2012).

The plans of the *tasehs* are semi–circular, creating a triangular space similar to *shaparak*, between them and the wall corner. The details can be observed in Figure 5.16. This sample is more developed compared to Samanid Mausoleum dome chamber cornering. Figure 5.17 shows the position of this portal on the building plan, as well as its back view and Table 5.6 provides information about the constituent elements of the squinch.

Table 5.6: Constituent elements table, Jurjir Mosque.

Element (Tier)	*Toranj*	*Tee*	*Taseh*						*Shaparak*			*Parak*			*Shamseh*			*Espar*			*Takht*	
			Whole			Half				Type			Type			Type			Type			
			No	2D	3D	No	2D	3D	No	2D	3D	No	2D	3D	No	half	whole	No	half	whole	No	Type
One		2	√																			

The squinch of the portal in Jurjir Mosque is also a single–tier structure which has form–, size–and height–transition applications. Categorized as having a square lattice plan, the squinch has the support of ribs for transferring the load to the base and is structural. The decoration is one–sided and has similar construction material with the building.

5.3   Inception of *Muqarnas*

After the entrance of Jurjir Mosque, a remarkable change happened in the development of this ornament. For the first time, in Na'in Jame' Mosque the simple squinch deforms from a structure in a corner of a construction into a kind of veneer to decorate a vault. This is the first example in which the whole vault is covered with *tasehs* projecting over each other and is hence the beginning of more advanced decorations leading to the creation of *muqarnas*.

5.3.1   Na'in Jame' Mosque, I and II

Na'in Jame' mosque, shown in Figure 5.18, is in fact a pre–Seljuk mosque, constructed in 960 C.E., with a simple hypostyle plan as illustrated in Figure 5.19 (Michell & Grube, 1995; O'Kane, 1996; Pope, 1977b; Schroeder, 1977). Despite the changes, reconstructions and additions, the mosque experienced during Seljuk times, the plan has remained simple. The courtyard façade belongs most probably to the alternations of the Seljuk period but central nave, which is located along the axis of qibla, i.e. the southwest direction in Iran, dates back to the original construction period. To emphasize on the direction of qibla, two angled piers are flanking the main nave, which are slightly thicker and taller than the rest. As a result, a lip is formed around the central nave and hence, the ceiling rests over the neighbour arcades' roofline (Hillenbrand, 2004; Michell & Grube, 1995; O'Kane, 1996; Pope, 1977b; Schroeder, 1977).

Figure 5.18 Na'in Jame' Mosque (Source: Author).

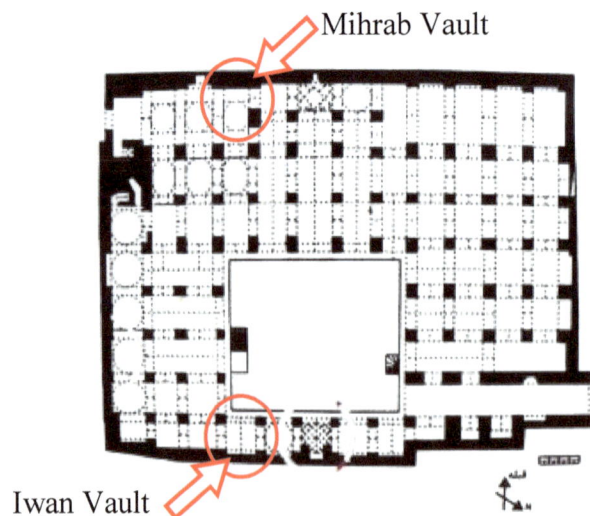

Mihrab Vault

Iwan Vault

Figure 5.19 Hypostyle plan of Na'in Jame' Mosque (Source: Iran Cultural Heritage and Tourism Organization).

As marked on the building plan, in Figure 5.19, there are two pseudo–*muqarnas* vaults in the building which are both very important and worth investigation. The south–western structure, marked as mihrab vault in Figure 5.19, is made from four tiers of *tasehs* over the sanctuary area of the mosque. The vault over the dome chamber of mihrab rests over the columns of the central nave on one side and on the brick wall behind mihrab on the other side. The geometry of the vault is not mature and ordered enough in this example and, unlike the rest of the samples that will be studied later in this manuscript, the *tasehs* are not located on the corners.

There are some angled steps between the four tiers to fulfil height transition needs and although there is no access to the hidden side of Mihrab vault, it is believed that the structure benefits from assistance of ribs to achieve its structural role. In order to simulate the structural characteristics of the vault, the two–dimensional plan of the vault obtained from inside the iwan was matched with the three–dimensional pattern of its ceiling, obtained from satellite photography of archaeological reports (Karimi, 2005). Based on this comparison, it was concluded that there should be four hidden major ribs and four minor assistive ribs that create and hold the *tasehs* of the vault. Figure 5.20 shows the major and minor hidden ribs, as well as the plan of the vault obtained from the ceiling.

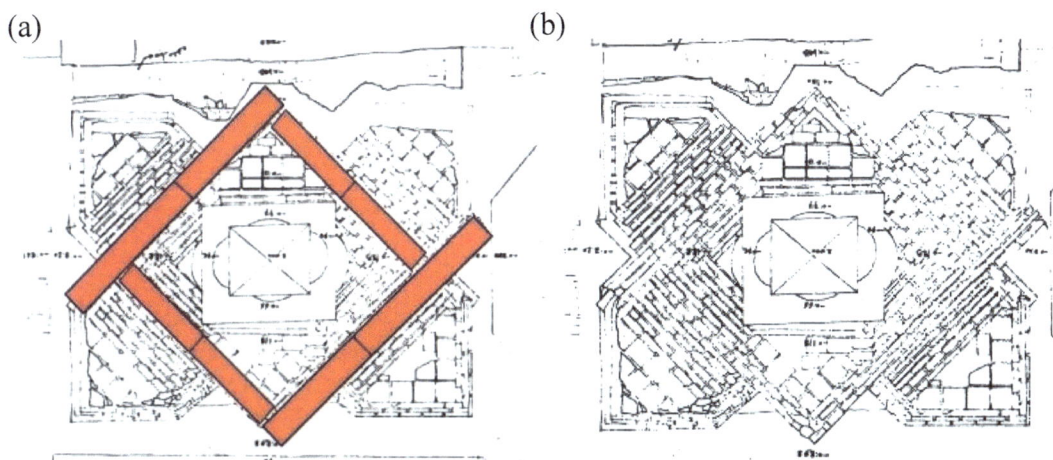

Figure 5.20 (a) The probable location of the hidden ribs of Mihrab vault (Safaeipour, 2009), (b) Plan of ceiling over sanctuary area in Na'in Jame' Mosque (Karimi, 2005).

Other than the mentioned characteristics, Mihrab vault does not have identical construction material with the building and has been made by plaster and its plan as shown in Figure 5.21 is a square lattice plan, having six *tasehs* in each tier, except the last tier which has four *tasehs*, as described in Table 5.8. More detailed information about the building is provided in Table 5.7.

Table 5.7:   Na'in Jame' Mosque building information, Mihrab.

Na'in Jame' Mosque (31.901389°N, 54.368611°E)	
Variant Names	Friday Mosque of Na'in, Masjid Jami' of Na'in, Masjid–e Jame, Congregational Mosque, Masjid–i Jami' of Na'in, Masjid Jami' and Minaret of Na'in, Masjid Jami Nain
Location	Na'in, Iran
Position	Mihrab
Date	960/348 AH
Period	Buyid
Century	10th
Building Type	Religious
Building Usage	Mosque
Copyright	Hamidreza Kazempour and Dr Gholam Hossein Memarian

Isometric view of 3D model    **Isometric view of *muqarnas***

2D plan    **Bottom view of *muqarnas***

Figure 5.21   View of pseudo-*muqarnas* in Na'in Jame' Mosque, Mihrab dome chamber (Source: Author and Memarian).

Table 5.8:   Constituent elements table, Na'in Mosque, Mihrab vault.

Element → Tier	Toranj	Tee	Taseh Whole No	Taseh Whole 2D	Taseh Whole 3D	Taseh Half No	Taseh Half 2D	Taseh Half 3D	Shaparak No	Shaparak Type 2D	Shaparak Type 3D	Parak No	Parak Type 2D	Parak Type 3D	Shamseh No	Shamseh Type half	Shamseh Type whole	Espar No	Espar Type half	Espar Type whole	Takht No	Takht Type
One			6		√				2	√		8	√									
Two			6		√																	
Three			6		√				6	√												
Four			4		√				4	√												

There is another significant vault, introduced in Table 5.9, in the same mosque which is worth considering. This vault, shown in Figure 5.22, is located on the North–Eastern iwan of the mosque, and has a square plan. It has a more ordered arrangement in terms of geometry of constituent element and the fact that makes it significant is the initial appearance of *shamseh*, i.e. the central sunburst medallion, as a constituent element in it.

The Iwan vault has two tiers of *tasehs* and as just mentioned a *shamseh* on the apex. The two tiers of the vault, as described in Table 5.10, have eight *tasehs* each, with 12 plumb *shaparak* elements filling the space between them in the first tier, and moving upward to the apex, i.e. an eight–pointed whole *shamseh*.

Table 5.9:   Na'in Jame' Mosque building information, Iwan.

Na'in Jame' Mosque (31.901389°N, 54.368611°E)	
Variant Names	Friday Mosque of Na'in, Masjid Jami' of Na'in, Masjid–e Jame, Congregational Mosque, Masjid–i Jami' of Na'in, Masjid Jami' and Minaret of Na'in, Masjid Jami Nain
Location	Na'in, Iran
Position	Iwan
Date	960/348 AH
Period	Buyid
Century	10th
Building Type	Religious
Building Usage	Mosque
Copyright	Hamidreza Kazempour

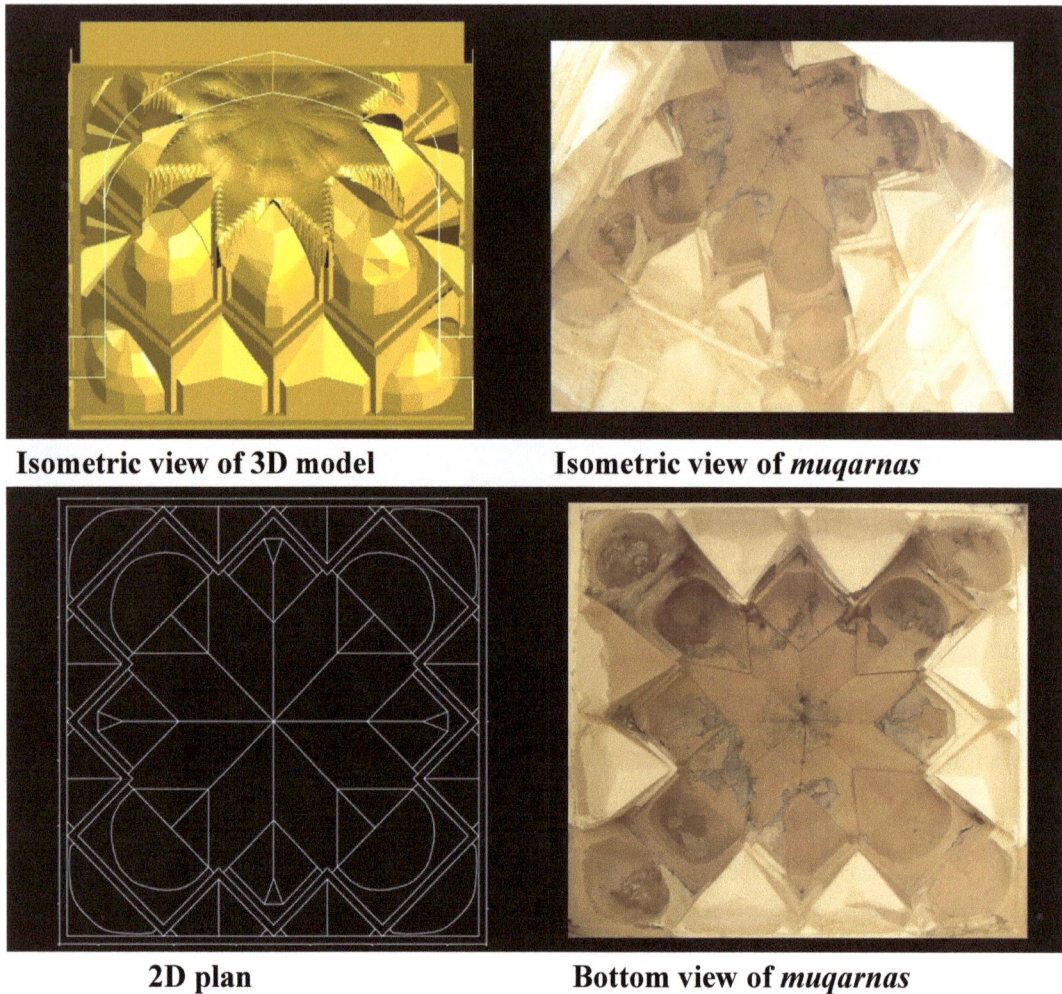

Isometric view of 3D model   **Isometric view of *muqarnas***

2D plan   **Bottom view of *muqarnas***

Figure 5.22   View of pseudo-*muqarnas* in Na'in Jame' Mosque, Mihrab dome chamber (Based on (Kazempourfard, 2006)).

Table 5.10:   Constituent elements table, Na'in Mosque, Iwan.

Element	*Toranj*	*Tee*	*Taseh*						*Shaparak*			*Parak*			*Shamseh*			*Espar*			*Takht*	
↓Tier			No	Whole		No	Half		No	Type		No	Type		No	Type		No	Type		No	Type
				2D	3D		2D	3D		2D	3D		2D	3D		half	whole		half	whole		
One			8		√				12	√												
Two			8		√										1		√					

The matching of two–dimensional plans of Iwan vaults from inside the iwan, and the plan obtained from the three–dimensional satellite pattern, results in the fact that four hidden ribs were used on the four corners of the vault of North–Eastern iwan as well. Figure 5.23 shows the schematic view of the vault's hidden ribs.

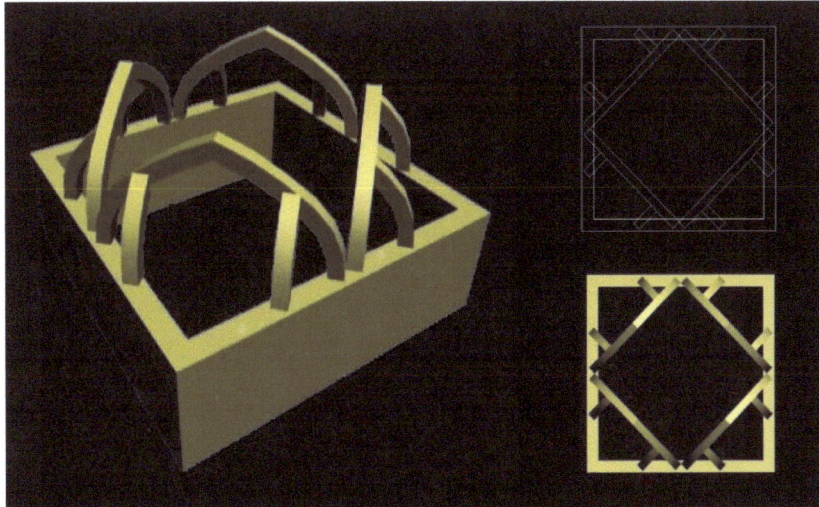

Figure 5.23   Hidden ribs of iwan, Na'in Jame' Mosque (Source: Author).

Compared to the mihrab vault, the shape of the constituent elements are more mature, in terms of shape and geometry. It is as if there are four squinches on the corners and well-established *shaparaks* can be observed clearly. Furthermore, there is a whole *shamseh* on the apex that altogether makes the north–eastern iwan very special and remarkable. As it can be concluded from the two–and three–dimensional figures that the pseudo–*muqarnas* vault of the North–Eastern iwan of Na'in Jame' Mosque, assists in form–, height and size-transition. The finishing material is plaster which is different from the structure. With the help of vault ribs, it has a structural role and the one–sided decoration follows the governing rules of a square lattice plan.

5.3.2   Davazdah Imam Shrine

After the Samanid Mausoleum, the transition technique used in Davazdah Imam Shrine dome chamber is the second remarkable Islamic squinch which guides us towards seeking the development path of *muqarnas*. Although the double–tier cornering squinch of this ancient mosque is not the first and is erected chronologically 30 years after the small double–tier squinch of Gonbad–e Qabus, 1006–7 C.E., shown in Figure 5.24, it was chosen as one of the main guiding samples for its very much developed and significant form, compared to that of Gonbad Qabus squinch, as well other significant characteristics of it, such as the vertical growth of *tasehs*, which will be elaborated below.

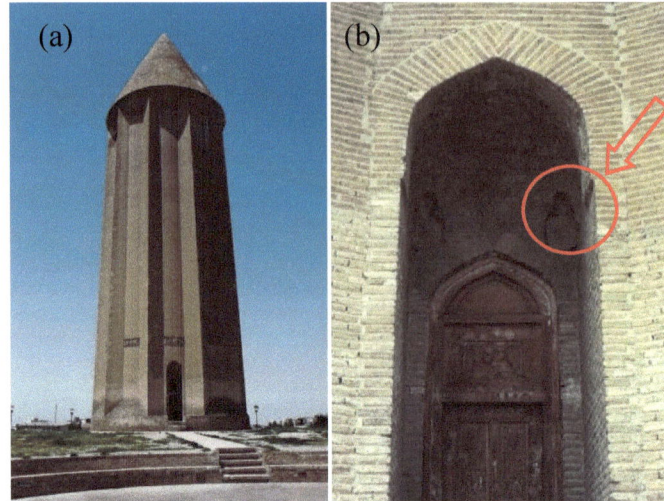

Figure 5.24   (a) Gonbad Qabus tower, Gorgan, Iran (Source: Panoramio.com), (b) Double–tier squinch over the entrance (Safaeipour, 2009)

The shrine of Davazdah Imam, shown in Figure 5.25, is a building with religious applications in the Fahadan quarter, Yazd. Davazdah Imam which literary means 12 Imams is the name given to the building as a memorial of the Shi'a 12 Imams. The shrine was constructed 1036–7 but the decorative pattern of the front portal was later added during Seljuk period (Pope, 1977b; Schroeder, 1977; Hillenbrand, 2004; Hutt & Harrow, 1978).

The squinch of this shrine, which is erected inside a shouldered arch, as shown in Figure 5.26, is famous as being one of the oldest available squinches. The Based on investigations of many scholars, the double-tier squinch of this building is considered as one of the main steps towards the creation of *muqarnas* (Pope, 1977b; Hillenbrand, 2004; Hutt & Harrow, 1978; Edwards & Edwards, 1999). The structure principally consists of a dome rising over an octagonal base created by a number of tri–lobed squinches, which itself rests over a square room. Figure 5.27 shows the section and the plan of the building. The building is erected by bricks and a plaster finishing over the brick. The double–tier squinches in the middle octagonal space are structural and provide a size–, form–and height–transition between the square walls and the heavy dome. The structural role of one–sided squinches is the result of their supporting ribs.

More details about the building can be found in Table 5.11 and Figure 5.28 shows the isometric three–dimensional view of the vault, as well as its two–dimensional pattern plan. The vertical expansion of *tasehs* can be observed in the squinches of Davazdah Imam and their significantly more ordered geometry and form can be clearly seen. The *parak* and *shaparak* elements can also be distinguished as part of the squinch, yet their form is not complete.

Figure 5.25   Davazdah Imam Shrine, Yazd, Iran (Source: Dome.mit.edu).

Figure 5.26   The double–tier Squinch of Davazdah Imam (Source: Archnet.org).

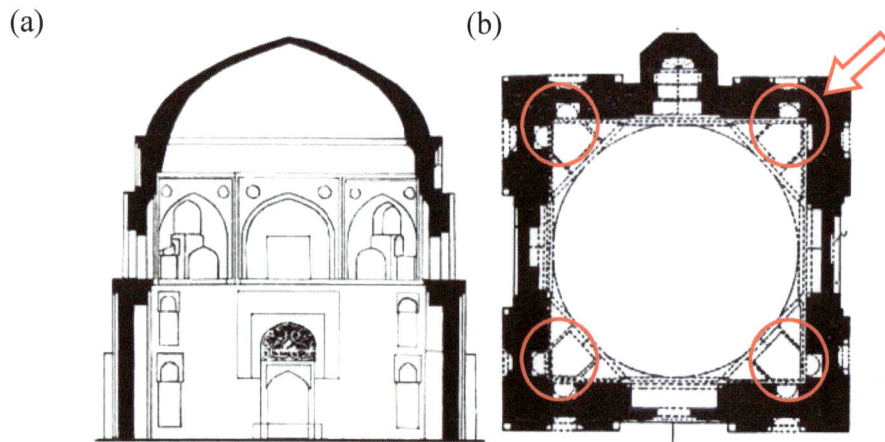

Figure 5.27 (a) Schematic section of the Shrine of Davazdah Imam, (b) Building plan showing the position of studied squinches (Pope & Ackerman, 1977).

Table 5. 11: Davazdah Imam Shrine building information.

Mashhad Davazdah Imam (31.897222°N, 54.367778°E)	
Variant Names	Davazdeh Imam Shrine, Mausoleum of Davazda Imam, Mashhad Davazdah Imam, Davazdah Imam Shrine, Shrine of the 12 Imams; Davazdeh Imam Mausoleum, Davuzdah Imam, Davazadah Imam, Maghbareh–ye Davazdah Emam, Mausoleum of Davazdah Imam
Location	Yazd, Iran
Position	Cornering
Date	1036–37/427–28 AH
Period	Buyid
Century	11th
Building Type	Funerary, Religious
Building Usage	Tomb, Shrine
Copyright	Hamidreza Kazempour (3D and Plan)

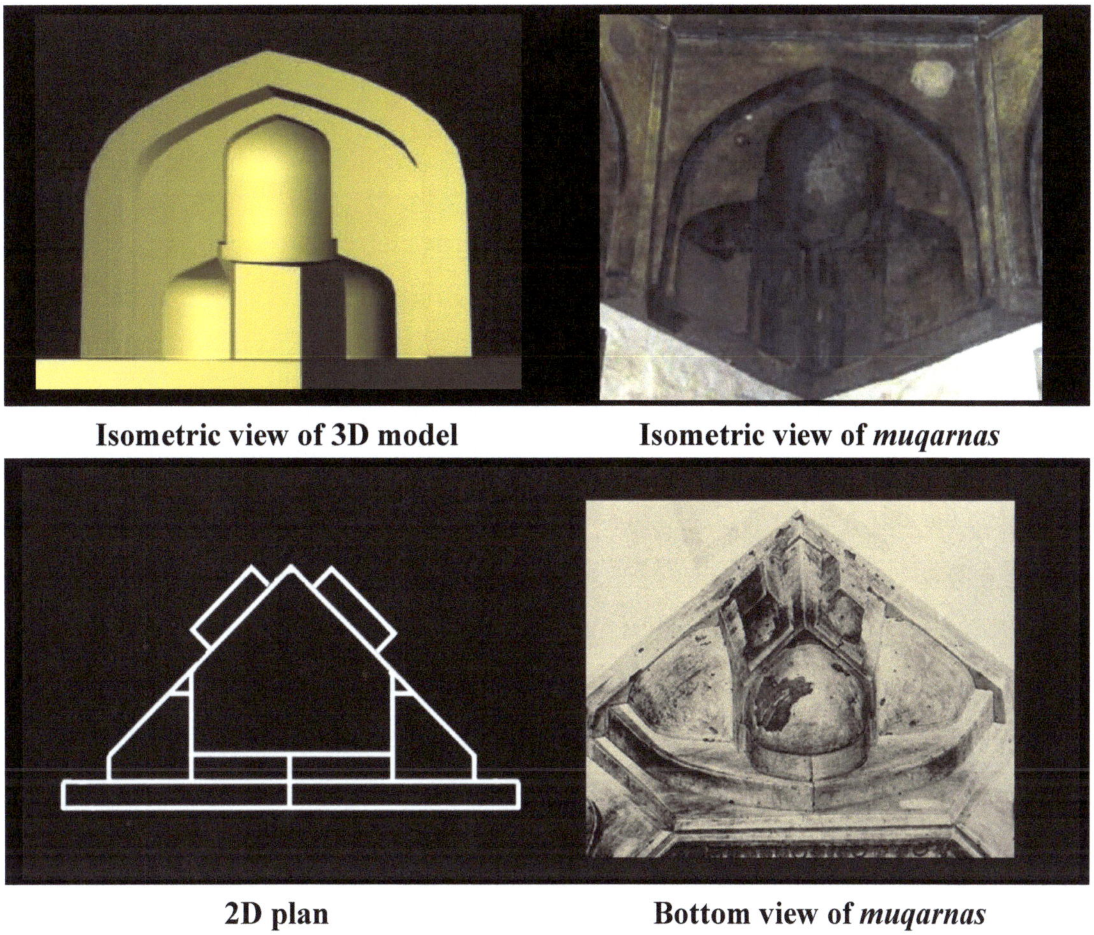

Isometric view of 3D model **Isometric view of *muqarnas***

2D plan **Bottom view of *muqarnas***

Figure 5.28   View of pseudo-*muqarnas* in Davazdah Imam (Source: Author).

Returning back to the relationship of Samanid Mausoleum and Davazdah Imam Shrine, although both have squinches to provide cornering in a dome chamber, and the main rib is not hidden in both structures, the number of tiers and *tasehs* are more advanced in Davazdah Imam. Table 5.12 provides information about the constituent elements of the latter.

Table 5.12:   Constituent elements table, Davazdah Imam Shrine.

Element ↓ Tier	*Toranj*	*Tee*	Taseh							Shaparak			Parak			Shamseh			Espar			Takht	
			No	Whole		No	Half			No	Type		No	Type		No	Type		No	Type		No	Type
				2D	3D		2D	3D			2D	3D			2D	3D		half	whole		half	whole		
One			4	√						3	√													
Two			1	√																				

5.3.3 Barsian Jame' Mosque

Barsian is a village located 42 kilometres north–east of Isfahan; a village in which there is a nice historical site consisting of a minaret, a mosque and a Caravanserai. The locals pronounce the name of the village as "Bersioon" that is probably an equivalent of "Parsian", which in turn literary means "Persian", indicating that the village was of significant reputation in earlier times. Caravanserais were normally attached to mosques and this mosque is specifically interesting because of its old minaret and its very fine brickworks, as well as its exceptional mihrab inside the prayer hall, having unusual brick decorations (Pope, 1977b).

The old minaret was built first in 1097 C.E but the mosque was added later to the complex in 1114. Therefore, the mosque does not follow the more typical "four iwan style" of Iranian architecture. The prayer hall of the mosque has four double–tier squinches to provide a smooth transition from the octagon base to the circular dome (Hillenbrand, 2004; Pope, 1977b). Although the building went through a long historical development and change, the Seljuk identity of its architecture can still be felt and the squinches are similar to the one in the Shrine of the Davazdah Imams, Yazd (Grabar, 1992; Pope & Ackerman, 2005).

The pseudo–*muqarnas* structures show variety of different characteristics in various buildings, making it hard to discover their nature and architectural role. In Jame' Mosque of Barsian, shown in Figure 5.29, two dissimilar pseudo–*muqarnas* structures are erected on two different positions in the building, i.e. on the corners of dome chamber and in the mihrab. The one on the corner, as can be seen in Figure 5.30, is a double–tier squinch which is, as mentioned before, very similar to that of Davazdah Imam in terms of structure, form and plan; although it was built 60 years after the shrine of Davazdah Imam. Furthermore, the squinches of both buildings benefit from the support of ribs.

The pseudo–*muqarnas* structure of Mihrab is another significant step forward in the evolution of *muqarnas* which will be discussed here. Figure 5.31 shows the section and plan of Barsian Jame' Mosque and shows the position of mihrab on the plan with Table 5.13 providing some information about the building.

The mihrab, as shown in Figure 5.32, is erected in three tiers, using brick. There are five *tasehs* in the first tier and two semi–*tasehs* on the sides. Second tier has six similar *tasehs* in terms of size and shape, whereas as the last tier consists of five similar *tasehs* and two semi–*tasehs* on the sides. The *shaparaks*, located between each two *tasehs*, are plumb and hence, they cannot be seen in the plan. There is a six–pointed *shamseh* on the top. Table 5.14 provides the details of this structure's constituent elements.

Figure 5.29   Barsian Mosque, Barsian, Isfahan (Source: Marematgar.ir).

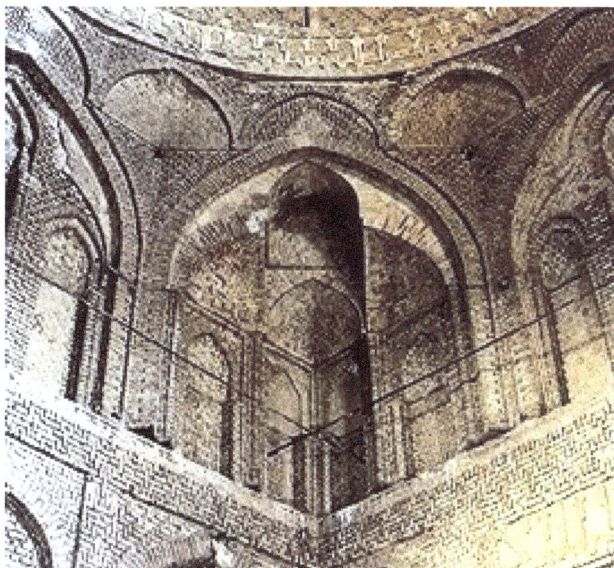

Figure 5.30   Double–tier squinch of Barsian Jame' Mosque (Source: Author).

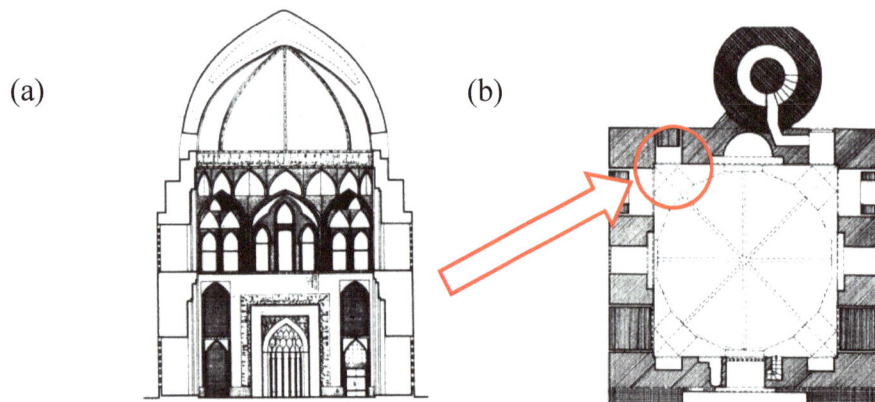

Figure 5.31   (a) Schematic section of the Barsian Mosque, (b) Building plan showing the position of studied mihrab (Source: Iran Cultural Heritage and Tourism Organization).

Table 5.13:   Barsian Jame' Mosque building information.

Masjid Jame' Barsian (32.600000°N, 51.983333°E)	
Variant Names	Friday Mosque of Barsian, Masjid–i–Jami' Bersian, Masjid–i Jami, Masjid–i Jomeh, Masjid–i–Jomeh, Bersian Mosque
Location	Barsian, Iran
Position	Mihrab
Date	1114/514 AH
Period	Seljuk
Century	12th
Building Type	Religious
Building Usage	Mosque
Copyright	Hamidreza Kazempour

Table 5.14:   Constituent elements table, Barsian Mosque.

Element / Tier	Toranj	Tee	Taseh Whole No	Taseh Whole 2D	Taseh Whole 3D	Taseh Half No	Taseh Half 2D	Taseh Half 3D	Shaparak No	Shaparak Type 2D	Shaparak Type 3D	Parak No	Parak Type 2D	Parak Type 3D	Shamseh No	Shamseh Type half	Shamseh Type whole	Espar No	Espar Type half	Espar Type whole	Takht No	Takht Type
One			5		√	2		√	6	√												
Two			6	√					5	√												
Three			5	√		2									1	√						

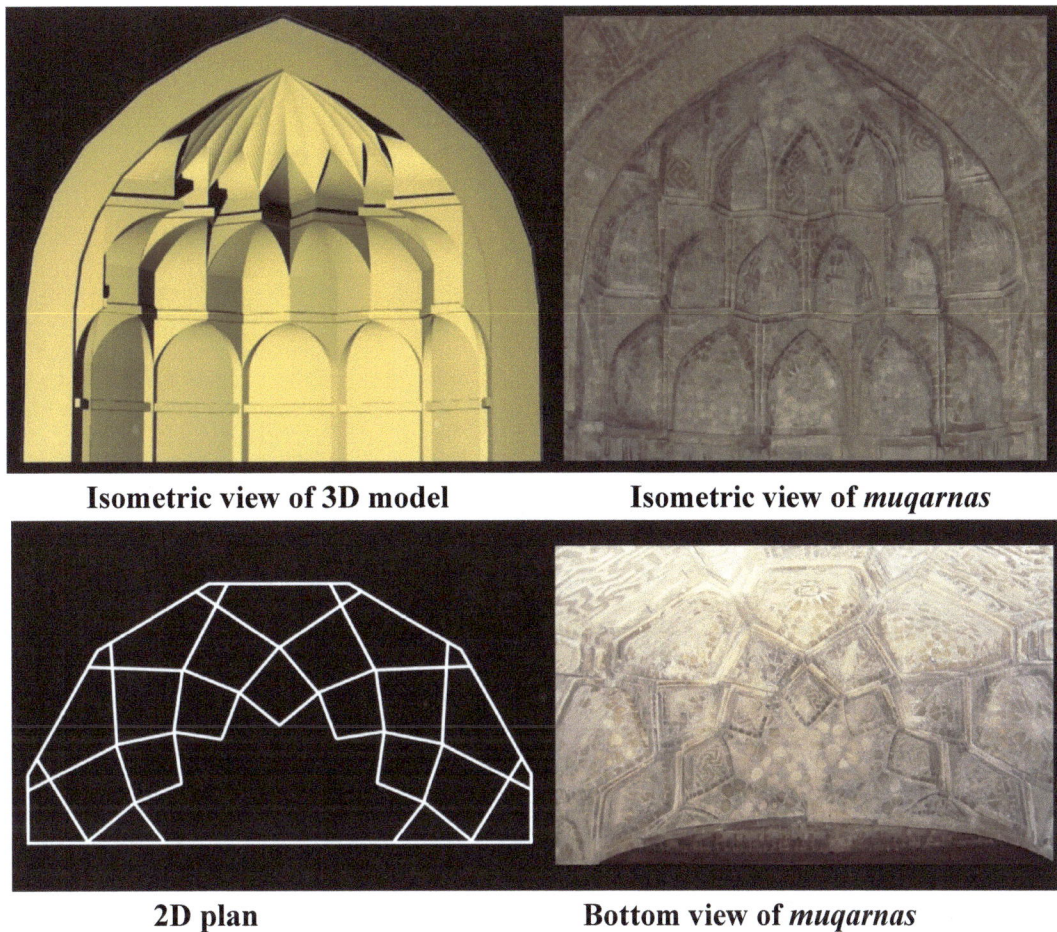

Isometric view of 3D model   **Isometric view of *muqarnas***

2D plan   **Bottom view of *muqarnas***

Figure 5.32   View of pseudo-*muqarnas* in Barsian Jame' Mosque (Based on (Kazempourfard, 2006)).

There are six pilasters, as shown in Figure 5.33, which are used to transfer the load from the six hidden ribs. The ribs are at the back and they are stretched from top of *shamseh* to the top of the columns. The one–sided pseudo–*muqarnas* structure of the mihrab in Barsian Jame' Mosque, like the rest of the samples, provides height–, size–and form–transition and has a structural role with the assistance of the ribs. The erection material is identical to that of the building although they were not constructed simultaneously. The plan is a square lattice plan but compared to previous squinches and vaultings, mihrab of Jame' Mosque of Barsian is more developed in terms of the number of elements and number of tiers.

Figure 5.33   The thin columns of mihrab, Jame' of Barsian (Based on (Kazempourfard, 2006))

5.3.4 Bastam Minaret

Located in the town of Bastam, there are two historical complexes which seem to be originally connected as one group. The larger building is the mausoleum of Bayazid, including the fragments of a Seljuk wall, which is a part of the mosque, and the famous Seljuk minaret of Bastam. Figure 5.34 shows the historical complex of Bastam. The Minaret and the wall fragment of mosque are from the oldest parts of the building, constructed between 1114 to 1120 C.E., based on the inscription band in the mosque (Wilber, 1969; Pope & Ackerman, 2005; Taboroff, 1981).

The construction of the shrine of Bayazid Bastami (d.874 or 877) started only a few years after his death but continued during centuries then after. Many parts were added to the structure and it was expanded to make today's historical complex. Two of these parts will be of significant importance in this investigation. One is the 14.5m high minaret, known as Bastam Minaret, which belongs to the Seljuk period and the other one is the main entrance portal, known as Ghazan Khan Entrance and Portal, built about 200 years after the minaret, which will be studied later in this chapter. Figure 5.35 illustrates the position of the minaret on the plan of the historical complex and building information is provided in Table 5.15.

Figure 5.34   Bayazid Bastami historical complex (Source: Wikipedia).

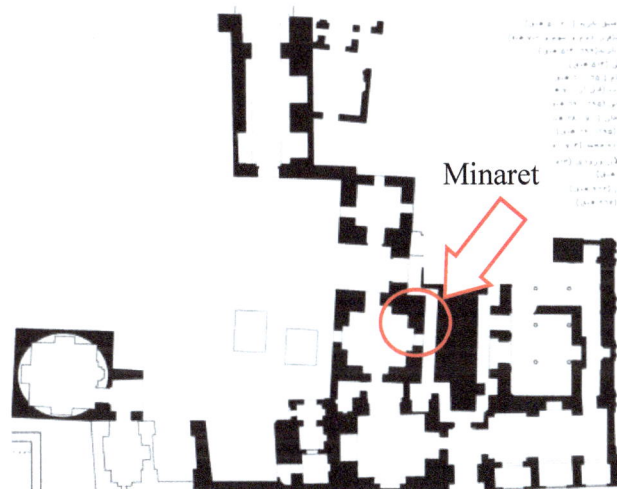

Minaret

Figure 5.35   Plan of Bayazid Bastami historical complex (Source: Iran Cultural Heritage and Tourism Organization).

The pseudo–*muqarnas* structure of Bastam Minaret, shown in Figure 5.36 is important for two main reasons. First, it has a circular plan around the minaret, making it the first pole-table type of plan studied and second, this is the only sample among all 100 samples which has been constructed by a mixture of "suspended unit" and "amalgamated construction" techniques. The suspended unit method is a construction technique which is rarely used in Iran. In this method, the constituent elements are first erected separately and then, they are attached one by one on the wall or ceiling. The spaces between these elements are then filled with decorative brick and plaster. The suspended units of the pseudo-*muqarnas* structure of the minaret, shown in Figure 5.37, are placed over erected bricks which were assigned for this purpose, to point out of the minaret's cylinder when building it.

Table 5.15:   Bastam Minaret building information.

Bastam Minaret (36.485278°N, 54.999722°E)	
Variant Names	Minar Bastam; Bayazid Mausoleum; Mashhad Bayazid; Bayazid Shrine
Location	Bastam, Iran
Position	Bastam Minaret
Date	1120/514 AH
Period	Seljuk
Century	12th
Building Type	Educational, Funerary, Religious
Building Usage	Madrasa, Mausoleum, Shrine
Copyright	Hamidreza Kazempour

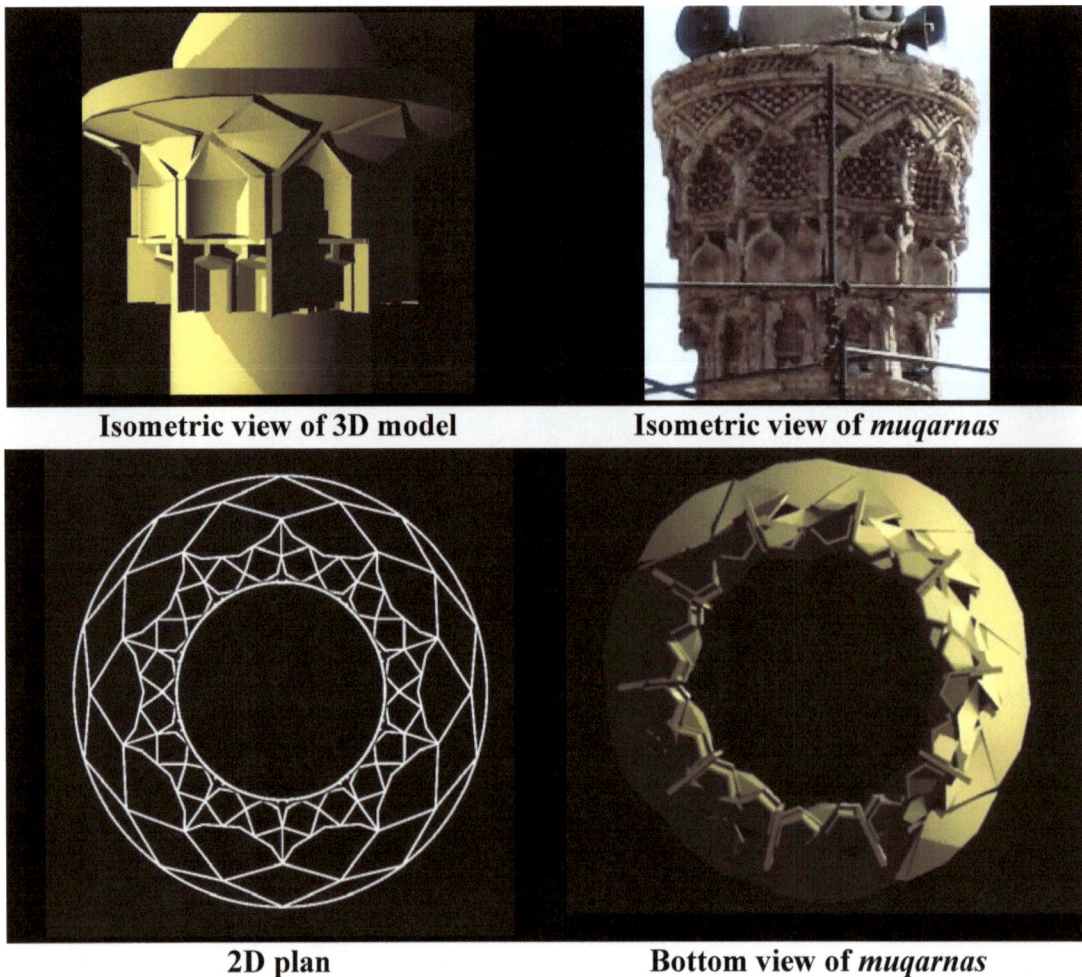

Isometric view of 3D model   **Isometric view of *muqarnas***

2D plan   **Bottom view of *muqarnas***

Figure 5.36   View of pseudo-*muqarnas* in Bastam Minaret (Based on (Kazempourfard, 2006)).

Figure 5.37 Suspended units in Bastam Minaret (Based on (Kazempourfard, 2006)).

This sample is built in three tiers. Table 5.16 provides the information about the sample's constituent elements. As the construction method of this pseudo–*muqarnas* is the "suspended unit", the shapes of the elements are not smooth, especially in the first two tiers. In the last tier, however, *tasehs* and *toranjes* were manipulated more carefully using brick. The plan of this structure has a radial 10-fold symmetry in a full circle. It has been confined between the circular section of the minaret and a bigger circle created according to the outward movement of the elements. Figure 5.38 shows the radial symmetry of the pole table plan. The sample of Bastam Minaret is structural and it is erected with similar material as the building itself.

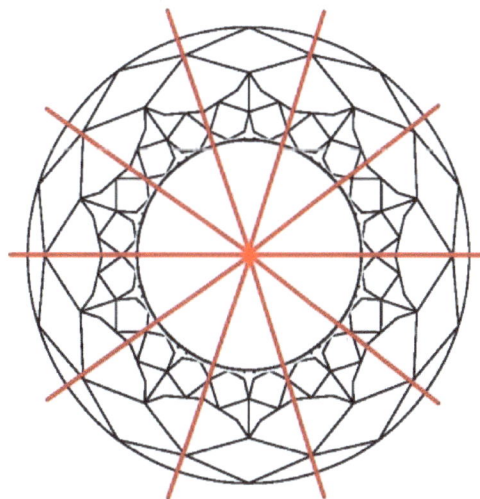

Figure 5.38 Radial symmetry of the plan in Bastam Minaret.

Table 5.16:   Constituent elements table, Bastam Minaret.

Element / Tier	Toranj	Tee	Taseh No	Whole 2D	Whole 3D	No	Half 2D	Half 3D	Shaparak No	Type 2D	Type 3D	Parak No	Type 2D	Type 3D	Shamseh No	Type half	Type whole	Espar No	Type half	Type whole	Takht No	Type
One						10	√		20	√												
Two			20		√				10	√		20	√									
Three	10		10		√																	

5.3.5 Sin Jame' Mosque

The architectural characteristics of domes are very complex from structural and stability points of view. In Jame' Mosque of Sin, as illustrated in Figure 5.40, a pseudo–*muqarnas* structure is used for the first time as a veneer to cover a whole interior dome chamber. Sin Jame' Mosque, shown in Figure 5.39, is located 24 Kilometres on North of Isfahan. According to its inscription, the minaret and the mosque of the Seljuk construction were built in 525 and 528 AH, respectively (Oghabi, 1999; Mehrabadi, 1973; Godard, 1965; Honarfar, 1995). Table 5.17 provides key information about the building.

Figure 5.39   Sin Jame' Mosque (Source: Author).

Table 5.17: Sin Jame' Mosque building information.

Masjid Jame' Sin (32.802222°N, 51.621111°E)	
Variant Names	Friday Mosque of Sin, Masjid–i Jami of Sin, Masjid–e Jomeh, Great Mosque, Masjid–e Jame, Congregational Mosque
Location	Sin, Iran
Position	Dome Chamber
Date	1134/528 AH
Period	Seljuk
Century	12th
Building Type	Religious
Building Usage	Mosque
Copyright	Hamidreza Kazempour

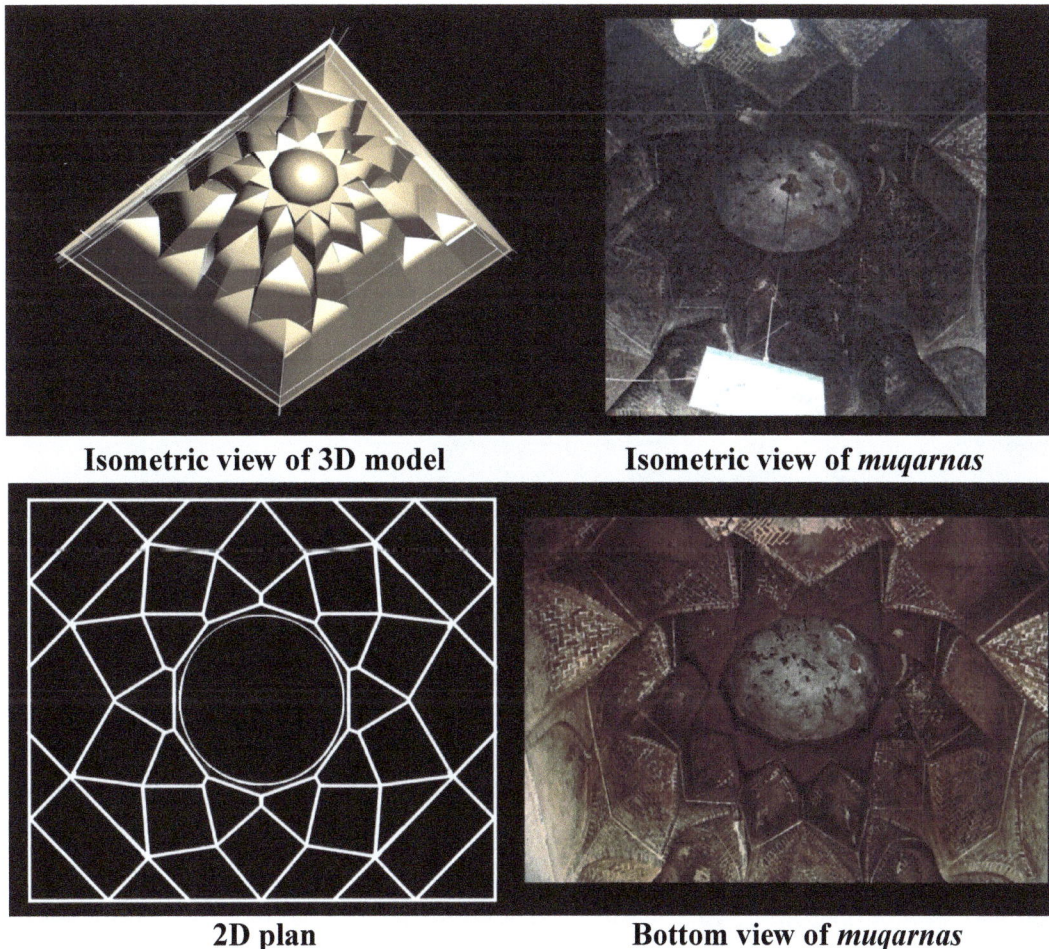

Isometric view of 3D model · Isometric view of *muqarnas*

2D plan · Bottom view of *muqarnas*

Figure 5.40 View of pseudo-*muqarnas* in Sin Jame' Mosque (Based on (Kazempourfard, 2006)).

The mosque has a 7×7 m^2 apron but the prayer hall is constructed by brick in a 4x5.6 m^2 rectangular space. Figure 5.41 shows the schematic plan of the building. This pseudo–*muqarnas* structure, as brought in Table 5.18, has three tiers of *tasehs* and plumb *shaparaks* and the mason has been able to cover the rectangularity of the base very elegantly by creating four squinches on the four corners and then filling up the rest of the space with *tasehs* and *shaparaks*, using radial arrangements which start from *shamseh* on the apex, in a way that the observer may hardly notice the base is not square.

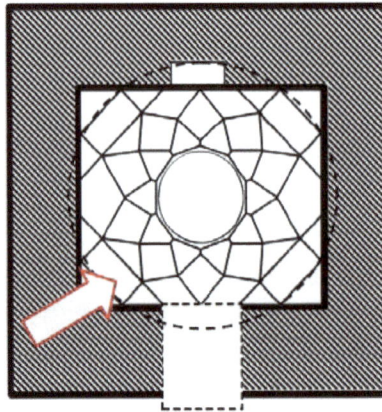

Figure 5.41   Plan of dome chamber, Sin Mosque (Source: Author).

Table 5.18:   Constituent elements table, Sin Jame' Mosque.

Element / Tier	Toranj	Tee	Taseh						Shaparak			Parak			Shamseh			Espar			Takht	
			No	Whole		No	Half		No	Type		No	Type		No	Type		No	Type		No	Type
				2D	3D		2D	3D		2D	3D		2D	3D		half	whole		half	whole		
One			14	2	12				16	√												
Two			10		√				10	√		8		√								
Three			10		√							10		√	1	√						

It is noteworthy that the squinches of this structure are very similar to those of Davazdah Imam and Barsian. Figure 5.42 shows the plan of the pseudo–*muqarnas* decoration, in which the squinches are highlighted. This plan shows the two distinctive parts of the structure, i.e. the squinches which are complete structures on themselves and the radial set of *tasehs* which have been added professionally to them to cover the whole ceiling. The plan of Sin Mosque's dome chamber is also of square lattice type and the plumb *shapakars* cannot still be identified in it.

Figure 5.42 Two distinctive parts of plan of Sin Jame' Mosque, i.e. the squinches and the radial set of *tasehs* (Source: Author).

In order to judge the construction technique of this sample, it is necessary to access the space between the dome and the pseudo–*muqarnas* veneer, which was impossible. Hence, based on the plan geometry and analyzing it, it can be concluded that similar to Davazdah Imam and Barsian cases, the cornering ornaments benefit from the help of the ribs to achieve structural role in the constructions, and the rest of the structure is erected by *tasehs* projecting over the squinches. The ribs are responsible for transferring the load of *tasehs* to the squinches and from there to the walls. Figure 5.43 illustrates the ribs that use squinches to transfer the load of the veneer to the walls. This pseudo–*muqarnas* structure is also one–sided and provides size–, height–and form–transition. The building and the veneer are both built by brick but not at the same time.

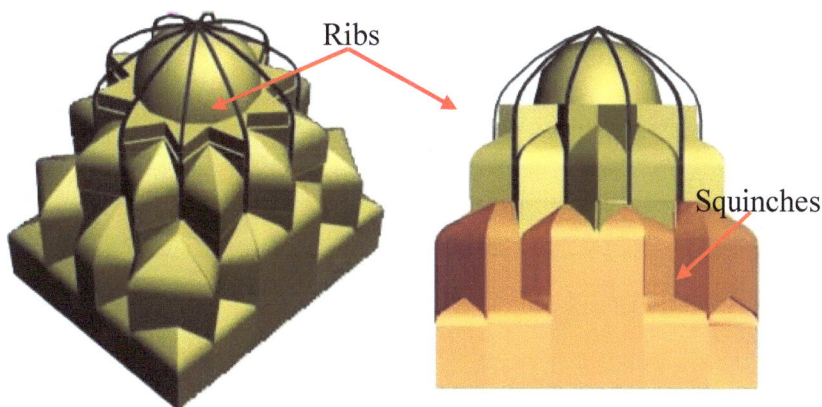

Figure 5.43 Ribs responsible for transferring the load from *tasehs* to squinches and from there to walls in Sin Jame' Mosque (Source: Author).

5.3.6 Isfahan Jame' Mosque, I and II

Many outstanding samples of creative initiation of traditional arts can be observed in decorations of Isfahan Jame' Mosque, shown in Figure 5.44. For instance the pseudo–*muqarnas* structure used to cover the skylight in the nave is one of them. This nave, along with the other iwans and portals, makes a total of 484 different vaults, making the mosque a unique museum of curved and arched vaults and veneers (Galdieri, 1984).

Figure 5.44   Isfahan Jame' Mosque (Source: iranboom.ir).

Together with Mozaffarieh School, inside Isfahan Jame' mosque, a total of 16 vaults out of the 484 vaults have been recorded and studied for this manuscript. Two of these vaults will be studied in detail below and the rest are available for further investigation in Appendix A of the manuscript. The first specific vault of interest is the 47th vault based on Galdieri's classifications, marked in Figure 5.45. As mentioned before, the vault, shown in Figure 5.46 is in fact a skylight provided inside one of the main naves. Table 5.19 provides some information about the building.

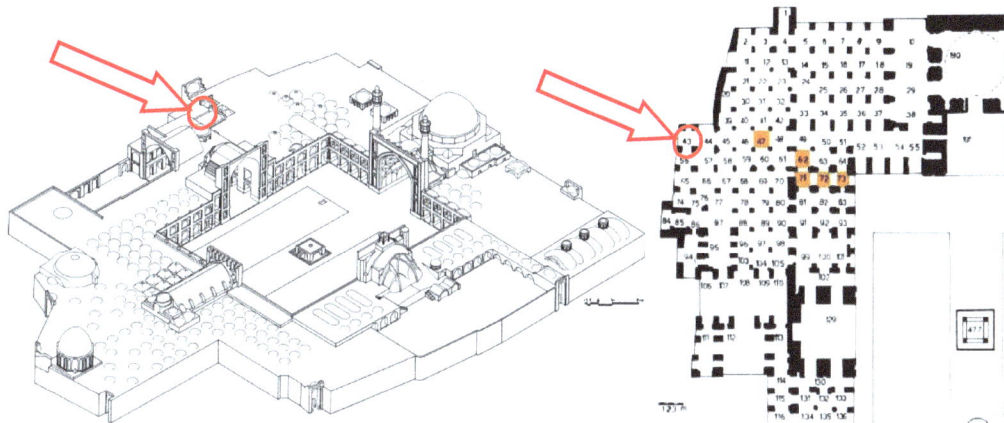

Figure 5.45   Sketch of Isfahan Jame' Mosque's and plan of its nave, showing vault number 47 (Galdieri, 1984).

Vault number 47, other than assisting in from–, size–and height–transition is important because of its diversity in terms of combination of forms and spaces among the rest of the vaults. The vault has 16 *tasehs* in two octagonal tiers, explained in Table 5.20, creating eight complete two–dimensional *shaparaks* between the two tiers. After the two tiers, there are layers of bricks coming forward to close the gap towards the octagonal skylight opening. The outward movement of brick layers in this vault are in fact the beginning for the creation of *toranj* elements, although the shape is not yet completely formed. Furthermore, the innovative replacement of an octagonal skylight with *shamseh* is another significance of vault number 47.

Table 5.19:   Isfahan Jame' Mosque building information, Nave.

Masjid Jame' Isfahan (32.671111°N, 51.685556°E)	
Variant Names	Friday Mosque of Isfahan, Masjid–i Jami' Isfahan, Masjid–i Jami of Isfahan, Masjid–e Jomeh, Great Mosque, Masjid–e Jame, Masjid–e Jameh, al–Jami' al–Kabir, Masjid–i Jami' Isfahan, Masdjid–i Djum'a, Congregational Mosque
Location	Isfahan, Iran
Position	Nave
Date	1118–1157
Period	Seljuk
Century	12th
Building Tpe	Educational, Religious
Building Usage	Madrasa, Mosque
Copyright	Hamidreza Kazempour

Isometric view of 3D model   **Isometric view of *muqarnas***

2D plan   **Bottom view of *muqarnas***

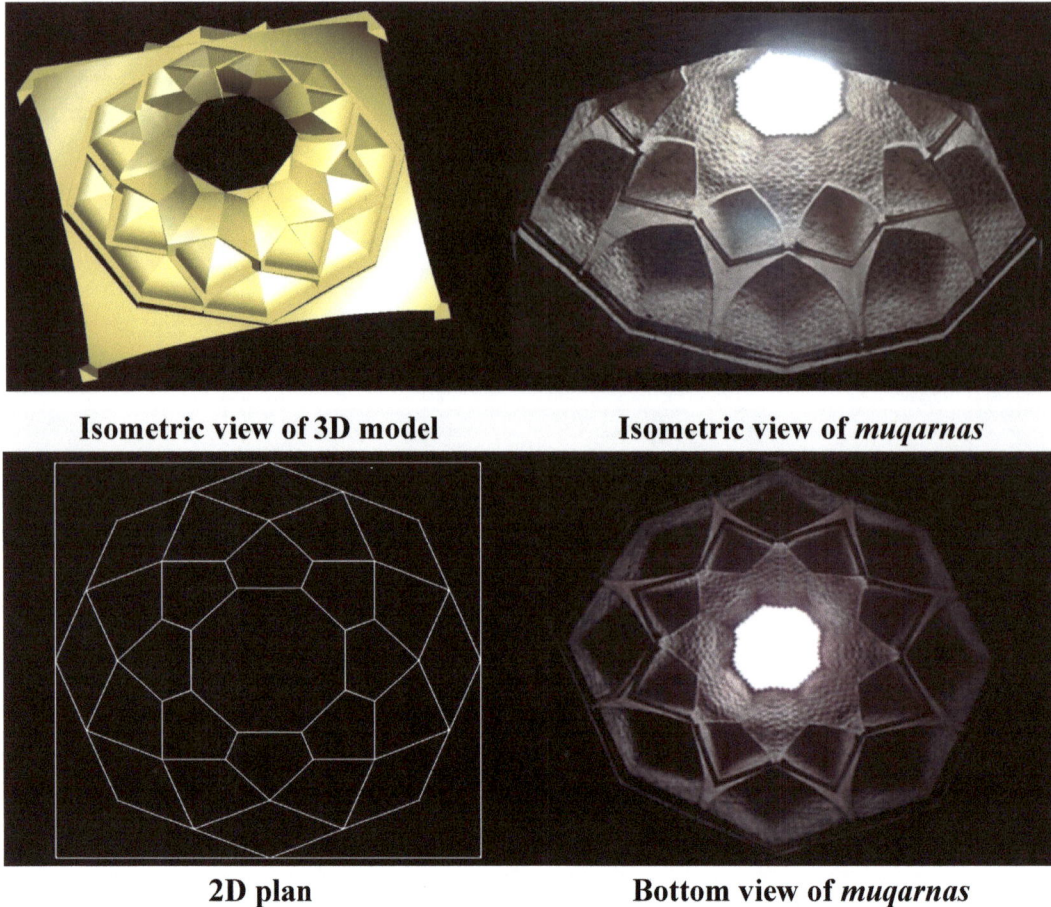

Figure 5.46   View of pseudo-*muqarnas* in Isfahan Jame' Mosque, Nave (Based on (Kazempourfard, 2006)).

Table 5. 20:   Constituent elements table, Isfahan Jame' Mosque, Vault number 47.

Element → / ↓Tier	Toranj	Tee	Taseh						Shaparak			Parak			Shamseh			Espar			Takht	
			No	Whole		No	Half		No	Type		No	Type		No	Type		No	Type		No	Type
				2D	3D		2D	3D		2D	3D		2D	3D		half	whole		half	whole		
One			8	√					8	√												
Two			8	√											1		√					

The material of vault number 47 is brick, identical to that of the building and is hence structural. The two–dimensional plan of the vault is a square lattice plan type, based on Shiro Takahashi's classification. The extraordinary design and erection of the vaults in the iwans of Isfahan Jame' Mosque, makes them all something above being a simple iwan. They all have reputation of their own and each of them in unique in some way. The iwan which will be investigated here is the Western Iwan, known as the "Maestro Iwan" or "Soffeh Ostad". The Western Iwan was constructed in the early 12th century (Honarfar, 1965). Figure 5.47 is the full plan of Isfahan Jame' Mosque; indicating Western Iwan and Vault Number 47 on it. For more details on the other iwans, one may refer to Appendix A of this manuscript.

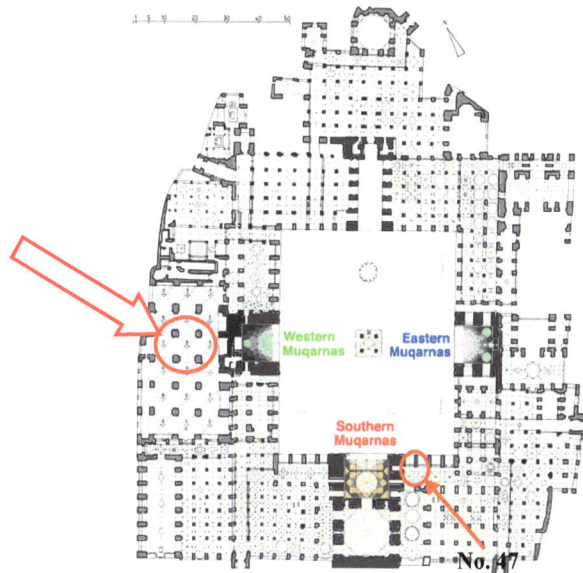

Figure 5.47   Full plan of Isfahan Jame' Mosque, indicating Western iwan and the vault number 47 (Galdieri, 1984).

Table 5. 21:   Isfahan Jame' Mosque building information, Iwan

Masjid Jamc' Isfahan (32.671111°N, 51.685556°E)	
Variant Names	Friday Mosque of Isfahan, Masjid–i Jami' Isfahan, Masjid–i Jami of Isfahan, Masjid–e Jomeh, Great Mosque, Masjid–e Jame, Masjid–e Jameh, al–Jami' al–Kabir, Masjid–i Jami' Isfahan, Masdjid–i Djum'a, Congregational Mosque
Location	Isfahan, Iran
Position	Western Iwan
Date	1118–1157
Period	Seljuk
Century	12th
Building Type	Educational, Religious
Building Usage	Madrasa, Mosque
Copyright	Hamidreza Kazempour

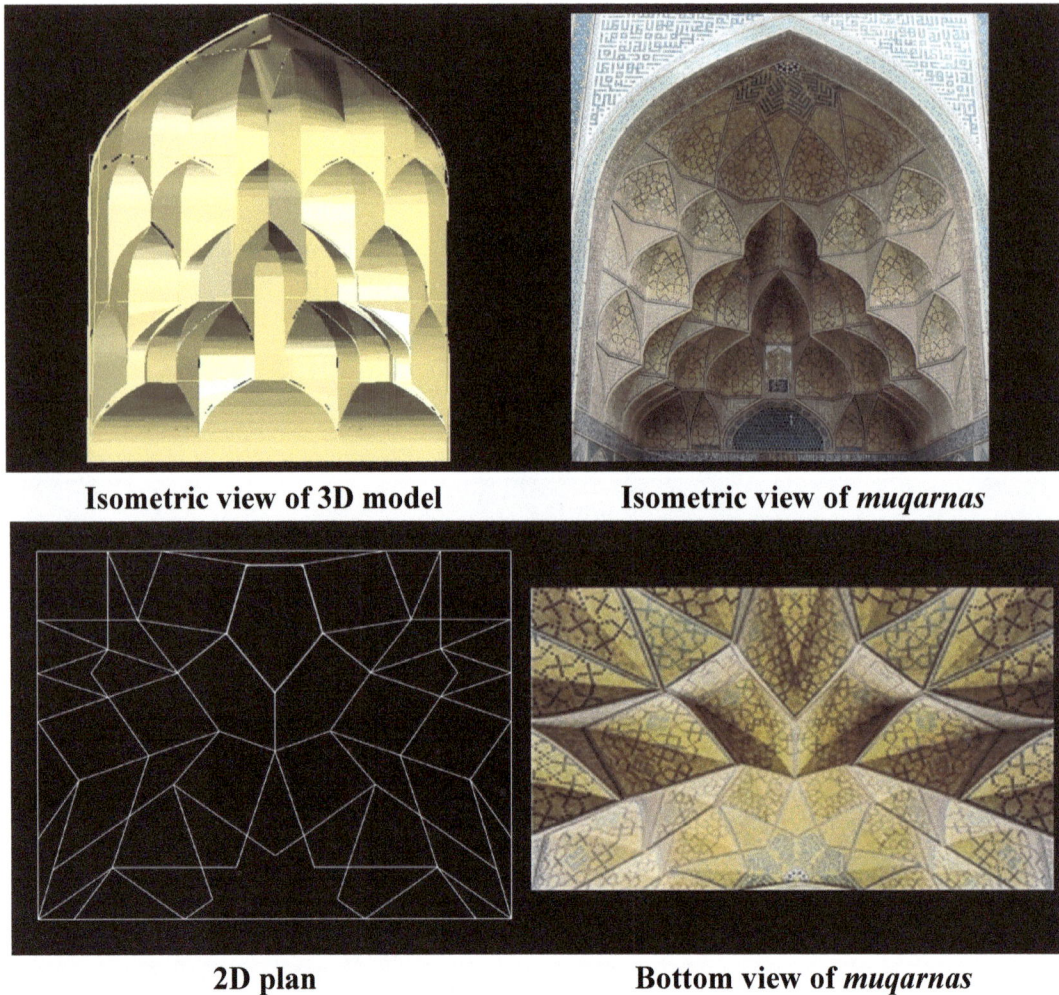

Isometric view of 3D model   **Isometric view of *muqarnas***

2D plan   **Bottom view of *muqarnas***

Figure 5.48   View of *muqarnas* in Isfahan Jame' Mosque, Western Iwan (Based on (Kazempourfard, 2006)).

The Western Iwan, as introduced in Table 5.21 and illustrated in Figure 5.48, is a turning point in terms of *muqarnas* constituent elements. There are many new elements used in this iwan and then vastly in later *muqarnas* and pseudo–*muqarnas* structures. Constructed in five tiers, the Western Iwan has 30 *tasehs*, *shamseh* and *toranjes*, as explained in Table 5.22.

Table 5.22: Constituent elements table, Isfahan Jame' Mosque, Western Iwan.

Element / Tier	Toranj	Tee	Taseh No	Whole 2D	Whole 3D	No	Half 2D	Half 3D	Shaparak No	Type 2D	Type 3D	Parak No	Type 2D	Type 3D	Shamseh No	Type half	Type whole	Espar No	Type half	Type whole	Takht No	Type
One			11	1	10	2	√		8	6	2	4		√				1		√		
Two		1	10		√				4	√		6	√									
Three			5		√	2	√		4	√		4	√									
Four			6		√				4	√		2	√									
Five	5		4		√	2		√							1	√						

For the first time, in this decorative construction, *Parak* and *shaparak* elements were erected two- and three–dimensionally. *T* element and *Espar* are the other innovative inventions of the maestro mason who built this iwan. Simultaneous use of *T* and *taseh*, and combining them results in the creation of *tanureh* in the masterpiece of Western Iwan, known as the Maestro Iwan. Figure 5.43 shows the two–dimensional pattern plan of Western Iwan and connects it with its three–dimensional figure. It is worth stating that all of these elements are supported by three major and nine minor ribs, as illustrated in Figure 5.49.

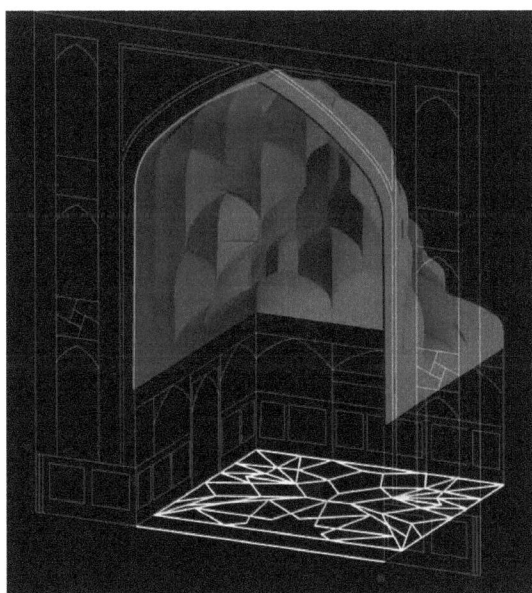

Figure 5.49 Two–dimensional plan of Western Iwan connected to its three–dimensional form (Based on (Kazempourfard, 2006)).

As there is no access to the space between the decorative veneer and exterior roof, the position of these ribs and their structural role were obtained by photographing their appearance from under the vault and over the roof and matching their geometry. Information about the connection between these two layers could only be achieved based on restoration reports of the walls. According to available evidence, there were two restorations during the last century, one was conducted by Ma'arefi, in 1950 and the other one was from 1972 to 1978 by Galdieri. The reports of the second restoration by Galdieri are published as a book, titled "Esfahan: Masgid–i Gum'a", where he just explains that some portions of the vault of the Western Iwan are load–bearing and have structural role but the rest are hanging to the ceiling (Galdieri, 1984). No more detailed information is provided about the structural behaviour of the vault.

The three main ribs are actually the arches of the main entrance, the small arch on the back of the iwan, and another curved arch constructed between them, numbered as 1, 2 and 3, respectively in Figure 5.50. The apex of the first main rib is attached to the second one by means of a minor rib, holding four *tasehs* of the uppermost tier, shown in Figure 5.51, between them.

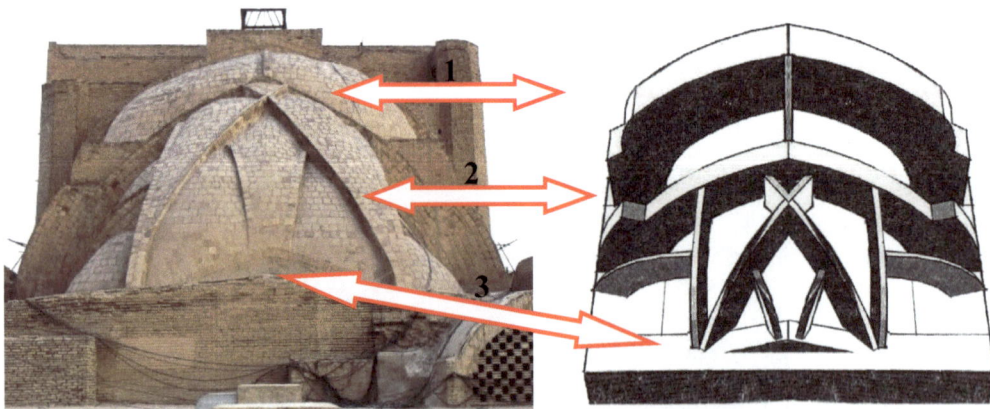

Figure 5.50   Major and minor ribs behind Western Iwan (Source: Author).

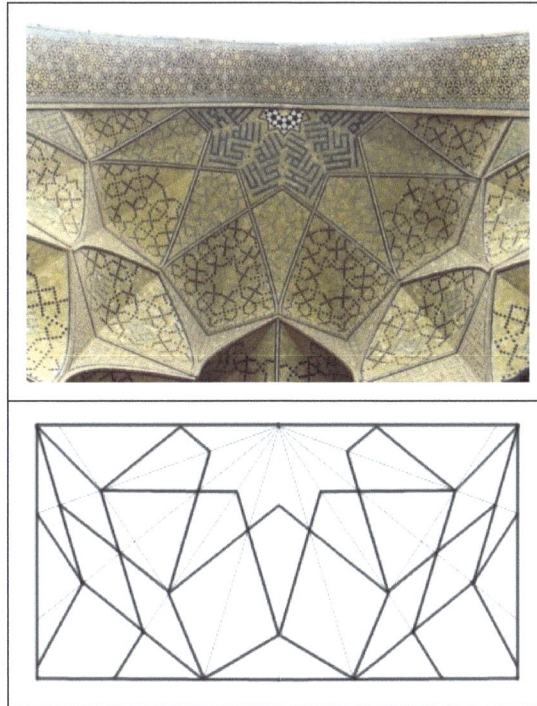

Figure 5.51   Plan and photo of the last tier of the Western Iwan, Isfahan Jame' Mosque (Source: Author).

As a result of the existence of the central main rib the geometry of the structure's two–dimensional plan seems to have two distinctive parts. The geometry of the plan between ribs 1 and 2 is in a radial regime, while the plan between the 2nd and the 3rd main ribs follow the geometry of the eight supportive minor ribs behind the work, holding other *tasehs* in between. Figure 5.52 shows the full view of the iwan and compares it with the plan, to show the mixed nature of it.

The finishing of the ornament is in bricks, as in the structure, and is decorated by glazed tiles, having masonry calligraphies. Halimi believes that the masonry calligraphies in Isfahan Jame' Mosque are mostly in harmony with the architectural elements to which they are attached. Especially about the Western Iwan, he mentions that there is a fine and unique geometrical harmony in the distribution of elements and the whole structure of the *muqarnas* of the iwan, making the combination a unique pattern for future ornaments in Iranian architecture (Halimi, 2011). As a proof of the unique harmony, as illustrated in Figure 5.53, Halimi shows that the centre of the decoration is projecting over the centre of the whole iwan.

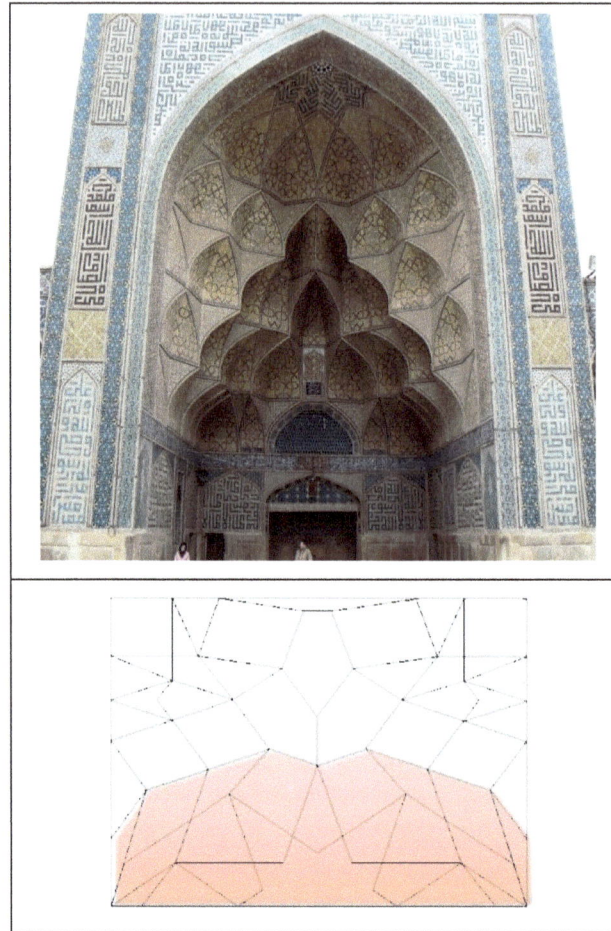

Figure 5.52   2DPP of Western Iwan, showing the two main sections of the plan. The highlighted section has a radial geometry (Source: Author).

Figure 5.53   The harmony between the decoration and the iwan (Halimi, 2011).

Although the structure belongs to the Seljuk period, it was renovated and decorated with the abovementioned tiles and masonry calligraphies during the time of Shah Sultan Hussein the Safavid. This type of decoration was not common during Seljuk period and is one of the most significant characteristics which were added to it later.

5.3.7 Bayazid Bastami Mausoleum

As mentioned before, other than the old minaret, the historical complex at Bastam has an entrance, known as Ghazan Khan Portal, shown in Figure 5.54. The portal is one of the oldest attachments of the building which was constructed about 200 years after the minaret by Ghazan Khan Uljaytu, in 698–712 AH / 1299–1313 C.E. (Wilber, 1969; Pope & Ackerman, 2005; Taboroff, 1981). There are also two other portals opposite each other and a school as parts of the historical complex.

Figure 5.54   Entrance of Bayazid Bastami historical complex (Persiatours.com).

The complex was under restoration twice. Once it was in 699 AH / 1299 C.E. and once during the time of the famous Ilkhanid ruler Uljaytu. During the first renovation, some carved stucco decoration was added to the mosque interior and a fine inscription was added to the mihrab. At the second reconstruction, the Eastern portal and corridor, as well as another iwan across the yard were added to the complex later during Ilkhanid dynasty (Wilber, 1969; Pope & Ackerman, 2005). The pseudo–*muqarnas* structure of entrance portal is marked in Figure 5.55.

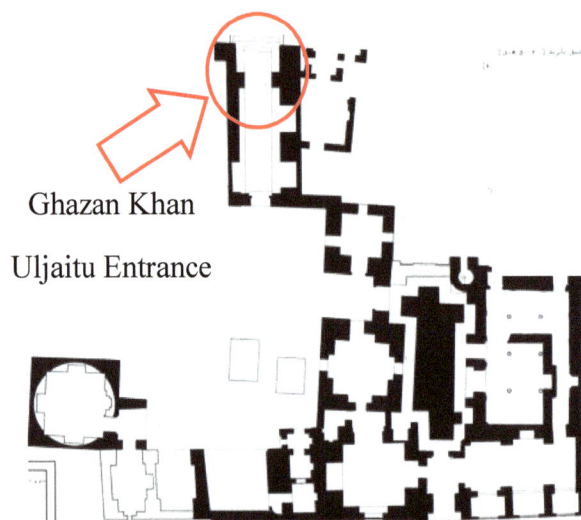

Ghazan Khan

Uljaitu Entrance

Figure 5.55　Plan of Bayazid Bastami historical complex (Source: Iran Cultural Heritage and Tourism Organization).

Table 5.23 provides key information about the structure and Figure 5.56 shows the pseudo–*muqarnas* decoration in entrance portal of Bayazid Bastami Mausoleum. The structure is unique in terms of the number of tiers. It has seven tiers for the first time and a half–*shamseh*, which is considered a lot at that period of time. Furthermore, as it will be explained, there is an innovative combination of *shaparak* elements in this work that creates some fan–like element and there are three indents on the lateral sides and on the centre where the structure penetrates the wall. This unique method is used to cover the angle difference between the walls and the structure itself.

This portal is the entrance of a corridor towards the Shrine and mosque of Bayazid. The iwan has fine stucco carving decorations with *gereh* patterns, which are in harmony with the design patterns of the surrounding walls. Furthermore, there is a noteworthy diversity in the patterns carved on various constituent elements of the pseudo–*muqarnas*. This structure is the first studied sample which is built by suspended layer method, i.e. some plaster layers are attached by means of rope and timber to the ceiling of the structure. These layers are responsible to play a structural role for the decorative structure. The ropes and timbers are then covered by plaster to be kept safe from being corroded quickly. Only edges of these layers can be seen later and the rest will be hidden behind the structure. Tier by tier, the space between these layers is filled later with constituent elements based on the corresponding plan. *Susani*, the combination of *espar* and *taseh* is observed in this iwan as one of the first samples.

Table 5.23: Bayazid Mausoleum building information.

Mashhad Bayazid Bastami (36.485278°N, 54.999722°E)	
Variant Names	Shrine of Bayazid Bastami, Mashhad–i Bayazid Bistami, Aramgah–I Bayazid Bastami, Bayazid Bistami Shrine Complex, Shrine of Bayazid Bastam, Bayazid Shrine Complex, Manar Mashhad–i Bayazid Bastami, Minaret of Shrine of Bayazid Bistami, Shrine of Bayazid Bistami: Minaret, Shrine of Bayazid Bistami, Shrine of Bayezid al–Bistami, Tomb of Bayazid Bastami, Khaneqah and Shrine of Tayfur Abu Yazid al–Bistamiz
Location	Bastam, Iran
Position	Iwan Portal
Date	1299–1313/699–712 AH
Period	Ilkhanid
Century	13th, 14th
Building Type	Educational, Funerary, Religious
Building Usage	School, Mausoleum, Shrine
Copyright	Hamidreza Kazempour

The structure is erected in eight tiers and has no *takht* elements. Table 5.24 gives information on the constituent elements of each tier. In the first tier, other than *spars*, the rest of elements are used to create a kind of penetration inside the wall. This penetration was for the purpose of covering the size–and form–difference between the radial regime of higher tiers and the square base of walls. For that, *toranj* elements were used on the corners of the ornament. The 3rd and 4th tiers are built with *tasehs*, *paraks* and *shaparaks*, but an innovative combination of elements can be observed in the forth tier. Two or three *shaparaks*, with various angles, were attached together to create some fan–like element as illustrated in Figure 5.57.

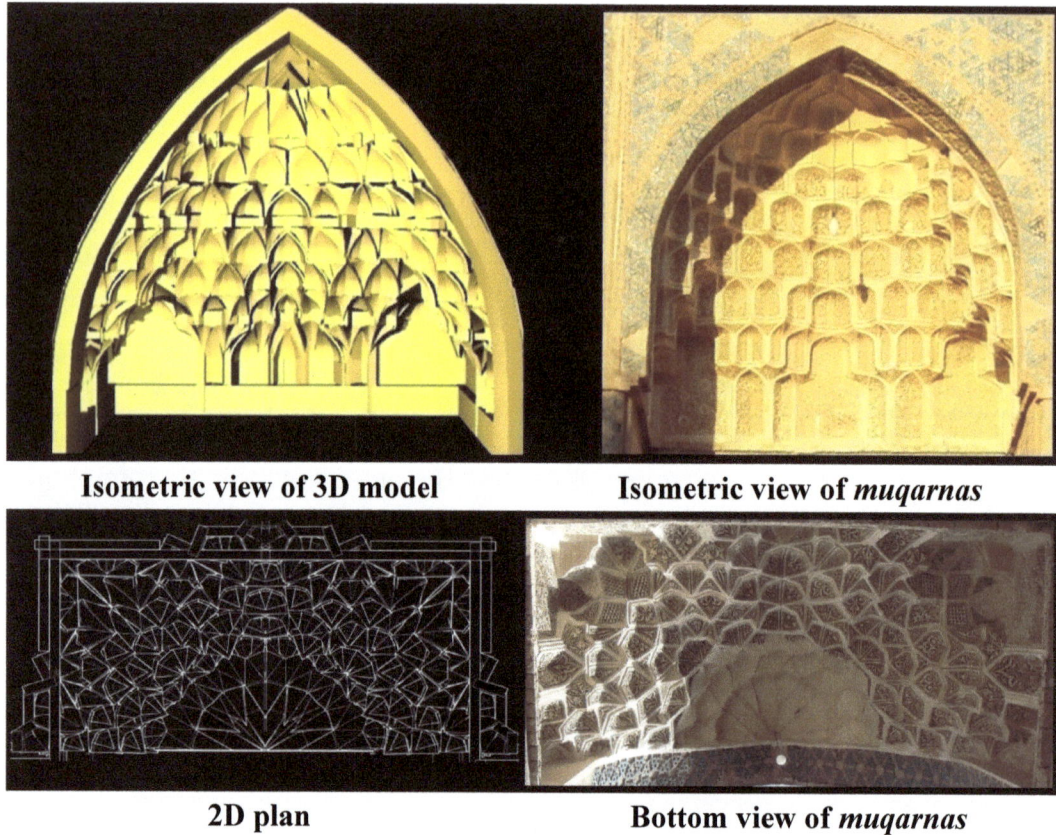

Isometric view of 3D model **Isometric view of *muqarnas***

2D plan **Bottom view of *muqarnas***

Figure 5.56 View of pseudo-*muqarnas* in Bayazid Mausoleum (Based on (Kazempourfard, 2006)).

Table 5.24: Constituent elements table, Bayazid Bastami Mausoleum.

Element ↓ Tier	Toranj	Tee	Taseh						Shaparak			Parak			Shamseh			Espar			Takht	
			Whole			Half			No	Type		No	Type		No	Type		No	Type		No	Type
			No	2D	3D	No	2D	3D		2D	3D		2D	3D		half	whole		half	whole		
One			8	√					32		√	4		√				7	2	5		
Two	6		33	√		2		√	8		√											
Three			17	√					12	4	8	4		√								
Four			13	√					20		√											
Five			9	√					8	√												
Six			8	√					9	√												
Seven			9	√					8		√	2		√								
Eight			5	√											1	√						

Figure 5.57 Combining various *shaparak* elements together (Based on (Kazempourfard, 2006)).

In some tiers, *tasehs* are expanded from two– to three–lobes. Even in some *tasehs*, two two–lobed *tasehs* were attached together to make a four–lobbed one. Unlike the combined *shaparaks* which should read as several distinctive *shaparak* elements, the combined *tasehs* should be read as a single *taseh*. Figure 5.58 illustrates the combination of several *tasehs* into a single one.

The combination of *toranjes* with *tasehs* has created a half–*shamseh* in the last tier of this *muqarnas*. In fact, the whole half–*shamseh* and a major part of the last tier were destroyed many years ago and were restored later. But unfortunately, the restoration was not accurate enough and it has produced some problem on the edges of the structure. Nevertheless, based on the rest of the decorative structure it can be concluded that the plan has a four–fold radial symmetry, shown in Figure 5.59, with three major guiding lines in each section that helps in coherent distribution of elements, making it a pole–table plan type.

Another remarkable point about this decoration is that it penetrates the wall on the lateral sides and in the middle of the first tier. This method, which is very rarely observed in other samples, is used to compensate the angle difference between the walls and the decorative one–sided pseudo–*muqarnas* structure. Figure 5.60 shows the structure's penetration into the walls. The structure provides form–, shape–and height–transition and the material of its finishing is different from that of the construction. Plaster plates was used for the purpose of erecting the decorative structure.

Figure 5.58 Several *tasehs* combined to create a single *taseh* in *muqarnas* of Bayazid Mausoleum (Source: Author).

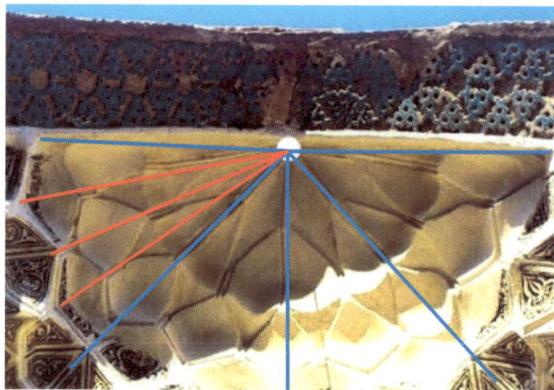

Figure 5.59 Symmetry in pattern of entrance portal, Bayazid Bastami Mausoleum (Based on (Kazempourfard, 2006)).

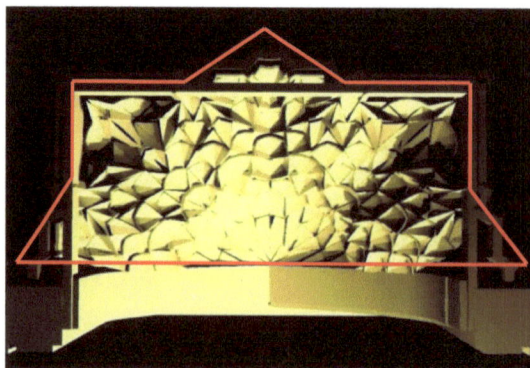

Figure 5.60 Penetrations of the structure into the walls (Based on (Kazempourfard, 2006)).

5.3.8 Pir Bakran Mausoleum

Pir Bakran Mausoleum, as shown in Figure 5.61, was constructed from 1303 to 1312, in Pir–i Bakran village, in the Linjan district, about 30 kilometres South–West of Isfahan. The building used to be the place of teaching and meditation of a famous Sufi of the era, Sheikh Mohammad ibn Bakran (d. 1303). The building was basically only a chamber behind its dome but later, after the death of the Sufi it was converted to his mausoleum. Based on the building's inscriptions, the constructions and enlargement was finally finished in 1312. Being constructed in early 14th century, the building is the first Ilkhanid building that is studied in this manuscript (Blunt, 1966; Grube, 1981; Javadi, 1984; Pope, 1930; Pope, 1965; Grube, 1974; Pope, 1934; Wilber, 1955).

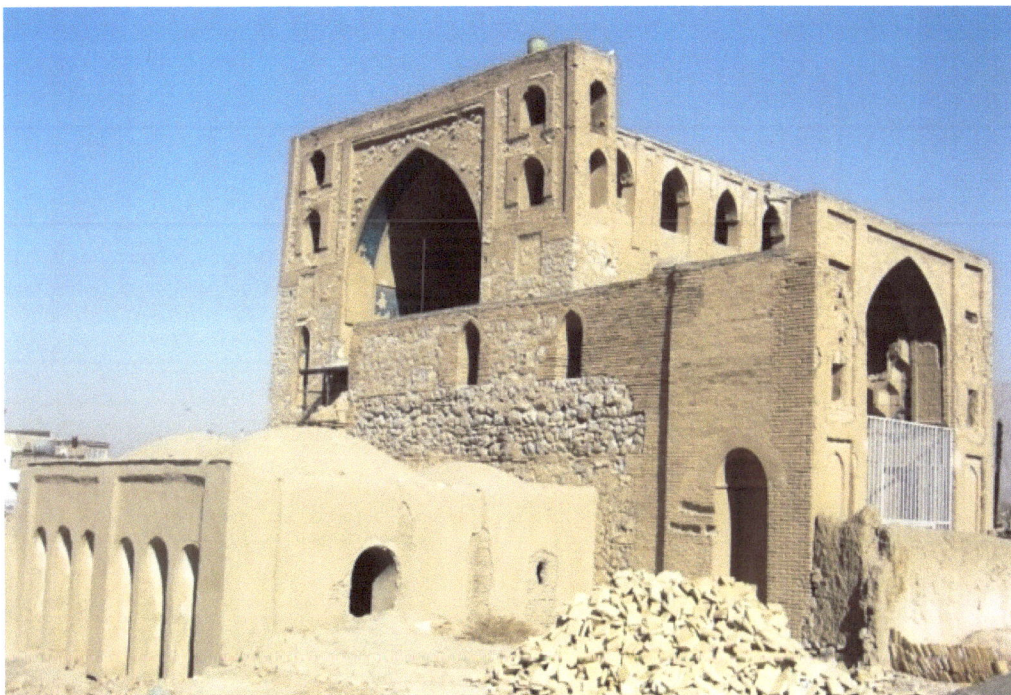

Figure 5.61   Pir Bakran Mausoleum (Based on (Kazempourfard, 2006)).

The building has three sections, namely, an undated small domed chamber, the main hall with a high iwan erected toward the direction of quibla covering one wall of the second and third floors as well as the shrine itself, with a low wall and a mihrab. There have been two rather similar *muqarnas* structures in the building, one of which, shown of Figure 5.62 is partially remained, providing evidences about the quality of the structure behind *muqarnas*, and useful information

166

about its construction technique. Figure 5.63 shows the building plan, with the studied decoration's location marked on it.

Figure 5.62   Pir Bakran Mausoleum, Fragmented *muqarnas* (Based on (Kazempourfard, 2006)).

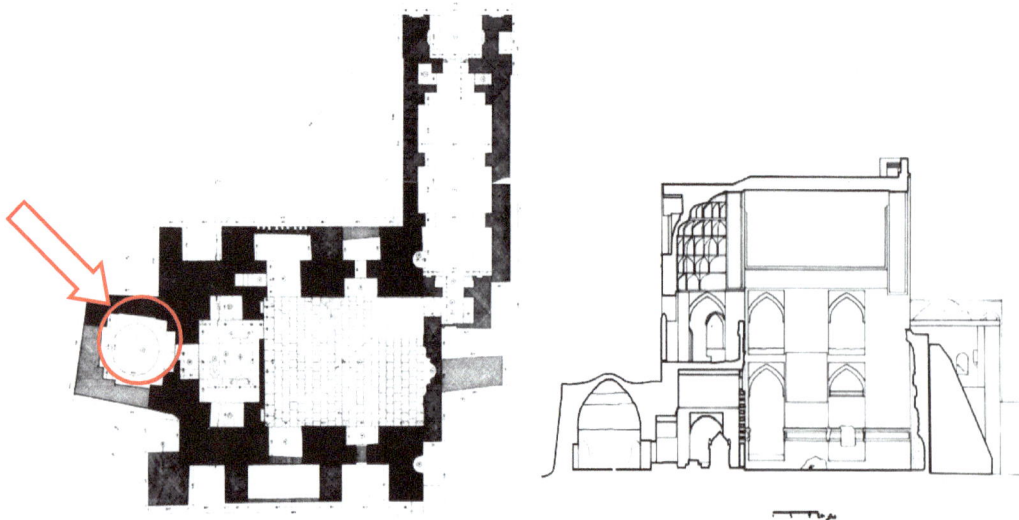

Figure 5.63   Pir Bakran Mausoleum, Building plan and section (Source: Iran Cultural Heritage and Tourism Organization).

The mausoleum of Pir Bakran is built with coarse stone and brick. It is one of the early examples in which star–and cross–shaped tiles were used on surfaces, as well as carved stucco decorations. Being also one of the earliest instances of buildings related to Sufism, and its innovative structural and functional features, the mausoleum of Pir Bakran is considered as one of the most significant examples of Iranian architecture during Ilkhanid period (Blunt, 1966; Grube, 1981; Javadi, 1984; Pope, 1930; Pope, 1965; Pope, 1934; Wilber, 1955).

The pseudo–*muqarnas* structure of Pir Bakran Mausoleum, as shown in Figure 5.65, was constructed over the iwan in the second floor of the building, using suspended layer technique. The decorative structure was erected at the time as the shrine and by the same mason. Table 5.25 provides some key information about the building.

For erecting the rectangular square lattice plan of the structure, specific proportions were considered by the mason, as displayed in Figure 5.64. These proportions were applied first to increase the width of the structure but in fact, they provided a base for the geometry as well. By adding a 45° line beside the corner square in the plan, two *tasehs* were created and the neighbour *tasehs* and *shaparaks* and *paraks* were made same way. The rest of the elements can be obtained by following simple geometric parallel lines, creating *tasehs* of the second and the third tiers. Table 5.26 provides details of the constituent elements.

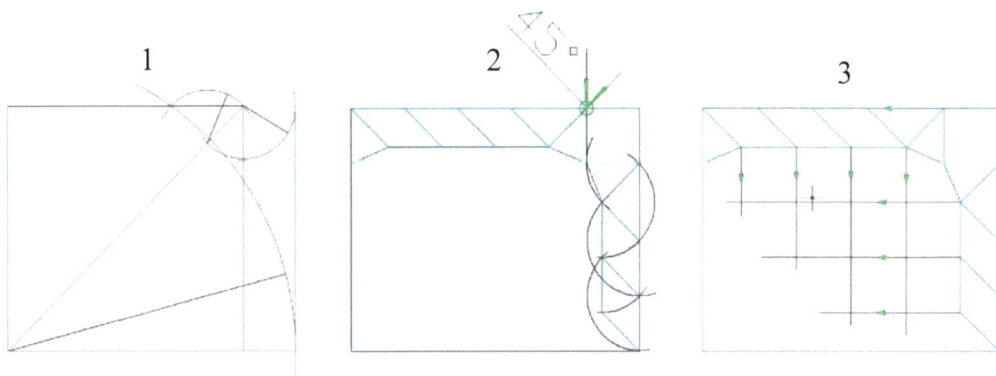

Figure 5.64   Steps of drawing the structure's 2DPP (Based on (Kazempourfard, 2006)).

Table 5.25:   Pir Bakran Mausoleum building information.

Gunbad Pir Bakran (32.468889°N, 51.557778°E)	
Variant Names	Pir–i Bakran Mausoleum, Gunbad Pir–i Bakran, Shrine of Pir–i Bakran, Pir–e Bakran, Shrine of Pir Bakran, Tomb of Sheikh Muhammad ibn Bakran, Shrine of Pir–i Bakran, Pir–e Bakran, Pir–i Baqran Mausoleum, Gunbad–i Muhammad ibn Bakran, Gunbad–e Pir–e Bakran
Location	Linjan, Iran
Position	Iwan
Date	1303–1312/702–711 AH
Period	Ilkhanid
Century	14th
Building Type	Funerary, Religious
Building Usage	Mausoleum, Shrine
Copyright	Hamidreza Kazempour

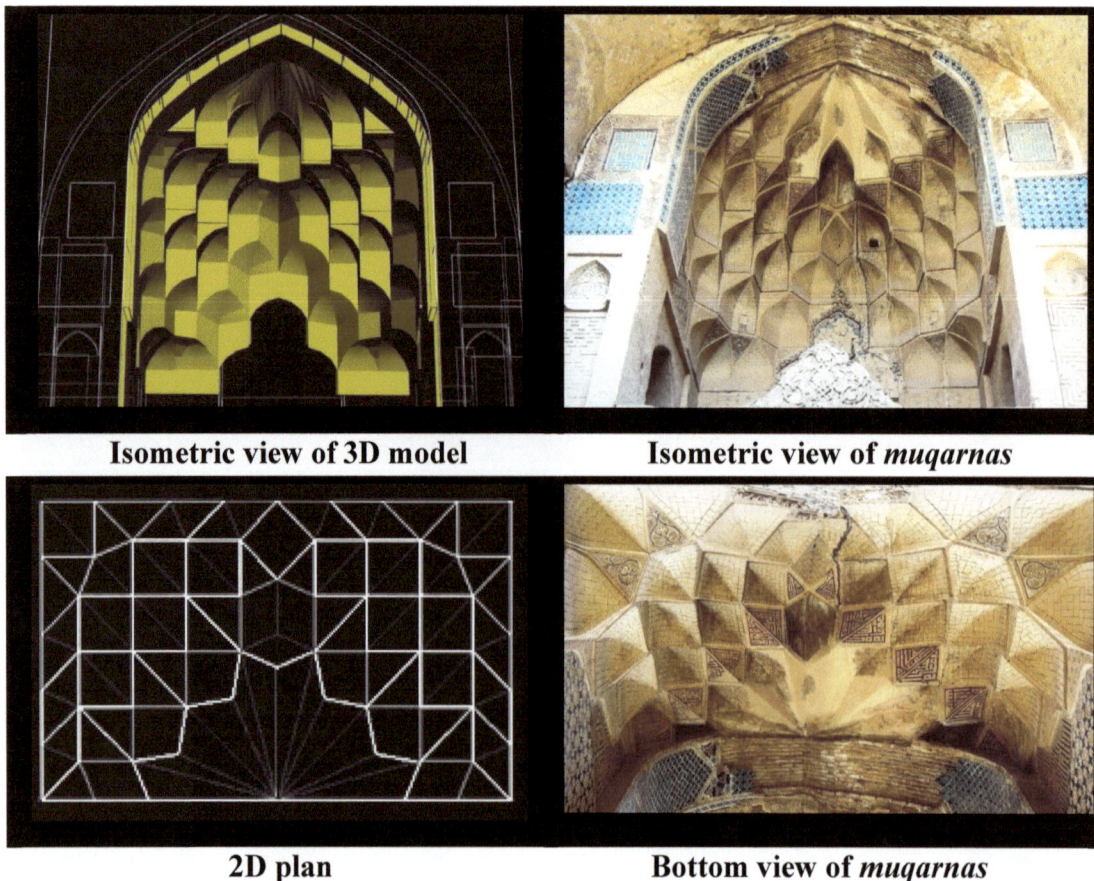

Isometric view of 3D model   **Isometric view of *muqarnas***

2D plan   **Bottom view of *muqarnas***

Figure 5.65   View of pseudo-*muqarnas* in Pir Bakran Mausoleum (Based on (Kazempourfard, 2006)).

Table 5.26: Constituent elements table, Pir Bakran Mausoleum.

Element / Tier	Toranj	Tee	Taseh Whole No	2D	3D	Taseh Half No	2D	3D	Shaparak No	Type 2D	3D	Parak No	Type 2D	3D	Shamseh No	Type half	whole	Espar No	Type half	whole	Takht No	Type
One			6		√	10	√		4	√		2		√				<u>3</u>	2	1		
Two			6	√		1	√		8	√								2	√			
Three			8	√		2	√		9	√												
Four	6		9	√								4		√								
Five			5	√		2		√							1	√						

Another innovative approach in this sample is the creation of a new element by composing one *taseh* and two *paraks*, leading to the creation of a four–lobed *taseh* which looks like a hexagon in the structure's two–dimensional pattern plan (2DPP). The element looks like as if it is stretching itself toward *shamseh*. Figure 5.66 shows the plan and three–dimensional pattern of the new element and Figure 5.67 shows how these elements are arranged in each tier.

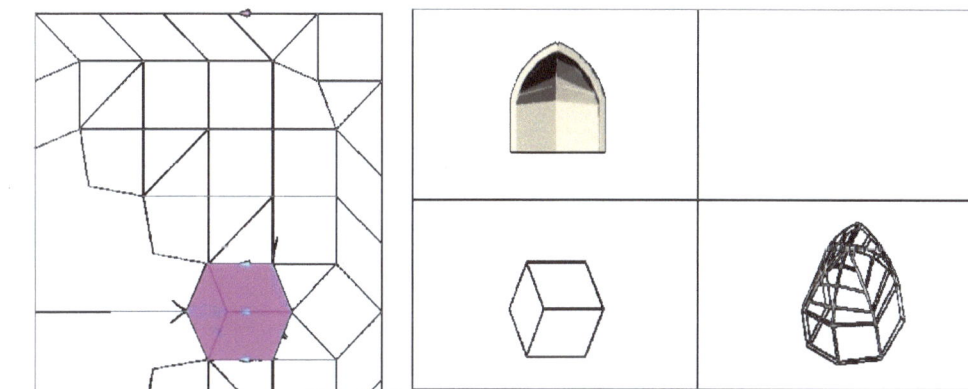

Figure 5.66 New four–lobed *taseh* element, Pir Bakran Mausoleum (Based on (Kazempourfard, 2006)).

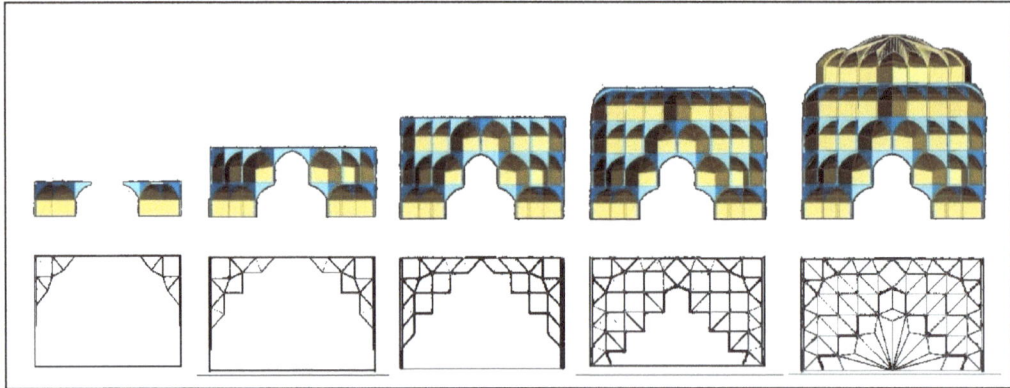

Figure 5.67 2DPP and 3D pattern of each tier, Pir Bakran Mausoleum (Based on (Kazempourfard, 2006)).

The one–sided pseudo–*muqarnas* structure of Pir Bakran is a five–tier decorative structure that provides size–, height–and form–transition. The finishing material of the sample is in tiles, bricks and plaster, different from that of the building. The structure is considered purely decorative with reference to the structure of the fragmented *muqarnas*.

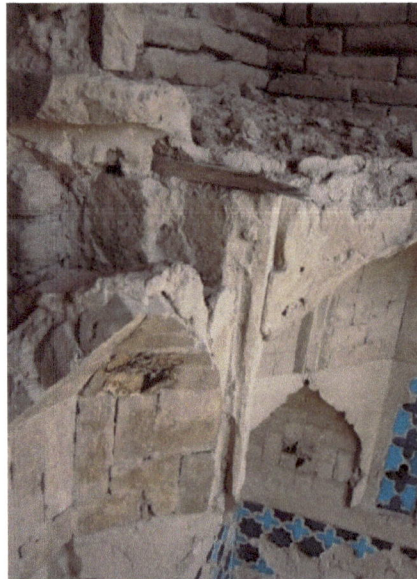

Figure 5.68 Fragments of the other *muqarnas* of Pir Bakran Mausoleum. A broken wood timber can be identified (Based on (Kazempourfard, 2006)).

Based on the available evidence, the load of its first two tiers is completely carried by the walls. On the top of the second tier, there is a wooden beam erected out of the wall, as illustrated in Figure 5.68, supported by mud and plaster. There are no other traces of timber for supporting upper tiers and hence, it can be concluded that the other three tiers used to hang on the ceiling in some way and probably their support was not strong enough to keep them.

5.3.9 Soltanieh Dome

Soltanieh is the name of an ancient city near Zanjan, Iran, dating back to the Ilkhanid dynasty, from which only an octagonal building with a huge double–shell brick dome has remained. The dome, shown in Figure 5.69, is covered entirely with turquoise glazed tiles and is still the third largest brick dome in the world, after the domes of Florence Cathedral and Hagia Sophia. It was started to be built in 1302 by Sultan Uljaytu Khudabanda, and was finished in 1312–1313. Uljaytu was planning to build the largest and most magnificent city in the world in that place but after he passed away, the building's construction scheme changed and it was converted to be Uljaytu's shrine (Godard, 1965; Godard, 1977; Hutt, 1978; Pope, 1977b; Wilber, 1955).

The plan of the building, as illustrated in Figure 5.70, is an octagon and the dome rests over eight terraces on each side of the octagon. There are also eight minarets on the upper sides of each gallery on the corners. The decorative structure which is studied here, as marked in Figure 5.70, is located in the entrance portal of the building. There have been other plaster pseudo–*muqarnas* structures in this building too, but most of them are damaged seriously and impossible to be restored.

Figure 5.69   Soltanieh Dome, Zanjan, Iran (Source: Panoramio.com).

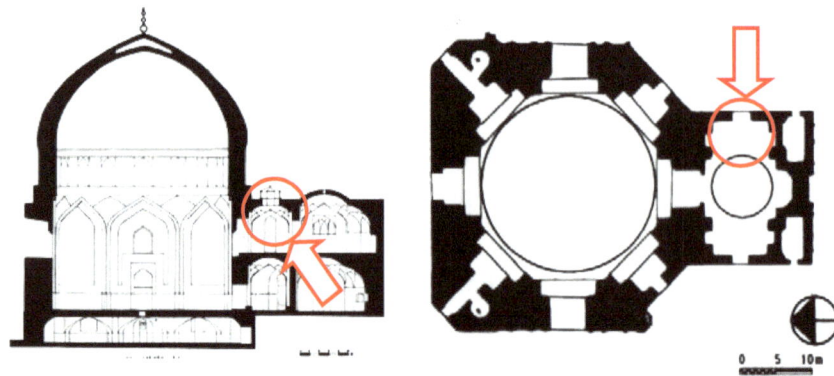

Figure 5.70 Plan of Soltanieh Dome (Source: Iran Cultural Heritage and Tourism Organization).

The recorded pseudo–*muqarnas* structure of Uljaytu or Soltanieh is shown in Figure 5.71 and Table 5.27 provides some important details about the building. The studied sample has nine tiers with 110 *tasehs*, seven three–dimensional *shaparaks* and 106 plumb two–dimensional *paraks* and *shaparaks*, as well as a half–*shamseh* which has been restored recently. These constituent elements as described in Table 5.28. From all *paraks* and *shaparaks*, only the seven three–dimensional ones can be distinguished in the plan and the rest can only be seen in the professional eyes of a master mason.

Table 5.27: Soltanieh Dome building information.

Gunbad–i Uljaytu (36.434722°N, 48.797500°E)	
Variant Names	Gunbad–i Uljaytu Khudabandah, Mausoleum of Oljaytu, Uljaytu Tomb, Tomb of Uljaytu, Oljeitu Tomb, Mausoleum of Sultan Muhammad Öljeitü Khudabanda, Uljaytu Khudabanda Gonbad, Oljeytu Gunbad, Gunbad–i Sultaniya, Gonbad–e Soltaniyeh, Tomb of Uljaytu, Gunbad–i Sultania, Gonbad–e Soltanieh, Gunbad–i Soltaniyeh
Location	Sultanieh, Iran
Position	Iwan
Date	1312–1313/711–712 AH
Period	Ilkhanid
Century	14th
Building Type	Funerary
Building Usage	Tomb
Copyright	Hamidreza Kazempour

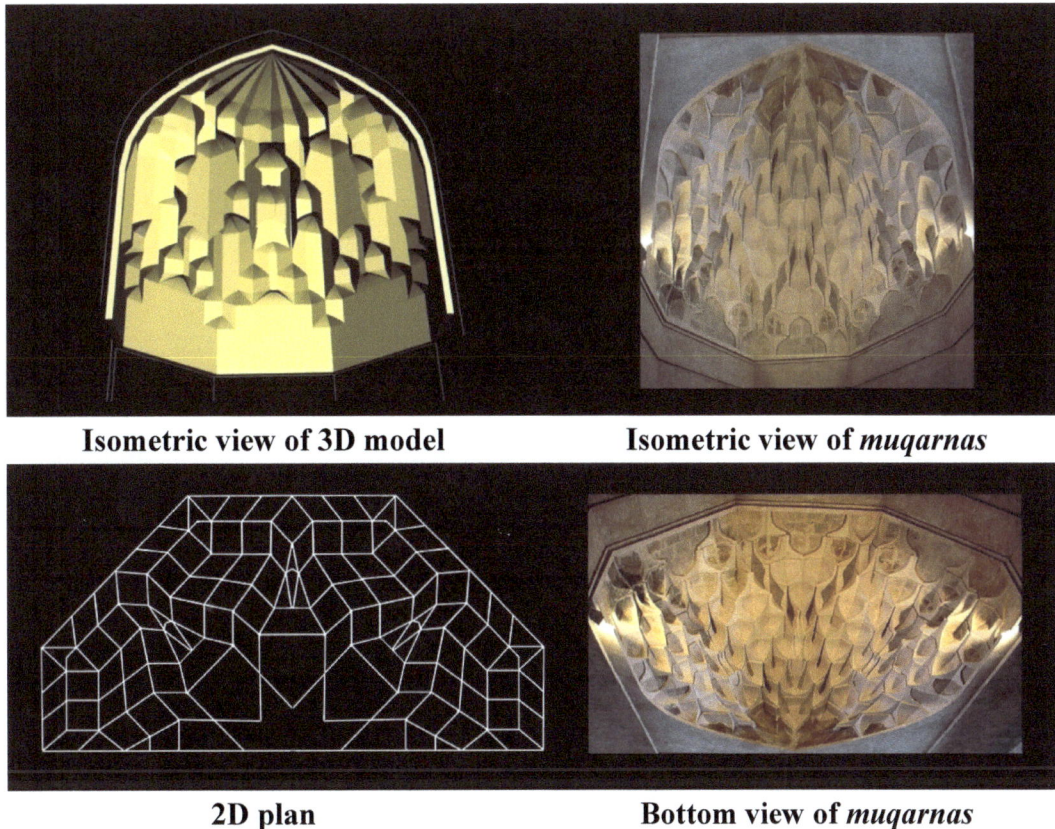

Isometric view of 3D model **Isometric view of *muqarnas***

2D plan **Bottom view of *muqarnas***

Figure 5.71 View of pseudo-*muqarnas* in Soltanieh Dome (Source: Author).

Table 5.28: Constituent elements table, Soltanieh Dome.

Element / Tier	Toranj	Tee	Taseh No	Whole 2D	Whole 3D	No	Half 2D	Half 3D	Shaparak No	Type 2D	Type 3D	Parak No	Type 2D	Type 3D	Shamseh No	Type half	Type whole	Espar No	Type half	Type whole	Takht No	Type
One			8		√				12	8	4							5	2	3		
Two			15	3	12	2		√	16	√												
Three			16		√				15	√		2	√									
Four			15		√	2		√	16	√												
Five			16	2	14				15	12	3	2	√									
Six			15	2	13				14	8	6	2	√									
Seven		3	10		√				12	√												
Eight			9		√				8	√												
Nine			6		√										1	√						

The plan of this structure is a half–octagon in which all the elements, except the ones on the two lateral sides, have radial symmetry, i.e. in general, the plan has an eight–fold symmetry and therefore it has a square lattice 2DPP. The one–sided structure is merely decorative and has no load–bearing function. It provides height–, form–and size–transition. The material used to erect the structure is plaster which is different from the building material, and the decorative structure is built in the suspended layer technique.

5.3.10 Natanz Jame' Mosque, I and II

Abd al–Samad was another Sufi Sheikh who died in Natanz, in 1299. A decade after his death, his shrine was developed into a complex, including Natanz Jame' Mosque and his tomb, which is still one of the best preserved examples of shrines in Ilkhanid period (Blair, 1986; Blair, 1986; Hutt, 1978; Pope, 1977). Construction material of the building, shown in Figure 5.72, is mainly baked brick, but in many parts, there is a white plaster coating over it. Furthermore, many facades in the building are decorated by glazed tiles, and patterned stucco. It has a four–iwan type plan and an octagonal sanctuary with a dome. There is also a single minaret next to the octagonal dome. Figure 5.73 illustrates the plan and section of the building, on which the studied pseudo–*muqarnas* decorations are marked. There is another pseudo–*muqarnas* decoration around the minaret of the building which is available for further reference in Appendix A.

There used to be a monastery in the current location of the mosque from which only the western iwan has remained. The early fourteenth century monastery or khanaqah was destroyed and reconstructed as a mosque during 1930s. The restorations revealed that the sanctuary area originally dates back to Buyid period and there was a stand–alone pavilion there which had been constructed in 999 C.E. / 389 AH (Blair S. S., The Ilkhanid Shrine Complex at Natanz, Iran, 1986; Blair S. S., 'The Octagonal Pavilion at Natanz: a Reexamination of Early Islamic Architecture in Iran, 1986; Hutt A. , 1978; Pope, The Fourteenth Century, 1977). There is however an inscription on the mosque and shrine's portal showing that the main building was erected in 1304–5 / 704 AH.

Figure 5.72   Abd Al–Samad Shrine Complex (Based on (Kazempourfard, 2006)).

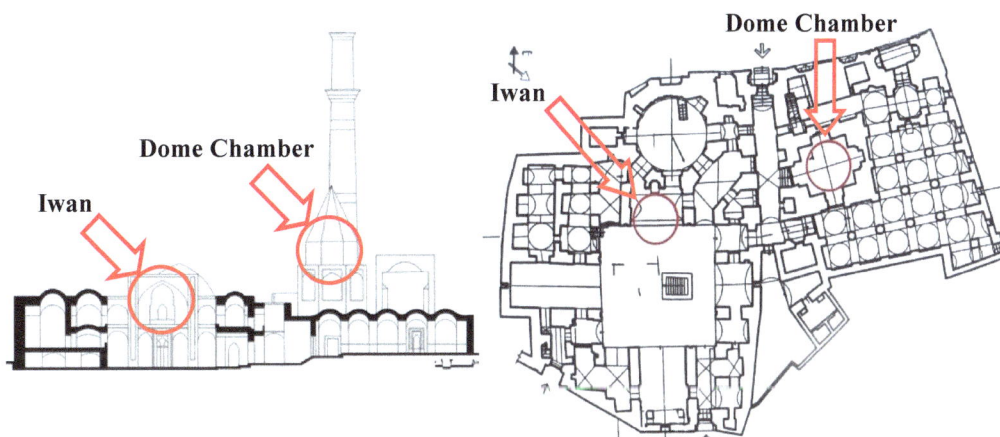

Figure 5.73   Plan and section, Natanz Jame' Mosque (Source: Iran Cultural Heritage Organization).

As mentioned before, two of the decorative structures are studied here. Table 5.29 provides some key information about the building. First, the decorative veneer of the sanctuary's octagonal dome is reviewed and then the pseudo–*muqarnas* structure of the Southern iwan.

The sample in dome chamber of the shrine, shown in Figure 5.74, is significant in terms of its rather complex geometry; it has an irregular polygon plan which is confined inside the regular octagonal plan of the dome. It is a complete and dense veneer covering the whole dome and finally the design of the decoration is a perfect and flawless one of its kind. The structure, like other ornamental structures studied here is provides form–, height– and size–transition. The material of the decoration is plaster, which is different from that of the building and the plan follows square lattice type rules.

This pseudo–*muqarnas* structure is erected in 10 tiers. Having one *shamseh* on the apex, resembling the sun, as well as some star like components, created by combination of some *toranjes* and *tasehs*, the whole veneer looks like a sky. Although these stars are not two–dimensional, as the form is common in later *muqarnas* structure, the mason elegantly pleased these 16 symmetric stars between the domes sky light chambers to give them a nice glow during daytime. Figure 5.75 shows one of these stars and the details of the structure's constituent elements are available in Table 5.30.

As mentioned before, the plan of the structure is an irregular polygon which is different from that of the base of dome. This variation is noticeable in Figure 5.76, in which the yellow colour shows the borders of the structure's plan and the red colour shows the dome's exterior plan and the white colour is the dome chamber's interior plan birders. The areas coloured with yellow are the skylight providing chambers.

Table 5.29:   Natanz Jame' Mosque building information, Dome.

Natanz Jame' Mosque (33.513333°N, 51.916389°E)	
Variant Names	Shrine of Shaykh Abd al–Samad, Shrine Complex of Shaikh 'Abd al–Samad: Octagonal Pavilion, Masjid–i Sheikh 'Abd al–Samad, Mosque of Shaykh 'Abd al–Samad, Majmu'ah–i Shaykh 'Abd al–Samad, Shrine Complex of Shaykh Abd al Samad, Shaykh 'Abd al–Samad Mosque, Masjid–i Jami Natanz, Shrine of Abd al–Samad, Abdol Samad Monastery
Location	Natanz, Iran
Position	Under dome
Date	1304–1325/703–725 AH
Period	Ilkhanid
Century	13th
Building Type	Funerary, Religious
Building Usage	Tomb, Shrine
Copyright	Hamidreza Kazempour

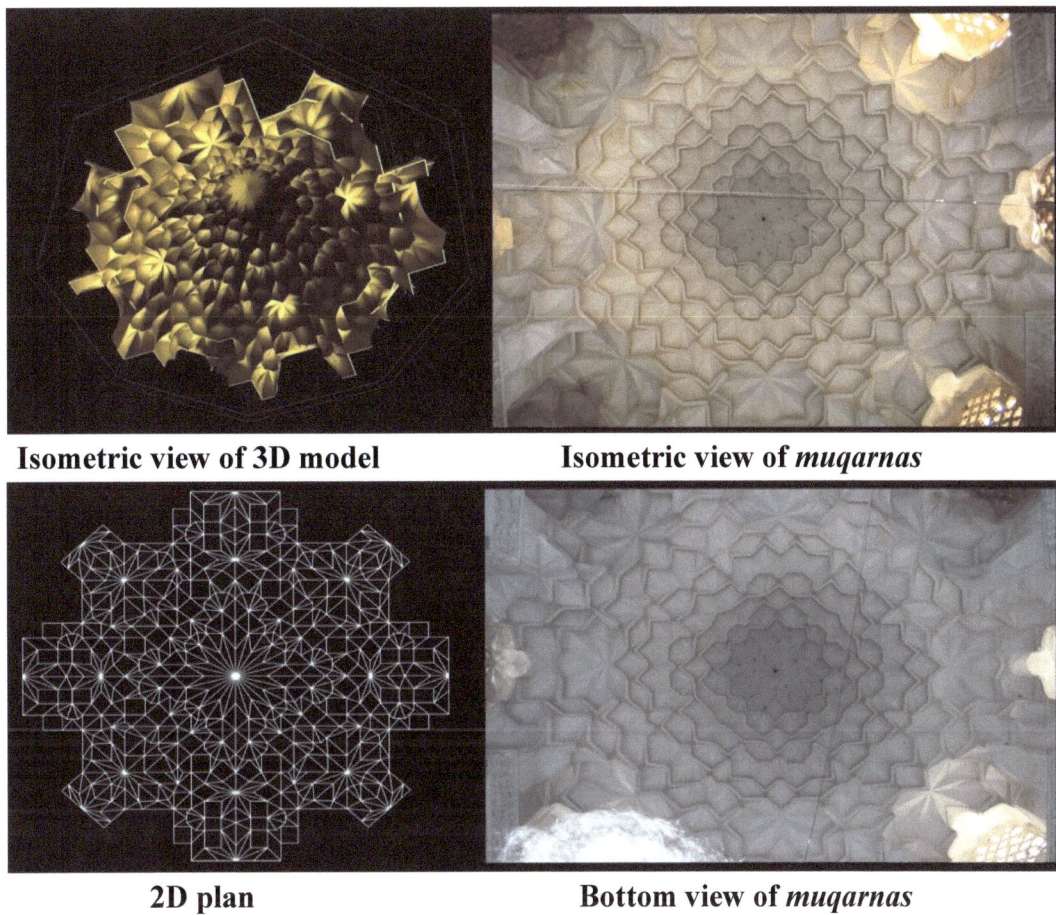

| Isometric view of 3D model | Isometric view of *muqarnas* |

| 2D plan | Bottom view of *muqarnas* |

Figure 5.74 View of pseudo-*muqarnas* in Natanz Jame' Mosque, Dome (Based on (Kazempourfard, 2006)).

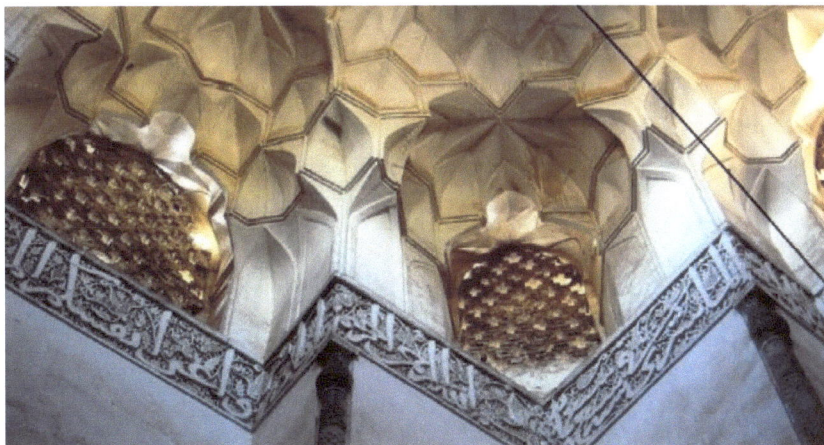

Figure 5.75 Three–dimensional stars of Natanz Jame' Mosque (Based on (Kazempourfard, 2006)).

Table 5.30: Constituent elements table, Natanz Jame' Mosque, Dome chamber.

Element / Tier	Toranj	Tee	Taseh No	Taseh Whole 2D	Taseh Whole 3D	Taseh No	Taseh Half 2D	Taseh Half 3D	Shaparak No	Shaparak Type 2D	Shaparak Type 3D	Parak No	Parak Type 2D	Parak Type 3D	Shamseh No	Shamseh Type half	Shamseh Type whole	Espar No	Espar Type half	Espar Type whole	Takht No	Takht Type
One			<u>32</u>	8	24				<u>56</u>	32	24	8	√					16	√			
Two			40		√				56	√								8	√			
Three	72		72		√				32	√		24		√								
Four			40		√				<u>32</u>	16	16											
Five			28		√				<u>32</u>	24	8											
Six			24		√				<u>36</u>	12	24											
Seven			20		√				12	√		8		√								
Eight			16		√				<u>48</u>	16	32											
Nine			12		√				12	√												
Ten			12		√										1		√					

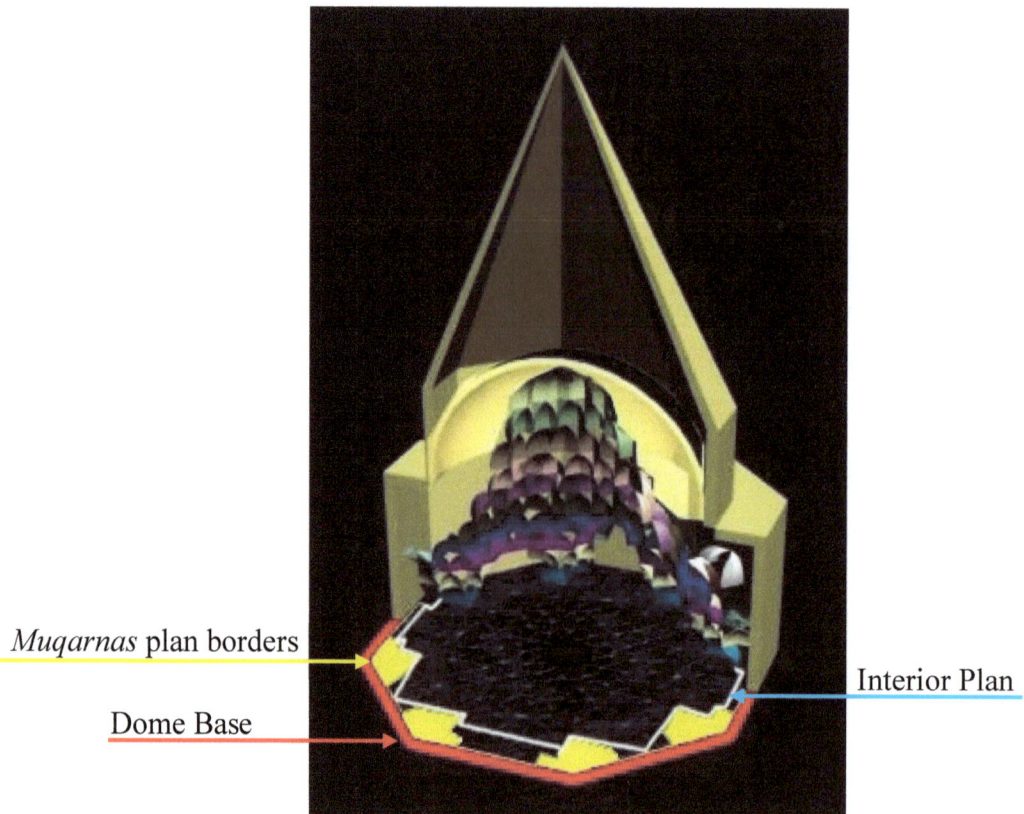

Figure 5.76 Comparison of base and plan of dome chamber decorative structure (Source: Author).

For constructing this decorative structure, the suspended layer method was applied. Figure 5.77(a) shows how the pseudo–*muqarnas* structure of dome chambers of Natanz Jame' Mosque is erected tier by tier and Figure 5.77(b) shows its plan on which the tiers are shown by different colours. Southern iwan, shown in Figure 5.78, has been chosen as the last sample before entering the *muqarnas* sub–section of this manuscript. Key information about the decorative pseudo–*muqarnas* veneer of Southern iwan at Natanz Jame' Mosque is provided in Table 5.31

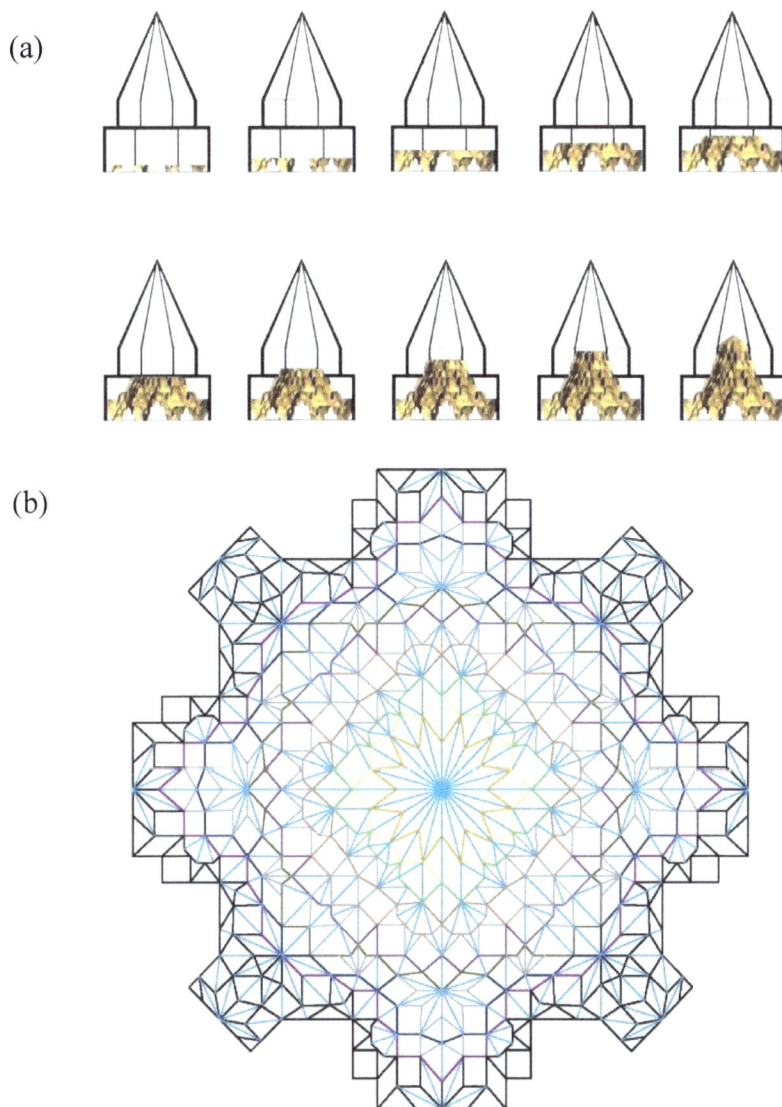

Figure 5.77   (a) Tier by tier construction of dome chamber ornament of Natanz Jame' Mosque, (b) Plan of the sample showing different tiers (Based on (Kazempourfard, 2006)).

Table 5.31:   Natanz Jame' Mosque building information, South iwan.

Natanz Jame' Mosque (33.513333°N, 51.916389°E)	
Variant Names	Shrine of Shaykh Abd al–Samad, Shrine Complex of Shaikh 'Abd al–Samad: Octagonal Pavilion, Masjid–i Sheikh 'Abd al–Samad, Mosque of Shaykh 'Abd al–Samad, Majmu'ah–i Shaykh 'Abd al–Samad, Shrine Complex of Shaykh Abd al Samad, Shaykh 'Abd al–Samad Mosque, Masjid–i Jami Natanz, Shrine of Abd al–Samad, Abdol Samad Monastery
Location	Natanz, Iran
Position	South Iwan
Date	1304–1325/703–725 AH
Period	Ilkhanid
Century	13th
Building Type	Funerary, Religious
Building Usage	Tomb, Shrine
Copyright	Hamidreza Kazempour

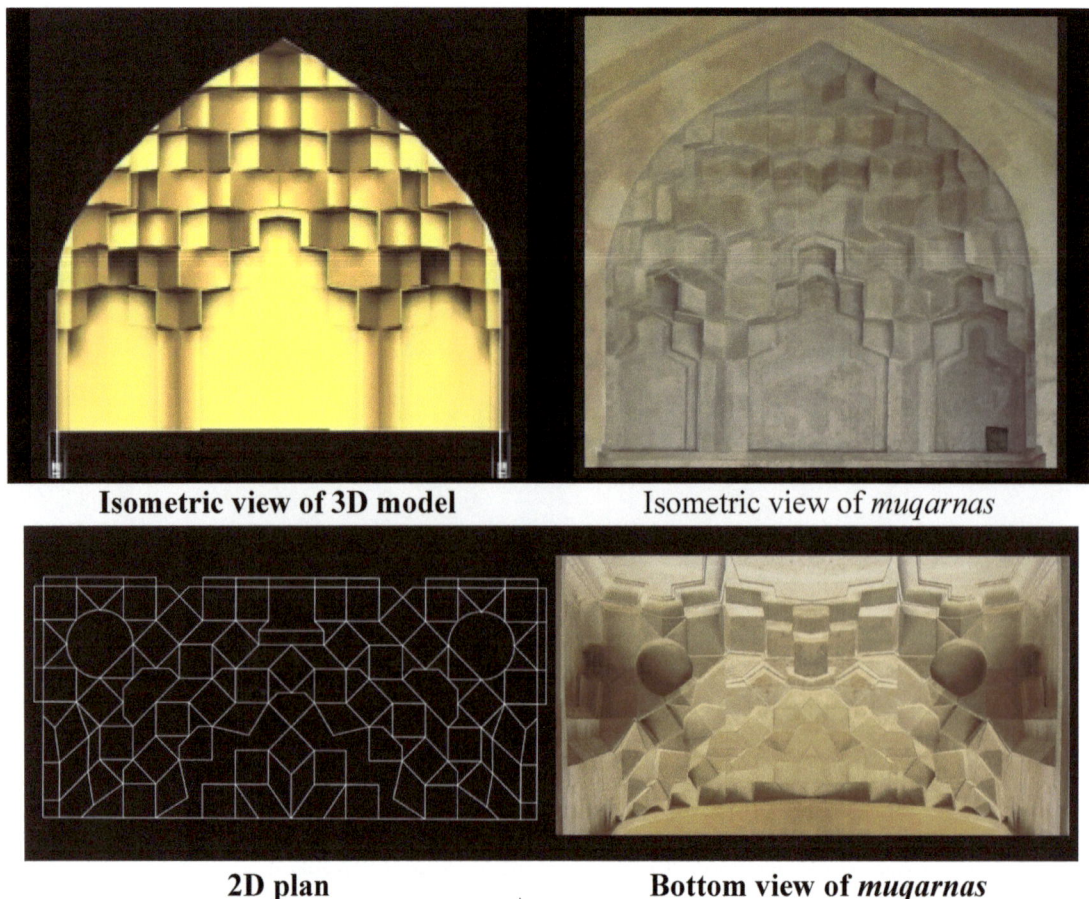

Isometric view of 3D model   Isometric view of *muqarnas*

2D plan   **Bottom view of *muqarnas***

Figure 5.78   View of pseudo-*muqarnas* in Natanz Jame' Mosque, South iwan

The constituent elements of this iwan are arranged in a way that an empty space, suitable for being filled with a *takht* element is created. This is the first and the only example among all collected samples that a half–cylindrical element is observed as a constituent element of structure. Although, the mason has tried to reshape the element in a way that it becomes similar to a *taseh*, because of having different architectural characteristics, the invented *takht* element cannot be categorized as a *taseh*. Figure 5.79 shows the plan and for of this anonymous constituent element.

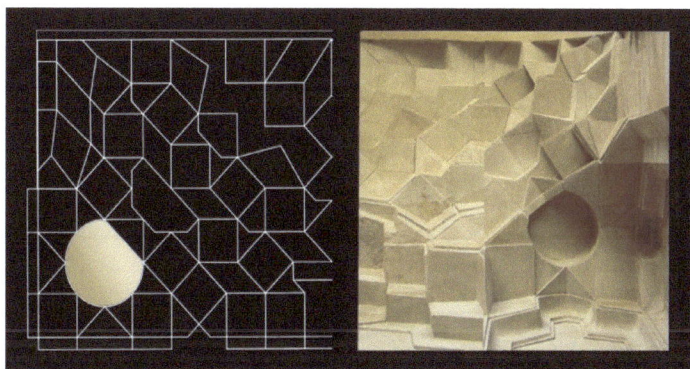

Figure 5.79   Plan and image of the anonymous element (Source: Author).

Another distinctive phenomenon, in terms of constituent elements, is also observed between the last two tiers of the structures. For the first and only time, there is an undefined flat area that can only be observed in the two–dimensional plan. This phenomenon which can be imagined as an anonymous flat tier is illustrated in Figure 5.80.

Figure 5.80   Anonymous flat tier, Natanz Jame' Mosque (Source: Author).

This pseudo–*muqarnas* structure is arranged in eight tiers and it has a half–*shamseh*. Table 5.32 provides information about the constituent elements of this structure.

Table 5.32: Constituent elements table, Natanz Jame' Mosque, Iwan.

Element / Tier	Toranj	Tee	Taseh No	Type 2D	Type 3D	Half–Taseh No	Type 2D	Type 3D	Shaparak No	Type 2D	Type 3D	Parak No	Type 2D	Type 3D	Shamseh No	Type half	Type whole	Espar No	Type half	Type whole	Takht No	Type
One			6		√													7	2	5		
Two			18		√							14		√				1		√		
Three			13		√	2		√	8		√	2		√								
Four			11		√				4		√	4		√								
Five			12		√				11	√												
Six			11		√				6		√											
Seven			7		√																	
Eight			5		√										1	√						

Although this flat tier cannot be categorized as any of the defined constituent elements, its existence is another clue towards the creation of *muqarnas* and the method of installing and attaching *takht* elements of *muqarnas* to the supporting arches or ribs. This is important because in *muqarnas* structures, the *takht* elements are installed first and the space between them is then filled with appropriate elements. This is why the geometry of elements near edges of the decorative structure does not normally follow that of the whole structure and it is more complex to construct a symmetric *muqarnas* over an arch, rather than erecting it in a whole space like a dome chamber. Figure 5.81 shows the borders of this decorative structure.

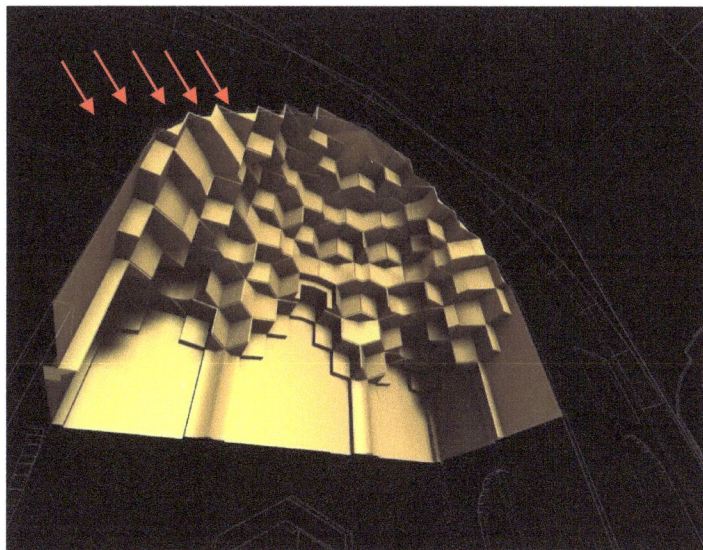

Figure 5.81 Three–dimensional isometric view of pseudo–*muqarnas* of Natanz Jame' Mosque (Based on (Kazempourfard, 2006)).

Pseudo–*muqarnas* of Natanz Jame' Mosque, like the rest of the samples studied so far is meant to provide a smooth form–, height– and size–transition. The finishing material of the decorative structure is from plaster, which is different from the building's construction material, i.e. brick and the plan of the structure can be categorized as square lattice.

5.4 *Muqarnas*

Since early 14th century, after the pseudo-*muqarnas* of the iwan of Natanz Jame' Mosque, in which some unusual and characterless constituent elements are observed, a new leap is observed in the development of *muqarnas*. The appearance of *takht*-elements in the structure of this ornament is the real beginning of the structure known as *muqarnas* and its major distinctive characteristic compared to pseudo-*muqarnas* structures.

5.4.1 Ashtarjan Mosque

The *muqarnas* structure at Ashtarjan Mosque is in fact the first *muqarnas* sample with horizontal *takht* elements and hence, one of the most important samples of this manuscript. Constructed in 1315 the Ilkhanid mosque and its decorations have survived very well and are in good condition (Golombek & Wilber, 1988; Pope, 1977; Wilber, 1969). Figure 5.82 shows the entrance portal of the mosque.

Figure 5.82   Entrance of Ashtarjan Mosque (Based on (Kazempourfard, 2006)).

The mosque has a rectangle courtyard surrounded by three prayer halls. On the Northern façade, there is a main entrance portal having two flanking minarets. The studied *muqarnas* is located inside this entrance portal. Figure 5.83 illustrates the plan of the building and its isometric view, on which the position of the studied *muqarnas* are marked.

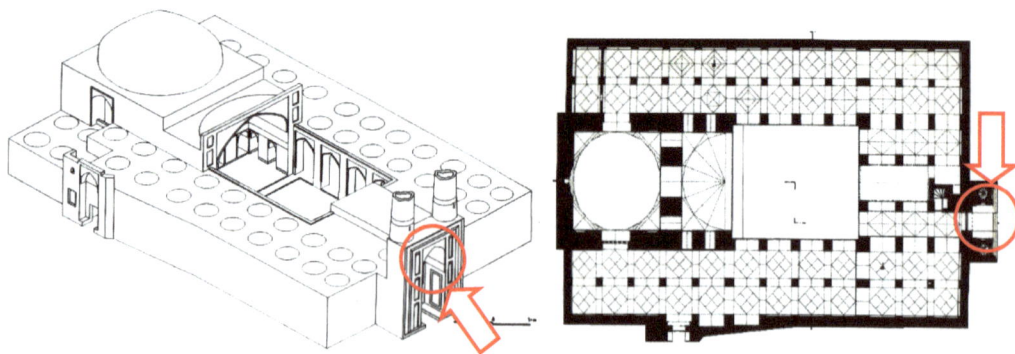

Figure 5.83   Schematic isometric view and plan of Ashtarjan Mosque, with marked entrance *muqarnas* (Source: Iran Cultural Heritage Organization).

The structure is in adobe which is unusual in the area but the outer layer of the building is made by fired brick. In a building erected totally with adobe and brick, the existence of an entrance *muqarnas*, which is covered with glazed tile and brick, catches the eyes of the observer. Figure

5.84 shows the *muqarnas* of the entrance portal and Table 5.33 provides some useful information about the building.

Table 5.33: Ashtarjan Jame' Mosque building information.

Masjid Jame' Ashtarjan (32.475833°N, 51.479167°E)	
Variant Names	Friday Mosque of Ashtarjan, Masjid–i Jami' Ashtarjan, Masjid–i Jami' of Ashtarjan, Masjid–e Jame, Masjid–i Jami, Masjed–e Jomeh, Masjed Jomeh–e Oshtorjan, Masjid–i Jami' at Ashtarjan
Location	Ashtarjan, Iran
Position	Entrance Portal
Date	1316/715 AH
Period	Ilkhanid
Century	14th
Building Type	Religious
Building Usage	Mosque
Copyright	Hamidreza Kazempour

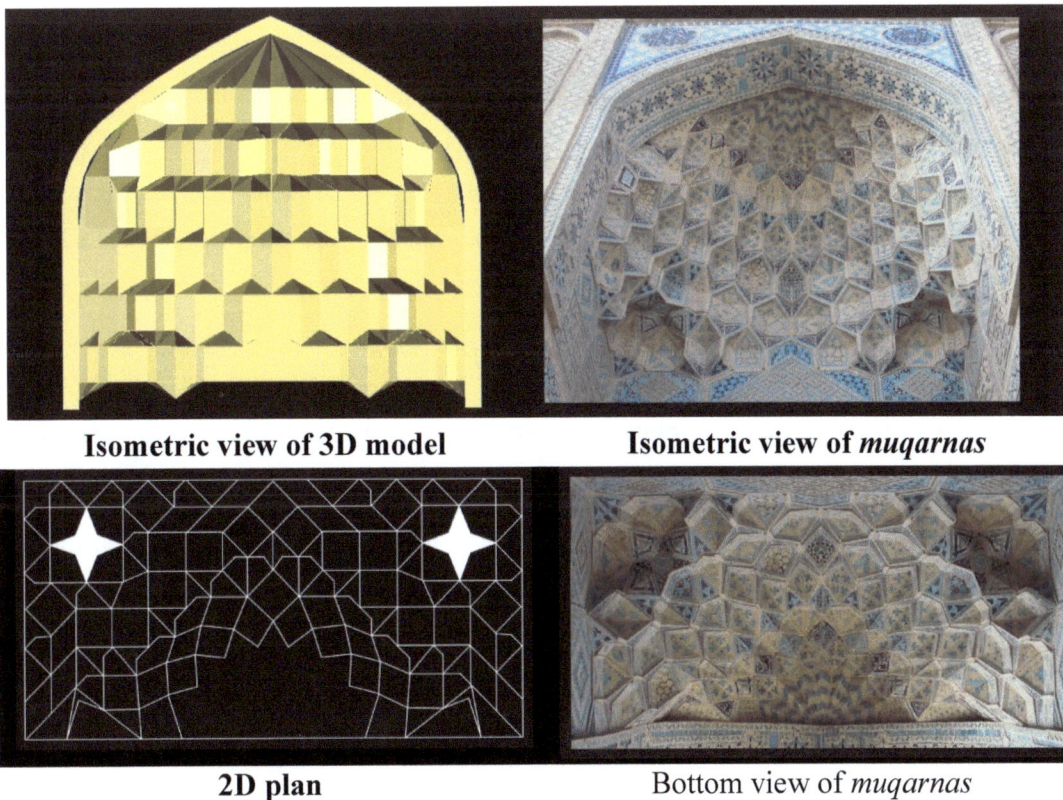

Isometric view of 3D model   **Isometric view of *muqarnas***

2D plan   Bottom view of *muqarnas*

Figure 5.84   View of *muqarnas* in Ashtarjan Mosque (Based on (Kazempourfard, 2006)).

The patterns and the materials applied in the decorative structures of the building are different from part to part. The exterior *muqarnas*, as mentioned above, is covered with glazed brick and tiles, while most of the interior is covered in bricks. Carved plaster veneer is extensively employed in many parts of the mosque, exhibiting unique geometrical and floral patterns and motifs and some Kufic inscriptions can also be observed in areas such as sanctuary and mihrab. The *muqarnas* of the entrance portal of Ashtarjan Mosque has seven tiers and a half–*shamseh*. Table 5.34 provides details of the structure's constituent elements.

Table 5.34: Constituent elements table, Ashtarjan Mosque.

Element / Tier	*Toranj*	*Tee*	Taseh No (Whole)	Taseh Whole 2D	Taseh Whole 3D	Taseh No (Half)	Taseh Half 2D	Taseh Half 3D	Shaparak No	Shaparak Type 2D	Shaparak Type 3D	Parak No	Parak Type 2D	Parak Type 3D	Shamseh No	Shamseh Type half	Shamseh Type whole	Espar No	Espar Type half	Espar Type whole	Takht No	Takht Type
One									16	4	12	2	√					7		√		
Two	6		13	√		2	v		12	8	4							5		√	2	4pointed
Three			14	√					10	√		6	2	4								
Four			11	√		2		√	14	10	4											
Five			14	2	12				18	2	16											
Six	7		10	√								2		v								
Seven			8	√											1	√						

The plan of the structure is a one by two rectangular with a main symmetry axis in the middle, as if the plan is made up of two identical squares. The diagonal lines of the squares are therefore the supportive lines in the plan of the structure. Figure 5.85 shows the main symmetry line and the auxiliary diagonal lines. As it can be observed, the plan is drawn based on some co–centric circles, having *shamseh* on the centre and two four–pointed *takht*–stars on the assistive diagonal lines. The structure, as explained above, has seven tiers. Figure 5.86 shows how these tiers were built one by one over each other.

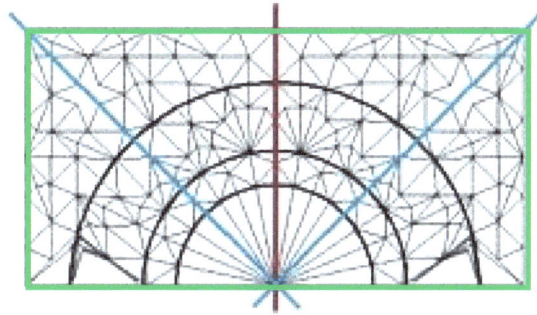

Figure 5.85   Main symmetry line of the *muqarnas* plan (Source: Author).

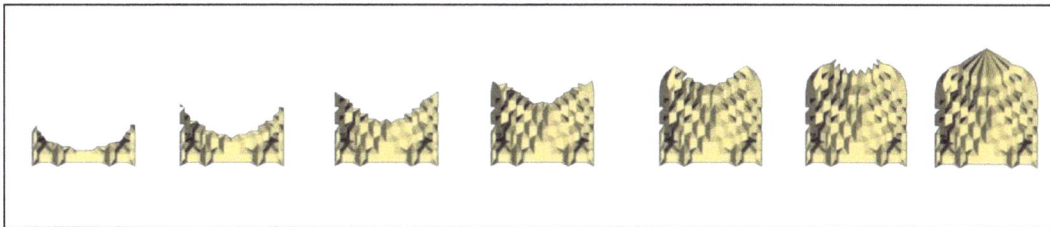

Figure 5.86   Tier by tier comparison of *muqarnas* structure (Based on (Kazempourfard, 2006)).

To conclude the investigation, it must be said that the *muqarnas* of the entrance portal of Ashtarjan Mosque is a one–sided purely decorative structure that provides form–, size– and height–transition. It is erected by plaster board technique and is hanged to the entrance ceiling by mud, rope and wood timber. The finishing material of this *muqarnas* is different from that of building and it was added to the structure as a decoration.

5.4.2   Varamin Jame' Mosque

Varamin Jame' mosque, shown in Figure 5.87, was built in 1322 in Varamin, 42 km South of Tehran. The mosque is famous as being the earliest surviving Ilkhanid mosque with a four–iwan plan in which the beauty of the ornaments of the mosque are enhanced by rich materials and the play of light and shadow in architectural elements. Furthermore, the building is famous for having relatively unusual aspect ratios. For instance, it has a small square courtyard, compared to the whole structure and as we will discuss below, the proportions of the dimensions of its *muqarnas* are also unique and uncommon. The western part of Varamin Jame' Mosque had been destroyed almost completely but the remaining parts of it, including the mosque's mihrab, main portal, Eastern portal and iwans, were restored and conserved (Blair & Bloom, 1994; Michell, 1978; Pope, 1977b; Pourjavady & Booth-Clibborn, 2001; Wilber, 1955).

Figure 5.87   Varamin Jame' Mosque (Source: Radiotehran.ir).

The exterior of the building is decorated by glazed and unglazed tiles, terracotta and plaster. The dome chamber was recently renovated and there are four *muqarnas* squinches on the corners of the main prayer chamber. The details of the squinches are available for further reference in Appendix A. The mosque has another *muqarnas* in the entrance iwan, shown in Figure 5.89. Table 5.35 provides key information about this *muqarnas*. There is an inscription band above this *muqarnas* which is erected by glazed tiles, but the units of *muqarnas* are made by bricks which are smaller in size, compared to that used as the building construction material. Figure 5.88 shows the plan and section of Varamin Jame' Mosque, on which the position of the studied *muqarnas* is marked with red.

Figure 5.88   Plan and section of Varamin Jame' Mosque, with marked *muqarnas* position (Source: Iran Cultural Heritage and Tourism Organization).

Table 5.35:   Varamin Jame' Mosque building information.

Masjid Jame' Varamin (35.320000°N, 51.650000°E)	
Variant Names	Friday Mosque of Varamin, Masjid–i Jami' Varamin, Masjid–e Jame, Congregational Mosque, Masjed–e Jameh
Location	Varamin, Iran
Position	Entrance
Date	1322/721–22 AH
Period	Ilkhanid
Century	14th
Building Type	Religious
Building Usage	Mosque
Copyright	Hamidreza Kazempour

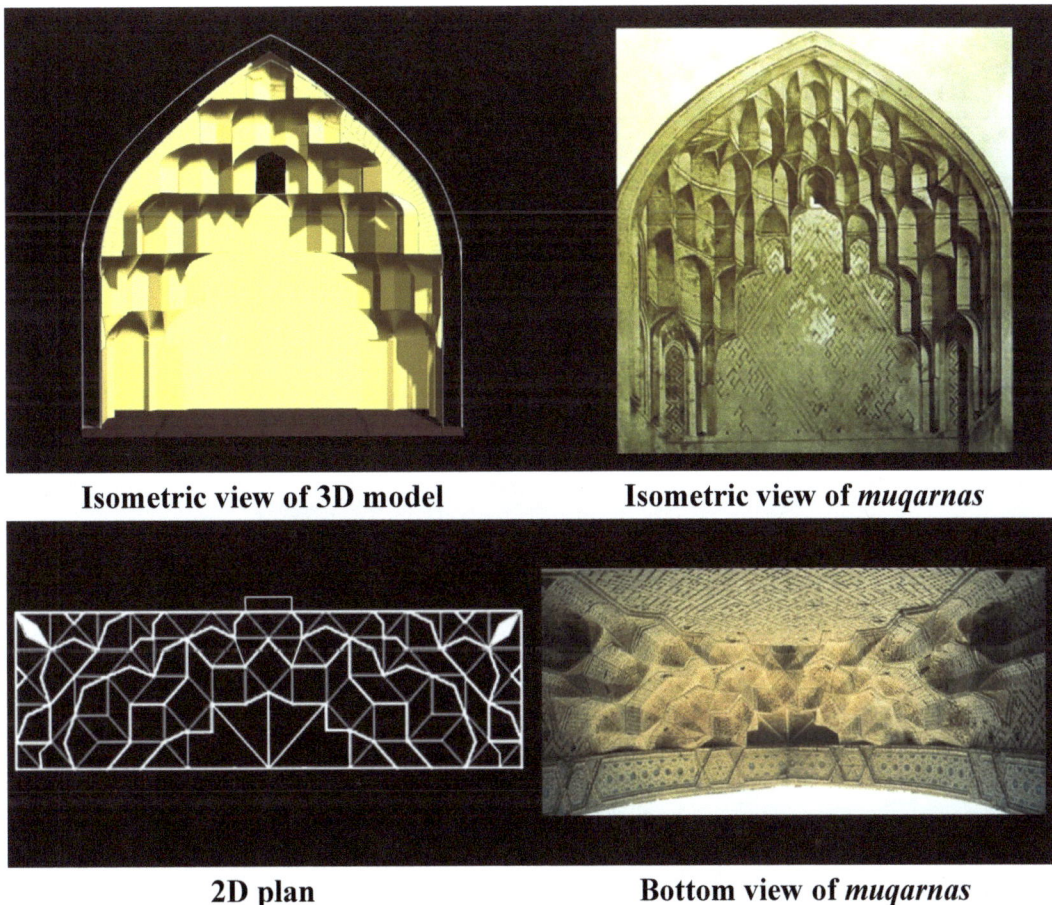

Isometric view of 3D model   **Isometric view of *muqarnas***

2D plan   **Bottom view of *muqarnas***

Figure 5.89   View of *muqarnas* in Varamin Jame' Mosque (Based on (Kazempourfard, 2006)).

As mentioned before, the proportions of the *muqarnas* of the entrance, in Varamin Jame' Mosque, are considered unusual compared to the rest of the recorded ornaments. The plan is a long rectangle which has almost a one-by-four aspect ratio. As result of the rather long and narrow plan of the structure, being comprised of four squares instead of typical two adjacent squares, the drawing of the pattern was a really hard task. The *muqarnas* in Jame' Mosque of Varamin, as explained in Table 5.36, is erected in six tiers and it has a fragmented half–*shamseh*. In the first tier of the *muqarnas*, one can observe a rhombic *takht* element, which has different construction material from that of the *muqarnas*. This *takht*, as shown in Figure 5.90, is made of glazed dark blue tiles which attracts the attention of the observer by glowing among unglazed brick structure of the ornament.

Table 5.36:   Constituent elements table, Varamin Mosque.

Element ↓ Tier	*Toranj*	*Tee*	*Taseh* Whole No	2D	3D	*Taseh* Half No	2D	3D	*Shaparak* No	Type 2D	3D	*Parak* No	Type 2D	3D	*Shamseh* No	Type half	whole	*Espar* No	Type half	whole	*Takht* No	Type
One			10	2	8				12		√							3	2	1	2	Rhombic
Two			6		√	2		√	10	4	6							1		√		
Three			12		√				11		√											
Four	8		13		√							8		√								
Five			9		√				2	√		8		√								
Six			2		√										1	√						

Figure 5.90   Glazed tiled rhombic *takht* element in *muqarnas* of Varamin Jame' Mosque (Source: Author).

As the building is very famous among archaeologists and architectural historians, the two–dimensional pattern plan of the ornamental structure was recorded by Takahashi. Referring to

Takahashi's, although most of the plans of his collection are accurate, as illustrated in Figure 5.91, some major mismatch can be identified in the case of Varamin Jame' Mosque, which can be listed as follows:

(1) There is no *takht* element in the first tier.

(2) The *shaparaks* of the third tier are not drawn.

(3) The indented window is not illustrated.

(4) Two *paraks* are drawn in the fourth tier, instead of one.

(5) *Shamseh* of this *muqarnas* had been damaged since at least a century ago (Pope, 1934). Hence, no *shamseh* should be drawn on the plan.

The one–sided *muqarnas* of Varamin Jame' Mosque plays the role of height-, form– and size–transition. The plan is of square lattice type. Based on Godard reports, the *muqarnas* of Varamin Jame' Mosque is absolutely decorative and it has no structural role (Godard, 1990). The ornamental structure is erected using suspended layer method.

5.4.3 Imamieh School

Madrasa Imami, or Imamieh School, shown in Figure 5.92, is a theological college in Isfahan. It is known as being one of the oldest extant theological colleges of Iran. Constructed in 1325, the brick building has a four iwan plan (Blunt, 1966; Hillenbrand, 2004). The plan and section of the building is shown in Figure 5.93. As the school is located attached to the shrine of a famous theologian, Baba Qasim, the school is sometimes being mentioned as Baba Qasim School.

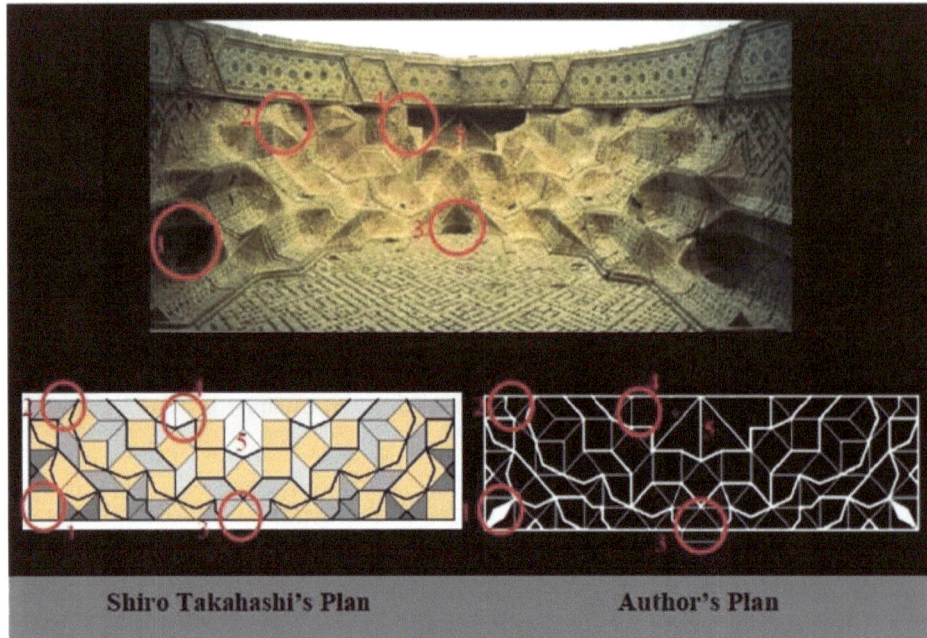

Figure 5.91 Comparing Takahashi's plan of Varamin Jame' Mosque with that of author's (Source: Author).

Figure 5.92 Imamieh School, Isfahan (Based on (Kazempourfard, 2006)).

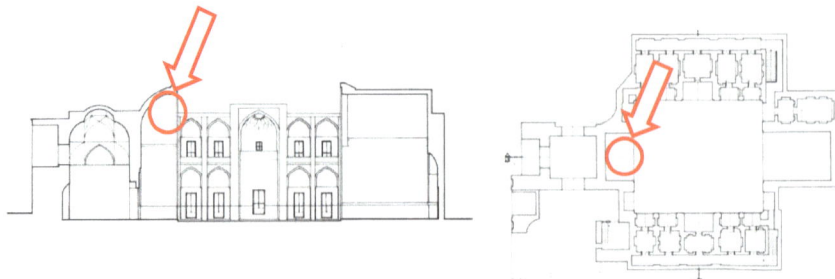

Figure 5.93 Plan and section, Imamieh School (Source: Iran Cultural Heritage and Tourism Organization).

The double–story school has a central courtyard with several small rooms around it to house students, and a main hall for prayers and study. The entrance iwans of each room and hall is decorated with a *muqarnas* structure. The school has therefore a sum of more than 20 diverse *muqarnas* and pseudo–*muqarnas* structures, with various geometries, from which only four samples have been recorded. The one studied here, is marked in Figure 5.93 and the other three are available for further reference in Appendix A.

The abovementioned *muqarnas* structures are mostly decorated by glazed tiles which have other significant characteristics. The patterns created by these decorative tiles are in one of the earliest phases of transformation of motifs from merely geometrical patterns to Arabesque floral designs. Floral motifs of Imamieh School are created by attaching small hand cut glazed tiles. The technique which is famous as *mu'araq* tiling became common later in Iran, during the Timurid dynasty in the 15th century scholars (Blunt, 1966; Hillenbrand, 2004; Golombek & Wilber, 1988). Figure 5.94 shows a sample of arabesque floral motifs created by *mu'araq* tile decoration in Imamieh School. More details about this method of tiling are provided in the next sub–section.

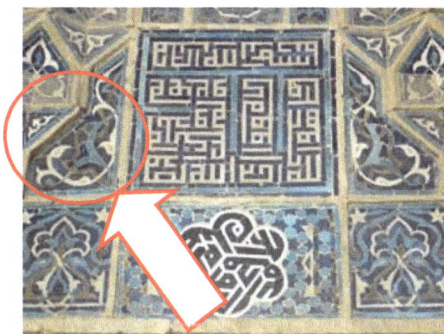

Figure 5.94   Non–geometric motif mu'araq tile decoration in Imamieh School (Source: Author).

The specific *muqarnas*, studied here, shown in Figure 5.95 is chosen from the other 20 samples inside Imamieh School, because it was a roof–patched *muqarnas*, having two four–pointed star *takhts*, in addition to its practical geometry which starts from a squinch and distributes radially towards the half–*shamseh*. Table 5.37 provides key information about the building

194

Table 5.37: Imamieh School building information.

Madrasa Imamieh (32.671111°N, 51.685556°E)	
Variant Names	Imami Madrasa, Madrasah–i Imami, Madrasa–yi Imami, Baba ghasem School, Emamieh school, Emamiah, Imamiah, Baba Qasim, Baba Qasem school, Madreseh
Location	Isfahan, Iran
Position	Iwan
Date	1325/726 AH
Period	Muzaffarid
Century	14th
Building Type	Educational
Building Usage	Madrasa
Copyright	Hamidreza Kazempour

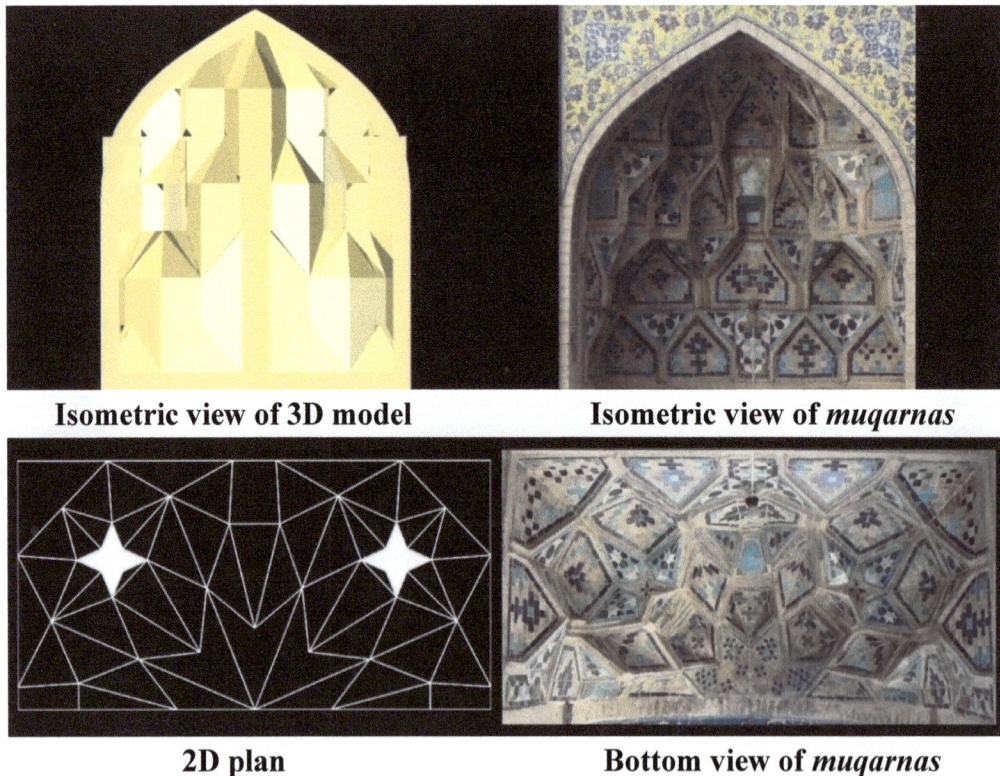

Isometric view of 3D model   Isometric view of *muqarnas*

2D plan   Bottom view of *muqarnas*

Figure 5. 95   View of *muqarnas* in Imamieh (Based on (Kazempourfard, 2006)).

The radial geometry of the structure is confined inside a one-by-two rectangular plan which can be divided into two identical squares by a main symmetry axis. Figure 5.96 shows the symmetrical plan of the studied *muqarnas*, which is of pole table type.

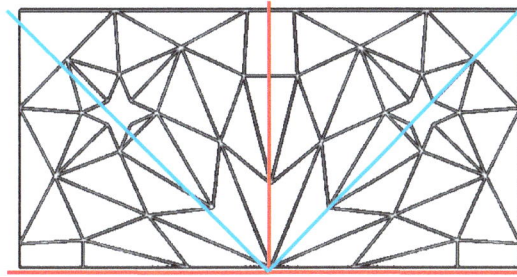

Figure 5.96   Symmetric plan of *muqarnas*, Imamieh School (Source: Author).

Furthermore, the *shaparaks* of the *muqarnas* are all plumb in this case and hence, they cannot be identified in the structure's two–dimensional pattern plan. It also contains *Tee*'s, *espars* and *toranjes*, as well as a half–*shamseh*. Details of the constituent elements of the four–tier *muqarnas* are available in Table 5.38.

Table 5.38:   Constituent elements table, Imamieh School.

Element → / Tier ↓	Toranj	Tee	Taseh						Shaparak			Parak			Shamseh			Espar			Takht	
			No	Whole		No	Half		No	Type		No	Type		No	Type		No	Type		No	Type
				2D	3D		2D	3D		2D	3D		2D	3D		half	whole		half	whole		
One			7	1	6	2	√		8	√												
Two	6		4		√				4		√	4		√				3	2	1	2	4point
Three			9		√				8	√												
Four			5		√										1	√						

The *Muqarnas* of Imamieh School is a one–sided decorative structure, erected by tile and brick. Although the building's construction material is also brick, the *muqarnas* was not erected at the same time with the building. It is absolutely non–structural and it was attached later to the building using suspended layer method.

5.4.4   Kerman Jame' Mosque

Kerman Jame' Mosque, shown in Figure 5.97, was built in the heart of the city's business district in 1349. It has an ideal four–iwan plan including a rectangular courtyard having four iwans the centre of each of its sides (Blair & Bloom, 1994; Golombek, 2003; Michell, 1978; Wilber, 1955). Although there are two other entrances on the South and West of Kerman Jame' Mosque, its main gateway, located to the North, is built in from of a 20–meter–high portal iwan, to be emphasised as the mosque's main entrance. This entrance which will be studied here is decorated with a flawless, complete *muqarnas* structure. Figure 5.98 shows the position of this entrance on the building's plan.

Figure 5.97   Kerman Jame' Mosque (Source: MehrNews.com).

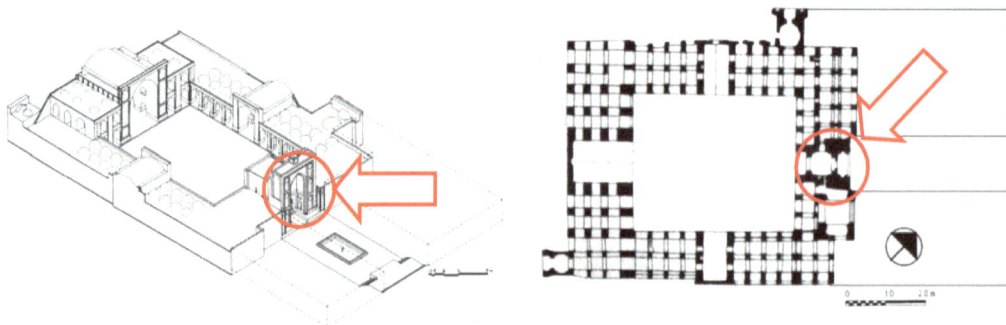

Figure 5.98   Plan and Section of Kerman Jame' Mosque, showing the position of the studied *muqarnas* (Source: Iran Cultural Heritage and Tourism Organization).

The *muqarnas* of the entrance portal is framed inside a Quranic inscription band. The exterior veneer of this ornamental structure is decorated by *mu'araq* tiling technique, with arabesque floral and geometrical motifs. Figure 5.99 shows the *muqarnas* of the entrance portal of Kerman Jame' Mosque and Table 5.39 provides some key information about the building.

The *muqarnas* sample of Kerman Jame' Mosque was chosen to be studied here because the sample can be considered one of the most complete *muqarnas* structures, in terms of form and structure, built by end of Ilkhanid period. It is built in nine tiers having one half–*shamseh* on its apex. With two eight–pointed *takht*–stars, the *muqarnas* of the entrance portal of Kerman Jame' Mosque has all the defined constituent elements of a *muqarnas* in it. Furthermore, the existence of eight–pointed *takht*–stars implies that the masons of the period were capable of creating mature and complex geometries within the structure of *muqarnas*. Table 5.40 provides the details of the structure's constituent elements.

Table 5.39: Kerman Jame' Mosque building information.

Masjid Jame' Kerman (30.283333°N, 57.083333°E)	
Variant Names	Friday Mosque of Kerman, Masjid–i Jami' of Kirman, Masjid–e Jame, Congregational Mosque
Location	Kerman, Iran
Position	Entrance Portal
Date	1349/749 AH
Period	Ilkhanid
Century	14th
Building Type	Religious
Building Usage	Mosque
Copyright	Hamidreza Kazempour

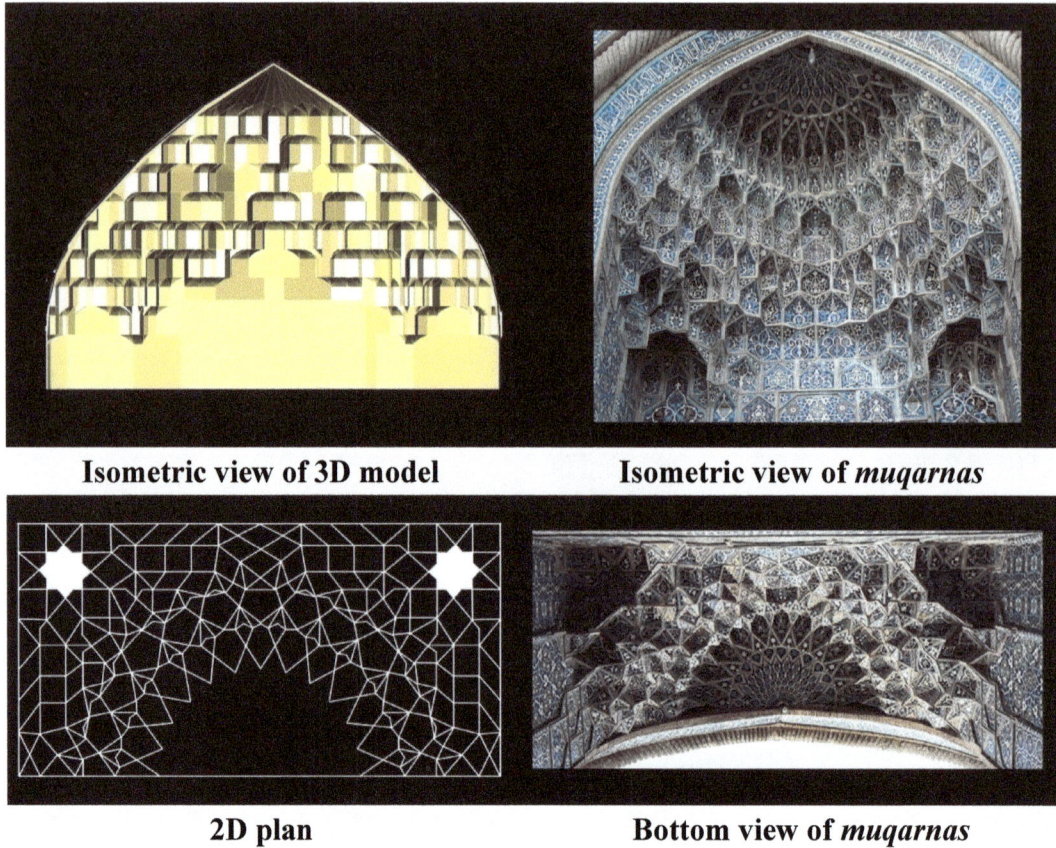

| | Isometric view of 3D model | | Isometric view of *muqarnas* |

| | 2D plan | | Bottom view of *muqarnas* |

Figure 5.99 View of *muqarnas* in Kerman Jame' Mosque (Source: Author).

Table 5. 40: Constituent elements table, Kerman Mosque.

Element → / ↓ Tier	*Toranj*	*Tee*	Taseh No	Whole 2D	3D	No	Half 2D	3D	Shaparak No	Type 2D	3D	Parak No	Type 2D	3D	Shamseh No	Type half	whole	Espar No	Type half	whole	Takht No	Type
One			<u>6</u>	4	2				<u>12</u>	10	2	2	√					7		√		
Two	12		<u>23</u>	11	12	2	√		16	√		12		√							2	8pointed
Three			<u>20</u>	2	18				<u>14</u>	2	12	<u>22</u>	2	20								
Four			24		√	2	√		19	√		6		√								
Five			25		√				12		√	24		√								
Six			25		√				24	√												
Seven			13		√				12		√	24		√								
Eight		13	12		√							24		√								
Nine			13		√										1	√						

The *muqarnas* of the entrance portal of Kerman Jame' Mosque is a one–sided decorative structure, built in the "suspended layer" method. As mentioned before, the finishing of the structure is *mu'araq* tiling, which means that the whole structure should have been built once by plaster. *Mu'araq* tiling is one of the most complex methods of decorating *muqarnas*. It is necessary to prepare moulds for every element. The small pieces of tile are then attached to the reverse mould and then a veneer is created for each element. The veneers are then mounted one by one on the tiers and are attached to the structure to cover up. Figure 5.100 shows the process of preparing *mu'araq* tiles, by the author under the supervision of Maestro Sha'rbaf, for the Faculty of Built Environment, Universiti Teknologi Malaysia. The mu'araq tile was cuted in Iran, but was then assembled inside UTM. This structure provides form–, height–, and size–transition. The plan of the structure can be divided into two sections of square lattice and pole table types and hence, it has a mixed plan type, based on Takahashi's categorizing of *muqarnas* plans. Figure 5.101 shows these two parts on the plan of the decoration. The highlighted part has pole table type and the rest has square lattice plan type.

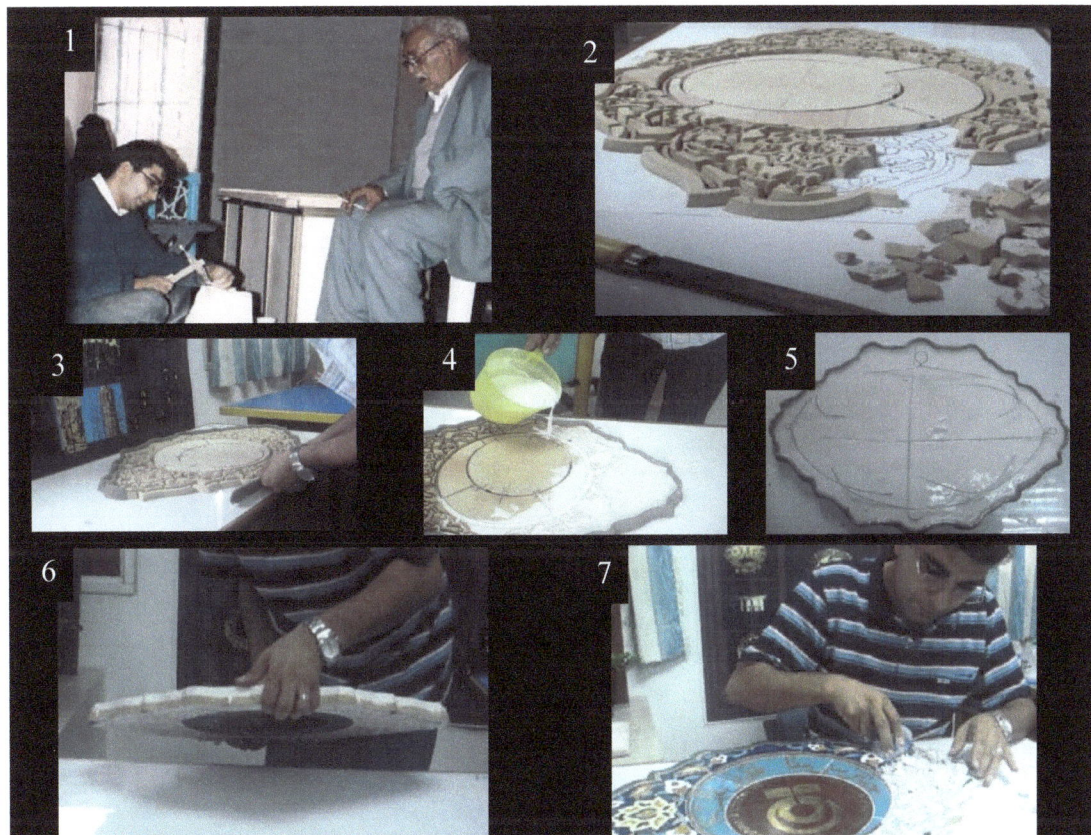

Figure 5.100 *Mu'araq* tile instalment process; (1) Cutting tiles in Iran, under supervision of Maestro Sha'rbaf, (2) through (7) Preparation and attaching in Malaysia

200

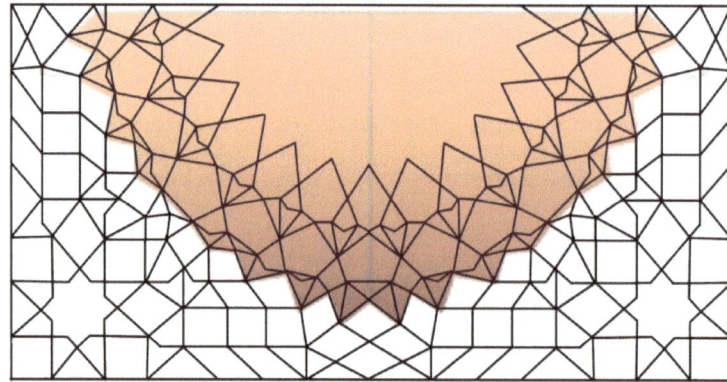

Figure 5.101 Square lattice and Pole table parts of the *muqarnas* plan of Kerman Jame' Mosque (Source: Author).

5.5   Matrix Analysis

As explained in Chapter 3, in order to obtain a comprehensive picture on the details of the studied samples, matrix analysis approach was selected as a method of systematic analysis. In Matrix 5.1, the selected samples were categorized based on the historical period to which they belong. As the purpose of this research is to study the evolution of *muqarnas* through time, the historical periods will be highlighted in the rest of the matrices in this chapter as well.

5.5.1   Chronology Matrix

As described before, samples were chosen in a way to include a comprehensive range, from the most basic squinches to a complete *muqarnas* structure. This range mainly covers a 450-year period, including two historical periods in Iran, i.e. Seljuk and Ilkhanid. Hence, Matrix 5.1, the chronology matrix, is arranged based on these historical periods.

Matrix 5.1   Characteristics Matrix, Chronology.

No	Date / Period → Sample ↓	Pre–Seljuk				Seljuk				Ilkhanid	
		< 900	900–950	950–1000	1000–1040	1040–1100	1100–1150	1150–1200	1200–1250	1250–1300	1300–1350
1	Sarvestan Palace	X									
2	Samanid Mausoleum		X								
3	Jurjir Mosque		X								
4	Nain Mosque, Mihrab			X							
5	Nain Mosque, Iwan			X							
6	Davazdah Imam Mosq.				X						
7	Barsian Mosque						X				
8	Bastam Minaret						X				
9	Sin Mosque						X				
10	Isfahan Jame', Nave							X			
11	Isfahan Jame', Iwan							X			
12	Bayazid Mausoleum									X	
13	Pir Bakran Mausoleum										X
14	Soltanieh Dome										X
15	Jame' of Natanz, Dome										X
16	Jame' of Natanz, Iwan										X
17	Ashtarjan Mosque										X
18	Varamin Jame'										X
19	Imamieh School										X
20	Kerman Jame' Mosque										X

5.5.2   Form and Structure Matrix

Form and structure of the studied samples were analysed in Matrix 5.2 and 5.3. In these characteristic matrices, different aspects in the form and structure of the samples were taken into account. The function of the samples in terms of form–, height– and size–transition, their application in terms of the position they were built in the building, including corners, dome chambers, Mihrab, Iwan, Entrance Portal, Minaret, etc., the number of tiers each sample has, as well as the construction dates and finishing materials are the different aspects studied in form and structure characteristics matrices.

Matrix 5.3 can be obtained based on analysing Matrix 5.2, which summarizes the gathered data in form of another matrix.

Matrix analysis of form and structure of the studied samples results in the following conclusions:

(a) The formal function of all *muqarnas* and pseudo-*muqarnas* structures is to provide a smooth transition of size, height and form between two different surfaces.

(b) *Muqarnas* and pseudo-*muqarnas* structures were most commonly built in iwans and entrance portals, in Iran.

(c) Pre-Seljuk *muqarnas* and pseudo-*muqarnas* structures have generally less than four tiers, whereas during Seljuk period, a maximum of five tiers is observed. During Ilkhanid period, the structures become more complex and they are built in more than five tiers.

(d) The finishing material has been normally either plaster or brick, while during Ilkhanid period glazed tiles started to become popular. The tiles that are normally installed on a plaster finishing, started to be used with arabesque floral and geometrical designs since late Ilkhanid period and reach the summit of their beauty during Safavid period. It is worth noting that during Seljuk period, using brick, an identical construction material with the buildings', as the finishing material was a common technique.

Matrix 5.2 Characteristics Matrix, Form and Structure.

			1	2	3	4	5	6	7	8	9	10	11	12	13	14	15	16	17	18	19	20
			Sarvestan Palace	Samanid Mausoleum	Jurjir Mosque	Na'in Msq., Mihrab	Na'in Msq., Iwan	Davazdah Imam Msq.	Barsian Mosque	Bastam Minaret	Sin Mosque	Isfahan Jame, Nave	Isfahan Jame', Iwan	Bayazid Mausoleum	Pir Bakran Mausoleum	Soltanieh Dome	Natanz Jame, Dome	Natanz Jame, Iwan	Ashtarjan Mosque	Varamin Mosque	Imamieh School	Kerman Mosque
Form and Structure	Formal Function	Height Transition	■	■	■	■	■	■	■	■	■	■	■	■	■	■	■	■	■	■	■	■
		Size Transition	■	■	■	■	■	■	■	■	■	■	■	■	■	■	■	■	■	■	■	■
		Form Transition	■	■	■	■	■	■	■	■	■	■	■	■	■	■	■	■	■	■	■	■
	Position in Building	Dome Chamber									■						■					
		Mihrab						■														
		Entrance/Portal			■														■	■		■
		Minaret								■												
		Iwan											■	■	■			■		■		
		Vault				■	■					■										
		Column																				
		Corner	■	■				■														
	Number of Tiers	1	■	■																		
		2				■						■										
		3							■	■												
		4					■														■	
		5											■		■							
		More than 5												■		■	■	■	■			
	Finishing Material	Brick		■				■	■	■	■									■		
		Stone/Rubble Stone	■																			
		Stucco/Plaster				■	■	■						■	■	■	■	■				
		Wood																				
		Tile & Ceramic																			■	■
		Glass & Mirror																				
		Mixed (Tile&Brick)																	■			

Matrix 5. 3 Form and Structure Analysis Matrix.

			In Percent
Form and Structure	**Formal Function**	Height Transition	100%
		Size Transition	100%
		Form Transition	100%
	Position in Building	Dome Chamber	10%
		Mihrab	5%
		Entrance / Portal	**20%**
		Minaret	5%
		Iwan	**30%**
		Vault	15%
		Corner	15%
	Number of Tiers	1	15%
		2	15%
		3	15%
		4	10%
		5	10%
		More than 5	35%
	Finishing	**Brick**	**40%**
		Stone/Rubble stone	5%
		Stucco/Plaster	**40%**
		Tile and Ceramic	10%
		Mixed (Tile&Brick)	5%

Pre–Seljuk	Seljuk	Ilkhanid

5.5.3 Construction Technique Matrix

The construction technique used for building the studied structures were explored in detail, using Matrix 5.4. The load–bearing role of the structure, as being merely a decorative veneer or a structural vault or squinch, the sample's construction material and investigating about it being either identical to that of building or different from it, the structure's plan type and method of building it are the details that are examined in these matrices. The glossary section can be referred to, as a quick reminder for the definition of the terms used in these matrices.

By a detailed review of Matrix 5.4 one can conclude that:

(a) Most buildings in Iran were constructed in bricks.

Construction Technique

Matrix 5.4   Characteristics Matrix, Construction Technique.

Category	Sub-category
Architectural Role	Decorative
	Structural
Building Construction Material	Brick
	Stone/Rubble stone
	Stucco and Plaster
	Adobe
Construction / Finishing Material	Identical
	Different
2DPP Type	Square Lattice Style
	Pole Table Style
	Mixed
Construction Method	One-Sided
	Two-Sided
	Suspended Layer
	Suspended Units
	Amalgamated Construction
	By means of Supporting Ribs

#	Building	Period
1	Sarvestan Palace	Pre-Seljuk
2	Samanid Mausoleum	Pre-Seljuk
3	Jurjir Mosque	Pre-Seljuk
4	Na'in Msq., Mihrab	Pre-Seljuk
5	Na'in Msq., Iwan	Pre-Seljuk
6	Davazdah Imam Msq.	Pre-Seljuk
7	Barsian Mosque	Seljuk
8	Bastam Minaret	Seljuk
9	Sin Mosque	Seljuk
10	Isfahan Jame, Nave	Seljuk
11	Isfahan Jame', Iwan	Seljuk
12	Bayazid Mausoleum	Ilkhanid
13	Pir Bakran Mausoleum	Ilkhanid
14	Soltanieh Dome	Ilkhanid
15	Natanz Jame, Dome	Ilkhanid
16	Natanz Jame, Iwan	Ilkhanid
17	Ashtarjan Mosque	Ilkhanid
18	Varamin Mosque	Ilkhanid
19	Imamieh School	Ilkhanid
20	Kerman Mosque	Ilkhanid

(b) Unlike the two-sided *muqarnas* structures of Iraq and Syria, all *muqarnas* and pseudo-*muqarnas* structures recorded in Iran are one-sided.

(c) All the pre-Seljuk pseudo-*muqarnas* structures had square lattice plans. The pole table plan was first introduced during Seljuk period in Iran, although it was not very well established and it was used mostly as a mixture with the common square lattice type.

(d) All of the studied Seljuk and pre-Seljuk *muqarnas* and pseudo-*muqarnas* structures were load-bearing, whereas during Ilkhanid period, they change to purely decorative veneers whose load should have been somehow transferred to the building.

(e) The importance of decorative veneer during the Ilkhanid period is so much that not only the dead load is added to the building, but also the load of more luxurious finishing, like glazed tiles, is added to cover the exterior surface of the decoration.

(f) Before Ilkhanid period, supportive ribs were the most commonly used technique in Iran for making load-bearing *muqarnas* and pseudo-*muqarnas* structures, while amalgamated construction of a *muqarnas* and pseudo-*muqarnas* structures can only be observed in pre-Islamic structures.

(g) Although suspended unit technique was a common technique outside the borders of the influence of Iran, the technique is not observed to be used in Iran except for one sample, i.e. Bastam Minaret, in which it is mixed with the amalgamated technique, making an innovative technique of its own; and finally,

(h) Suspended layer construction technique becomes popular in Iran since the Ilkhanid period with no record prior to that.

5.6 Evolution of Constituent Elements

Based on the main aim of this research, after a detailed review of the constituent elements of structures of the selected 20 samples and presenting them in form of tables, Matrix 5.5 was designed to illustrate the variety of constituent element used in the structure of studied samples. The evolution of *muqarnas* from squinch was a gradual process that took place in about five centuries. As the sequence of the samples is arranged chronologically and the historical periods are highlighted in this matrix, Matrix 5.5 can explain in which historical period and in which building decoration each element was invented and used.

Matrix 5.5   Constituent Elements Matrix.

No	Samples	Toranj	Tee	Taseh Type 2D	Taseh Type 3D	Half–Taseh Type 2D	Half–Taseh Type 3D	Shaparak Type Plumb	Shaparak Type 3D	Parak Type 2D	Parak Type 3D	Shamseh Type Half	Shamseh Type Whole	Espar Type Half	Espar Type Whole	Takht	
1	Sarvestan Palace				■												Pre-Seljuk
2	Samanid Mausoleum				■												
3	Jurjir Mosque				■												
4	Nain Mosque, Mihrab				■			■		■							
5	Nain Mosque, Iwan				■								■				
6	Davazdah Imam				■												
7	Barsian Mosque				■		■					■					Seljuk
8	Bastam Minaret	■			■	■		■		■							
9	Sin Mosque			■	■	■		■			■	■					
10	Isfahan Jame', Nave				■			■				■					
11	Isfahan Jame', Iwan	■	■	■	■	■	■	■	■	■	■			■			
12	Bayazid Mausoleum	■	■	■	■	■		■		■				■			Ilkhanid
13	Pir Bakran Shrine	■			■	■		■		■							
14	Soltanieh Dome	■	■	■	■	■		■		■			■				
15	Jame' Natanz, Dome	■	■	■	■	■		■		■		■					
16	Jame' Natanz, Iwan	■		■	■		■	■		■	■			■			
17	Ashtarjan Mosque	■	■	■	■			■		■				■		■	
18	Jame' of Varamin	■		■	■		■	■		■				■			
19	Imamieh School	■		■	■			■		■							
20	Jame' of Kerman	■	■	■	■		■	■		■		■		■			

Analysis of Matrix 5.5 results in the following conclusions:

(a) Three-dimensional *taseh* is the one and only element that survives the creation of squinches in all various *muqarnas* and pseudo-*muqarnas* structures. Since 12th century, in Sin Mosque, two-dimensional *taseh* has been observed along with three-dimensional ones.

(b) The *Parak* and *shaparak* that Al-Kashi labels as intermediate elements are the second created elements. These elements, which are first created in plumb form, in the ornaments of the mihrab and iwan vaults of Na'in Jame' Mosque, are the second most important constituent elements, in terms of application in the structure.

(c) With the masterpiece at Isfahan Jame' mosque, the Western Iwan, three-dimensional *shaparak* is introduced for the first time in 12th century and since, this element is observed in almost all built *muqarnas* and pseudo-*muqarnas* structures. The first three-dimensional *parak*, on the other hand, is also observed for the first time in Sin Jame' Mosque's pseudo-*shamseh*.

(d) Iwan vaults of Na'in Jame' Mosque has the first *shamseh* in it, in mid-10th century and since then, they are an inevitable part of all *muqarnas* and pseudo-*muqarnas* structures that built over iwans, portals and dome chambers.

(e) The Western Iwan of Isfahan Jame' Mosque, is the place of birth of another important constituent element of *muqarnas*, i.e. *espar*. Since 12th century this element is also added to almost all *muqarnas* and pseudo-*muqarnas* structures, as a major element.

(f) 12th century is the birth century of another significant *muqarnas* element. *Toranj* is introduced in Bastam Minaret for the first time. Although it is not used in some pseudo-*muqarnas* structures, the element is a part of all roof-patched *muqarnas* structures.

(g) *T* element is another plumb element which was first used in Western Iwan of Isfahan Jame' Mosque to connect several tiers and make a bigger element unit; and finally,

(h) The *takhts*, as the distinctive constituents of *muqarnas* are observed as an established element in the 14th century. The existence of this element is the key difference between *muqarnas* and pseudo-*muqarnas* structures.

5.7 Conclusion

Based on the clarified characteristics of the studied *muqarnas* and pseudo-*muqarnas* structures, their formal application in the building, and their complexity, in terms of form and structure, the studied *muqarnas* structures are categorized into four groups, defined as follows:

(1) Squinch, which from constituent elements point of view, can be considered identical to a single *taseh*, is in fact a cornering technique which consists of two concaved vault–sections intersecting each other at right angle, in a line. The intersecting of these vault–sections creates squinch.

(2) *Patkaneh* is a load-bearing structure dating back to the 10th century. The structure consists of two main parts, the ribs that provide the structural role for the ornament, and the *tasehs*, built between these ribs, in two to maximum five successive projecting tiers. Plumb *paraks*

and *shaparaks* are the intermediate elements observed in *patkanehs*, although exceptional samples had been recorded with three-dimensional *paraks* or *shaparaks* in late 12th century.

(3) Being the most similar ornament to *muqarnas*, decorative *patkaneh* consists of non-load-bearing decorative constituent elements. This ornament which was born in the 13th century is normally erected by suspended layer technique, in which the elements are attached to the ceiling and walls by means of timber and rope. Together with *taseh*, three-dimensional *paraks* or *shaparaks* should be considered as main elements of decorative *patkaneh*. Furthermore, *espar* and *shamseh* are also inevitable constituent elements of this type of ornament, except on minarets and columns, respectively; and,

(4) *Muqarnas* is the most advanced ornament studied in this manuscript. The appearance of *takts* is the main distinctive characteristic of this ornament compared to other pseudo-*muqarnas* structures. Furthermore, the existence of *toranj* in all *muqarnas* samples is another important characteristic of this ornament. The constituent elements of *muqarnas* are confined to the eight introduced elements and this structure is the invention of the early 14th century.

Hence, based on the definitions above, the studied samples can be categorized into the four main groups, namely, squinch, *patkaneh*, decorative *patkaneh* and *muqarnas*, as explained in Tables 5.41. Figure 5.102, shows an example for each type of ornament introduced above.

Table 5. 41: Categorization of the studied samples.

No	Name of Building	Date	Type of Decoration	
1	Sarvestan Palace	420–480	Squinch	Pre-Seljuk
2	Samanid Mausoleum	907	Squinch	
3	Jurjir Mosque	939	Squinch	
4	Nain Mosque, Mihrab	960	*Patkaneh*	
5	Nain Mosque, Iwan	960	*Patkaneh*	
6	Davazdah Imam Mosque	1037	*Patkaneh*	
7	Barsian Mosque	1114	*Patkaneh*	Seljuk
8	Bastam Minaret	1120	*Patkaneh*	
9	Sin Mosque	1134	*Patkaneh*	
10	Isfahan Jame' Mosque, Nave	1118–1157	*Patkaneh*	
11	Isfahan Jame' Mosque, Iwan	1118–1157	*Patkaneh*	
12	Bayazid Mausoleum	1299–1313	Decorative *Patkaneh*	Ilkhanid
13	Pir Bakran Mausoleum	1303–1312	Decorative *Patkaneh*	
14	Soltanieh Dome	1312–1313	Decorative *Patkaneh*	
15	Jame' of Natanz, Dome	1304–1325	Decorative *Patkaneh*	
16	Jame' of Natanz, Iwan	1304–1325	Decorative *Patkaneh*	
17	Ashtarjan Mosque	1316	*Muqarnas*	
18	Varamin Jame' Mosque	1322	*Muqarnas*	
19	Imamieh School	1325	*Muqarnas*	
20	Kerman Jame' Mosque	1349	*Muqarnas*	

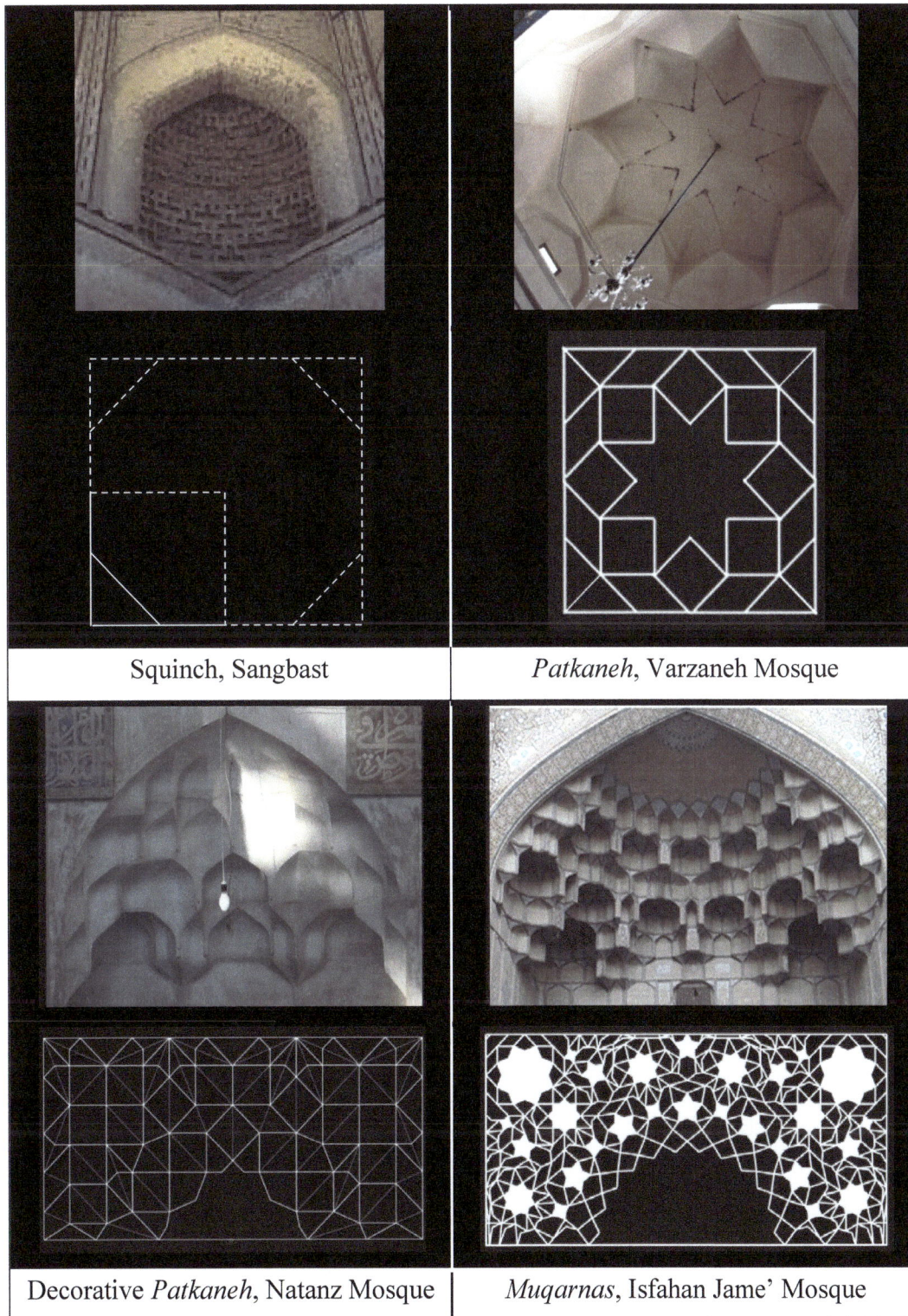

| Squinch, Sangbast | *Patkaneh*, Varzaneh Mosque |
| Decorative *Patkaneh*, Natanz Mosque | *Muqarnas*, Isfahan Jame' Mosque |

Figure 5.102 Four main types of studied ornaments, based on form and structure.

CHAPTER 6
Summary

6.1   Introduction

Referring back to the aim and objectives of this research and the selected methodology for achieving them, this chapter will provide an overall look to the whole process of research and its achievements.

The evolution of the phenomenon *muqarnas*, as an architectural element of Islamic architecture, was studied through a qualitative research approach. As the historical aspects of the decorative elements was an important factor to analyze the structure's path of development, a large amount of evidences was required to provide enough data for seeking the progress of the phenomenon. Hence, a comprehensive and worthy database of 100 *muqarnas* and pseudo-*muqarnas* samples were collected, measured and recorded for this research. The plan and three–dimensional patterns of all of 100 samples were created to assist in systematic analysis and evaluation of them, which involves different stages of data processing through an inductive approach.

At the first stage of research, open coding technique was accomplished with the purpose of reviewing the general characteristics of the gathered samples for identifying and categorizing them based on the variety of their constituent elements and probable coding concepts were extracted. Then, the samples were arranged in a chronological order for the second stage of analysis, i.e. axial coding. During the process of axial coding, the axis variable was chosen to be time and hence, after detailed investigations about the sample's date of construction, they were arranged based to their construction time. This was done in order to provide suitable leading concepts for the next step, i.e. important turning points in the evolution of *muqarnas*. Selective coding was accomplished later and referring to the achieved critical changes, 20 critical samples were chosen to be studied, based on the obtained concepts and variables, for demonstrating the phenomenon's path of development by detailed comparison of the characteristics, which was accomplished by matrix analysis.

6.2 Research Findings

The stated process was accomplished for achieving the research objectives, as explained below, which will in turn lead the author toward attainment of research aim and contributing it to the body of knowledge.

6.2.1 Historical Development of *Muqarnas*

"To seek the time period in which *muqarnas* structure is fully developed to its optimum form by recording existing *muqarnas* and pseudo-*muqarnas* samples in Iran."

Based on the axial coding of the gathered 100 samples of *muqarnas* and pseudo-*muqarnas* structures in Iran, the first transition from squinch toward a decorative veneer was recorded in 10th century, in Na'in Jame' Mosque. The ornaments were load-bearing until early 12th century, where after the iwan of Isfahan Jame' Mosque, they started to find a merely decorative role in the building and their construction technique change accordingly from using ribs to suspended layer technique. Samples of mature *muqarnas* which had *takht* elements, was found in the early 14th century in Iran and since then, the form was fully developed century and the composition reached maturity by the end of the Ilkhanid period and beginning of Timurid.

6.2.2 Constituent Elements of *Muqarnas*

"To investigate the constituent elements of *muqarnas*, with the purpose of developing a minimized but general set of basic constituent elements that could cover all elements of all *muqarnas* structures of all times."

Although Al–Kashi's described *muqarnas* cells are widely accepted as comprising elements of *muqarnas*, more advanced forms of *muqarnas* cannot be explain based on this definition. This is because no *takht* elements is considered in the definitions provided by Al–Kashi, whereas during Ilkhanid period, as provided in the samples studied in the scope of this research, the so–called roof–patched *muqarnas* structures had been invented. Therefore, the necessity of defining a general set of constituent element that could cover the components of all types of *muqarnas* structures urge the author to investigate about the issue by open coding technique of a 100 samples of *muqarnas* and pseudo-*muqarnas* structures in Iran.

This investigation was accomplished with the help of considering some important variables, namely the basic or intermediate role of the element, as well as its geometry as being an edge to point, point to edge or edge to edge element. Based on the above mentioned coding, as explained in Chapter 4, the elements of *muqarnas* were categorised to *shamseh*, *toranj*, *taseh*, *parak*, *shaparak*, *tee*, *espar*, and *takhts*, as shown below in Figures 6.1 and 6.2. Colours defined for each element type in Figure 6.1 is used in Figure 6.2 as well, to make it easier to recognize the constituent element. *Muqarnas* of the iwan of Qazvin Jame' Mosque, chosen here as an example, is simple but complete in terms of constituent elements.

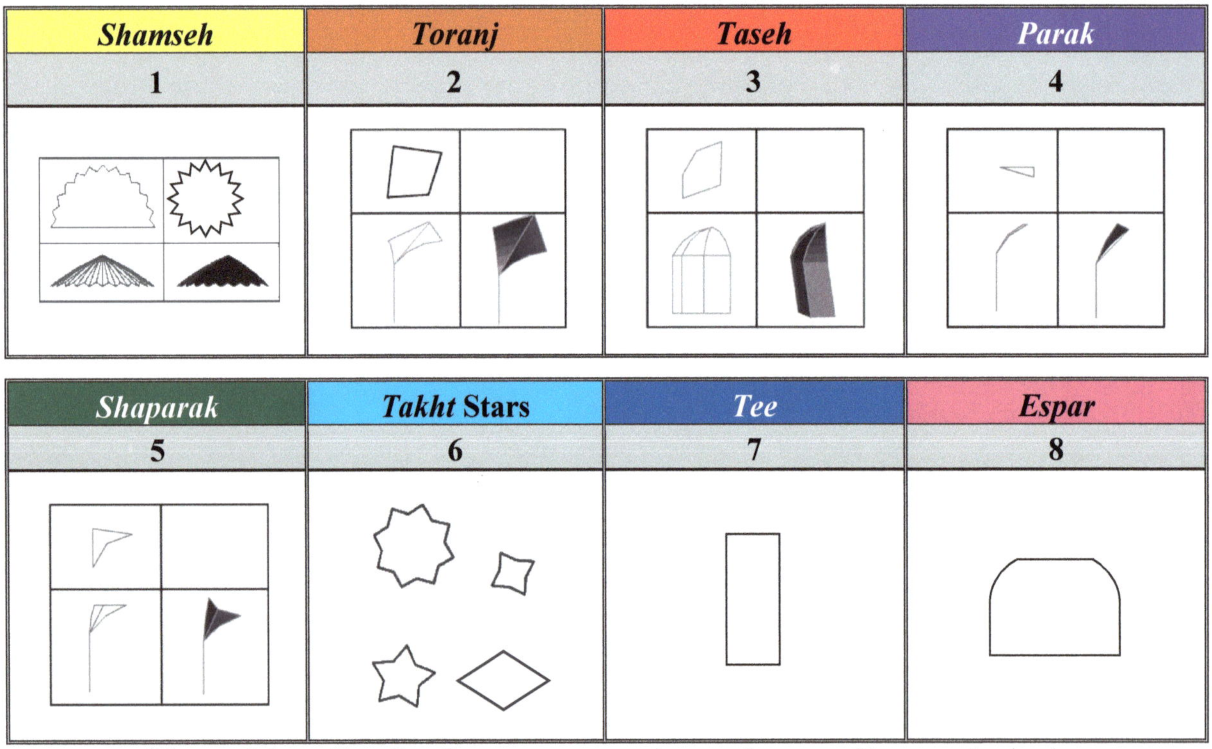

Shamseh	*Toranj*	*Taseh*	*Parak*
1	2	3	4

Shaparak	*Takht* Stars	*Tee*	*Espar*
5	6	7	8

Figure 6.1   Details of *muqarnas* constituent elements (Source: Author).

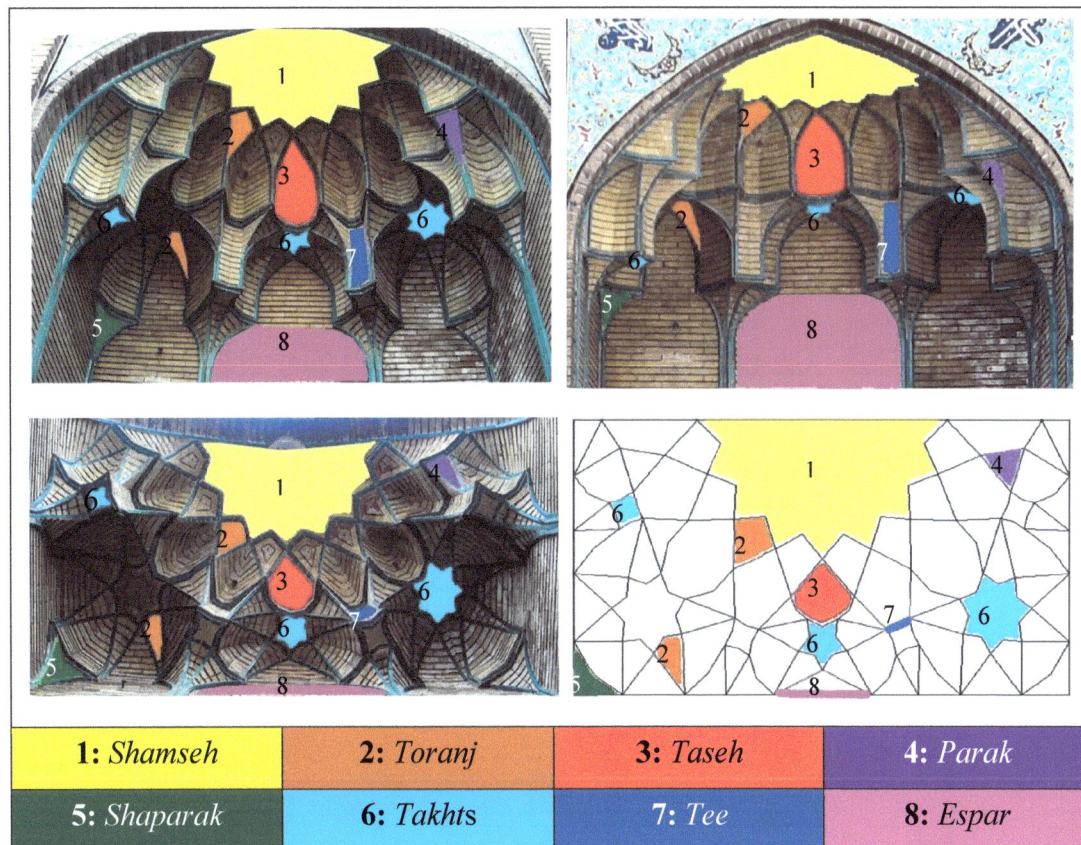

| 1: *Shamseh* | 2: *Toranj* | 3: *Taseh* | 4: *Parak* |
| 5: *Shaparak* | 6: *Takht*s | 7: *Tee* | 8: *Espar* |

Figure 6.2   The constituent elements of *muqarnas*, Qazvin Jame' Mosque (Source: Author).

Evolution of the constituent elements of *muqarnas* was found to be as described in Figure 6.3. *Taseh* was the first existing element, in form of squinch. The date of creation of *taseh* goes back to pre-Islamic history. The oldest recorded sample with squinch in this research was that of Ardeshir Palace in 224 C.E. Plumb *shaparak* and *parak*, as well as whole-*shamseh* were the next created elements in 960 C.E., whereas between 1100-1150 two-dimensional *taseh*, three-dimensional *parak*, *toranj* and half-*shamseh* were created. *Tee*, three-dimensional *shaparak* and *espar* were the next invented constituent elements in ornaments of Isfahan Jame' Mosque before 1200. Half-*espar* used for the first time before 1300 in *muqarnas* of Bayazid Mausoleum and *takht* was the last introduced element in *muqarnas* of Ashtarjan Jame' Mosque in 1316 .

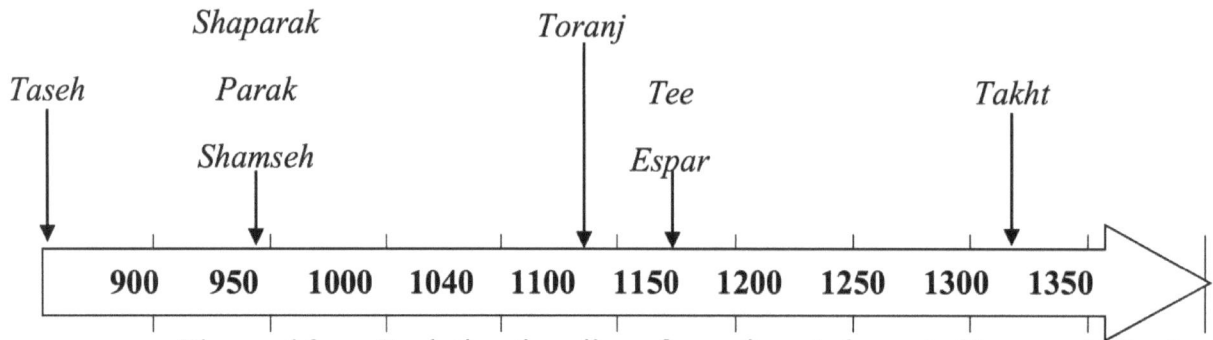

Figure 6.3   Evolution time-line of constituent elements (Source: Author)

6.2.3   *Muqarnas* vs. Pseudo-*Muqarnas*

"To clarify the similarities and differences between *muqarnas* and other pseudo-*muqarnas* structures."

The perception obtained by open and axial coding of the 100 sample database, as explained in Chapter 3, provided the author the ability of defining some practical criteria for selecting critical samples out of database. Twenty samples were selected based on the following criteria:

(a) As the chronological path of evolution of *muqarnas* structure was one of the key factors considered in this research, several samples were chosen from each historical period in *muqarnas* evolution timeline and from the main stages of the structure's development. It is worth considering that many of the studied buildings were renovated or even reconstructed several years after their date of construction. Hence, accurate investigation about the date of the specific part of the building under study was vital. Therefore, efforts have been made to find reliable dates for each building and the specific date of construction of that particular decoration. In traditional architecture of Iran, the date of erecting each specific part of the building and the mason's name used to be written in some form on an inscription band or among the patterns of the decorative veneer. Furthermore, dates provided and published by archaeologists and those involved in restoration of the buildings are another reliable source of seeking each sample's date of construction.

(b) The next criteria for choosing these samples were the importance of not missing any of the varieties of forms and structures of *muqarnas* during the studied eras.

(c) There have been various method for building *muqarnas* throughout its history in Iran, the construction technique of the *muqarnas* was also an important criteria in choosing the final *muqarnas* samples.

(d) Another very important criterion for choosing the samples was the first time that a constituent element of *muqarnas* was invented and used in a structure; and finally,

(e) The function of the structure in the building was also a factor that was considered for choosing the appropriate examples from the collected database.

The selected samples were studied in detail in Chapter 5 considering three major aspects, namely, chronology, form and structure, and construction technique. Referring back to the obtained results, the characteristics of squinch, *patkaneh*, decorative *patkaneh* and *muqarnas* can be explained as Tables 6.1 and 6.2 imply.

Table 6.1: Comparison of form and structure.

			Squinch	*Patkaneh*	Decorative *Patkaneh*	*Muqarnas*
Form and Structure	**Formal Function**	Height Transition	■	■	■	■
		Size Transition	■	■	■	■
		Form Transition	■	■	■	■
	Number of Tiers	1	■			
		2		Minimum	Minimum	
		3		↕	↓	Minimum
		4				↓
		5		Maximum		
		More than 5				
	Finishing	Brick	■	■		■
		Stone/Rubble stone	■			
		Stucco/Plaster		■		■
		Tile and Ceramic				■
		Glass and Mirror				■
		Mixed (Tile&Brick)			■	■

Table 6.2: Comparison of construction technique.

Construction Technique			Squinch	Patkaneh	Decorative Patkaneh	Muqarnas
	Architectural Role	Decorative			■	■
		Structural	■	■		
	Construction / Finishing Material	Identical	■	■		
		Different		■	■	■
	Construction Method	Suspended Layer			■	■
		By means of Supporting Ribs		■		
		Amalgamated Construction	■			

All mentioned ornaments are meant to provide a smooth size-, height- and form-transitions. Squinches were constructed in only one tier, whereas *patkanehs* could have a minimum of two to maximum four tiers. As mentioned before, there are two major techniques for constructing *muqarnas* and pseudo–*muqarnas* structures, namely suspended layer and ribs. The limit in the number of tiers is a consequence of the ornament's construction technique. Hence, by alerting the construction method from ribs to suspended layer technique, the limit in the maximum number of tiers is all disappeared, though the architectural role of the ornament also changes. Suspended layer technique, used for constructing decorative *patkaneh* and *muqarnas*, makes the ornament a purely decorative veneer, with a dead load that should be transferred to the walls and ceilings by means of ropes and timbers.

The preference of ornament to structure can be felt more when one looks at the difference between the building's material and the ornaments finishing. Although in *patkanehs* different finishing material is observed, in case of decorative *patkaneh* and *muqarnas* the difference becomes a rule. Glazed tiles are introduced as a luxurious finishing to be pasted over the plaster finishing of ornament which itself is another added load, showing the increasing importance of decorations by passage of time.

Table 6.3:   Comparison of constituent elements.

Elements ↓ → Date	Toranj	Tee	Taseh Whole		Taseh Half		Shaparak Type		Parak Type		Shamseh Type		Espar Type		Takht
			2D	3D	2D	3D	Plumb	3D	Plumb	3D	Half	Whole	Half	Whole	
Squinch				■											
Patkaneh	■	■	■	■	■	■	■	■	■	■	■	■		■	
Decorative Patkaneh	■	■	■	■	■	■	■	■	■	■	■	■	■	■	
Muqarnas	■	■	■	■	■	■	■	■	■	■	■	■	■	■	■

In terms of constituent elements, as shown in Table 6.3, squinches are made of *tasehs* only, where as in case of *patkaneh*, seven other basic elements are gradually added to the list during two centuries of development. *Takht* is the only specific element that is added in early 14th century to complete the list and to make *muqarnas* remarkably different from the rest of ornament. Although *takht* is the only added ornament that *muqarnas* has compared to *patkaneh*, it gives the ornament a significant difference. But the major difference between the two is in fact their construction technique and their structural role in the building.

Hence, *patkaneh* and decorative *patkaneh* have similar constituent elements, while *takht*, which was first invented in early 14th century, is the characteristically distinctive element, unique to *muqarnas* ornament. Further, *muqarnas* should be considered an absolutely different ornament from *patkaneh*, due to the fact that *muqarnas* is purely decorative but *patkaneh* is a load-bearing ornament. In addition, *muqarnas* is made by suspended layer technique whereas *patkaneh* is built by using rib. Hence, in spite of their similar appearance, the two structures should be differentiated.

6.2.4   Clarification of Definitions

"To define each ornament clearly, i.e. squinch, *patkaneh*, decorative *patkaneh* and *muqarnas*, from three aspects, namely chronology, form and structure, construction techniques."

Although *muqarnas* and squinch are considered known concepts, *patkanehs* are little introduced to the body of knowledge and the available mentions are mostly indirect and incomplete. Based on the clarified characteristics of the studied *muqarnas* and pseudo-*muqarnas* structures, their structural role in the building, and their complexity, in terms of form and structure, *patkaneh* should be introduced as a separate ornament from *muqarnas*.

Patkaneh is structurally different from *muqarnas*, i.e. it is load-bearing while *muqarnas* is merely decorative. *Patkaneh* is built by means of ribs, while suspended layer technique is used for constructing *muqarnas*. In terms of complexity of form, although the structures seem to have rather similar elements, *patkaneh* has no horizontal constituent element and the number tiers in *patkaneh* is limited to five, while no limit can be set for number of tiers in *muqarnas*.

Patkaneh which is invented for the first time in 960 C.E. in Na'in Jame' Mosque is introduced in this research as the missing link between squinch and *muqarnas*. The ornament developed gradually until 1316, when the first *takht* element appeared and *muqarnas* reached its fully developed form.

On the other hand, in the Western Iwan of Isfahan Jame' Mosque, built professionally in five tiers, between 1118 and 1157, *patkaneh* reaches its optimum point. Since then, the need for more decorated spaces urged the masons to use another construction technique which enables them to have more tiers of ornament. Introduction of suspended layer technique alerted the structural role of ornaments from load-bearing to purely decorative veneers, which is considered as the starting point of *muqarnas*.

The two ornaments overlap each other, making another category of ornamental veneers, which is named as "decorative *patkaneh*". *Patkaneh* and decorative *patkaneh* have identical constituent elements and can be considered as a single category, whereas in terms of structural role and construction technique, decorative *patkaneh* is similar to *muqarnas*. Hence, as shown schematically in Figure 6.4, the ornaments are categorized into four groups as described below:

Squinch is the oldest known transition technique, which can be considered a single *taseh*. It consists of two load-bearing concave vault–sections intersecting each other in a line at right angle.

Patkaneh is a set of niche–like components arranged in successive tiers in a definite geometrical regime, in order to cover vaults or ceilings. The structure is another load-bearing transition structure, dating back to the 10th century. The structure consists of two main parts, the ribs that provide the structural role for the ornament, and the *tasehs*, built between these ribs, in two to maximum five successive projecting tiers. Plumb *paraks* and *shaparaks* are the intermediate elements observed in *patkanehs*, although exceptional samples had been recorded with three-dimensional *paraks* or *shaparaks* in late 12ᵗʰ century.

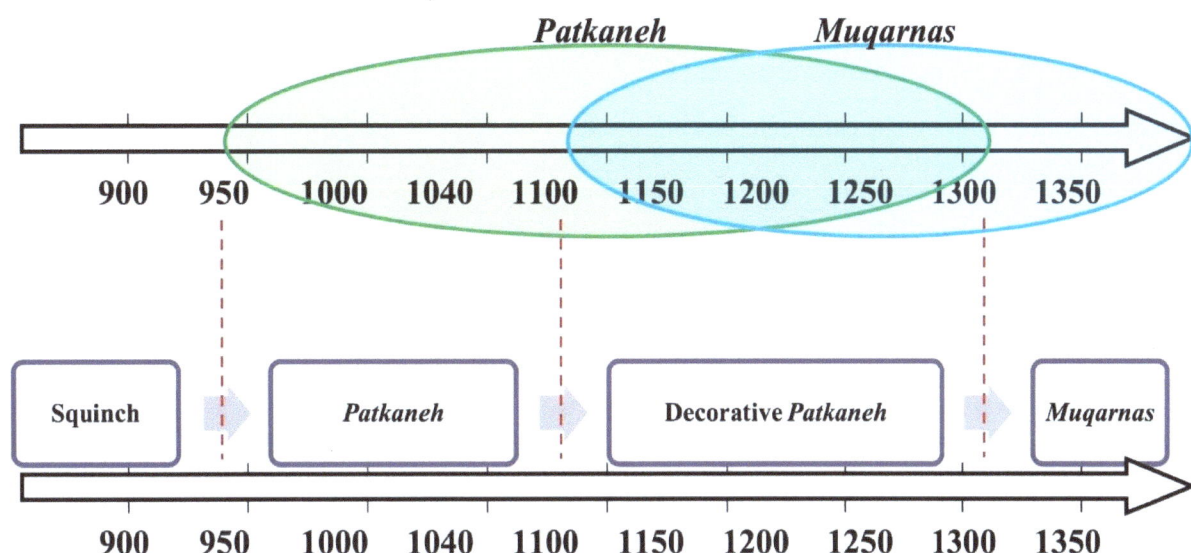

Figure 6.4   *Muqarnas* evolution time-line (Source: Author).

Being the most similar ornament to *muqarnas* in terms of structural role and construction technique, and having identical constituent elements with *patkaneh*, **decorative *patkaneh*** is a non-load-bearing decorative veneer which was born in the 13ᵗʰ century. Being erected by suspended layer technique, *taseh*, three-dimensional *paraks* or *shaparaks* should be considered as main elements of decorative *patkaneh*. Furthermore, *espar* and *shamseh* are also inevitable constituent elements of this type of ornament, except on minarets and columns; and finally,

Muqarnas is the most advanced ornament of this type invented in early 14ᵗʰ century. This structure is purely decorative and adds dead-load to the building. The constituent elements of *muqarnas* are confined to the eight basic elements, namely *shamseh*, *taseh*, *toranj*, *shaparak*, *parak*, *tee*, *espar* and *takht*. The introduction of *takht*-elements is the main distinctive characteristic of this ornament compared to other pseudo-*muqarnas* structures. Furthermore, due to the existence of star *takhts* and high-number pointed *shamsehs*, *toranj* is an inevitable element of all *muqarnas* samples.

6.3 Research Contribution

The contributions of this research is indebted to triangulation of these three sources, i.e. the two major theories by Michel Écochard and Henri Stierlin, positing that the origin of *muqarnas* should be sought in Iran and in squinch; the author's experience and knowledge about traditional and contemporary techniques of designing and building *muqarnas* in Iran; in association with the final achievements of this investigation in seeking the characteristics and definition of the missing link between the *muqarnas* and squinch. Hence, these findings are introduced as the main contributions of this research to the body of knowledge:

(a) *Muqarnas* is an ornamental structure that is originated in Iran. The structure roots can be found in the Iranian pre-Islamic transition technique, known as squinch, which was reformed in 10th century to another linking structure, named *patkaneh*.

(b) *Muqarnas* was fully developed to its optimum form in early 14th century by introduction of the *takht* element, which is the only horizontal element, specific to *muqarnas*.

(c) Constituent elements of *muqarnas* can be abstracted to a set of eight basic elements, namely *taseh, shaparak, parak, toranj, shamseh, Tee, espar* and *takht*. This set is extensible to all types of *muqarnas*, built anytime, anywhere in Iran.

(d) *Taseh* and *shaparak* and *parak* are the most extensively used elements of all *muqarnas* and pseudo-*muqarnas* structures, which can be observed in plumb or three-dimensional forms. *Tee* and *espar* are the two vertical elements and *takht* is the only horizontal one. *Shamseh*, which has mostly *toranjes* as its immediate neighbours, is a several-pointed three-dimensional star, usually located on the apex of the ornament.

(e) *Patkaneh*, the linking structure between squinch and *muqarnas*, is a load-bearing ornament consisting of two main parts, i.e. *tasehs* and ribs. The 10th century ornament, limited to maximum five successive tiers, reached its summit in late 12th century, when alternation of construction technique to suspended layer method changed the architectural role of the ornament to non-load-bearing.

(f) *Squinch, Patkaneh* and *muqarnas* are all transition techniques, built from bottom to top. Squinch is built by amalgamated construction technique, *patkaneh* is built by means of ribs but *muqarnas* is built on suspended plaster layers, and therefore, *muqarnas* is merely decorative.

Referring to the above mentioned contributions, the gradual development of *muqarnas* from squinch to *patkaneh* is illustrated in Figures 6.5 and 6.6, in which a quarter of the plan of several real world samples are shown with an image of the structure. The plans are arranged from left to right to show the deformation sequence, with previous step shown again in its next plan with same color to provide easier understanding. It is necessary to be mentioned that, as the selected shown samples are real world structures fitted to the specific defined dimension defined for illustration purpose, some minor alternation was imposed to the real size of constituent elements of the plans. The plan with correct dimension-ratios, though, can be found in the manuscript based on the references provided below each introduced sample.

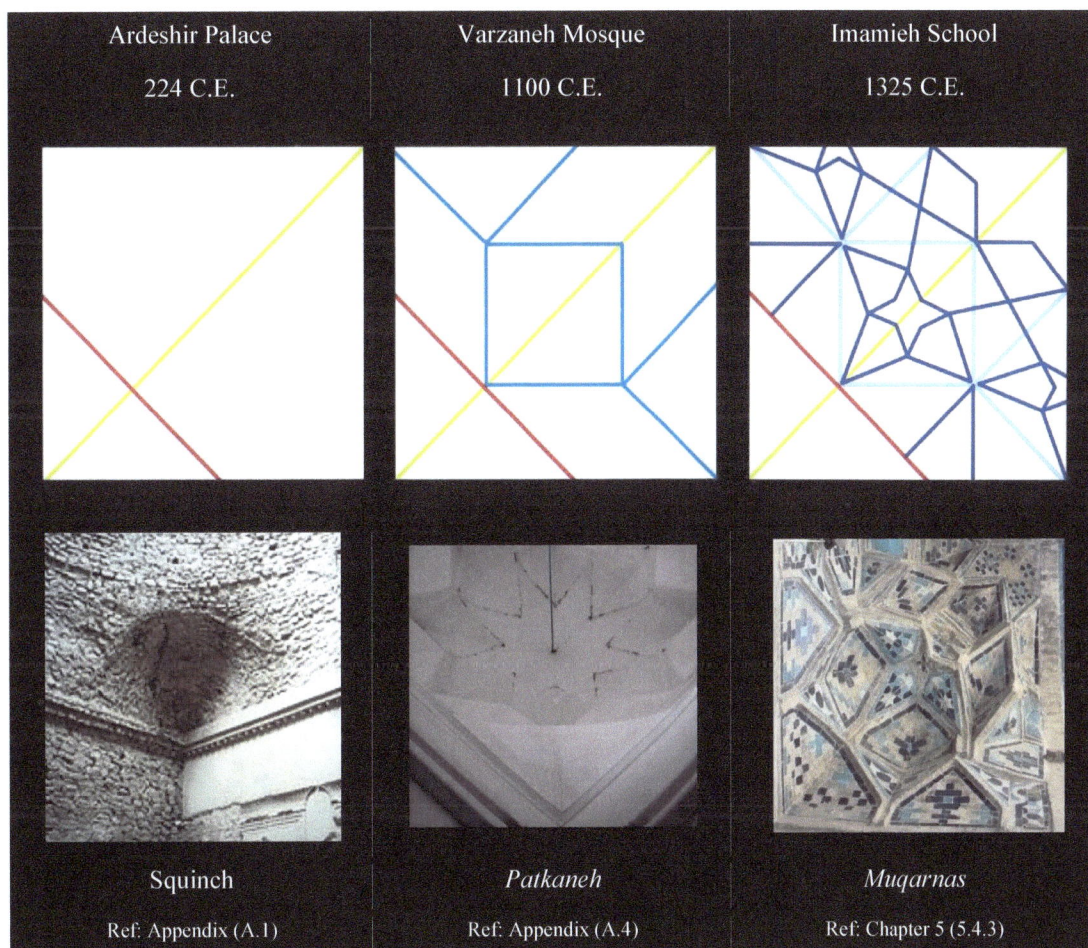

Figure 6.5 Demonstrating the evolution of *muqarnas*; from left to right, squinch (red), *patkaneh* (light blue), and *muqarnas* (dark blue) (Source: Author).

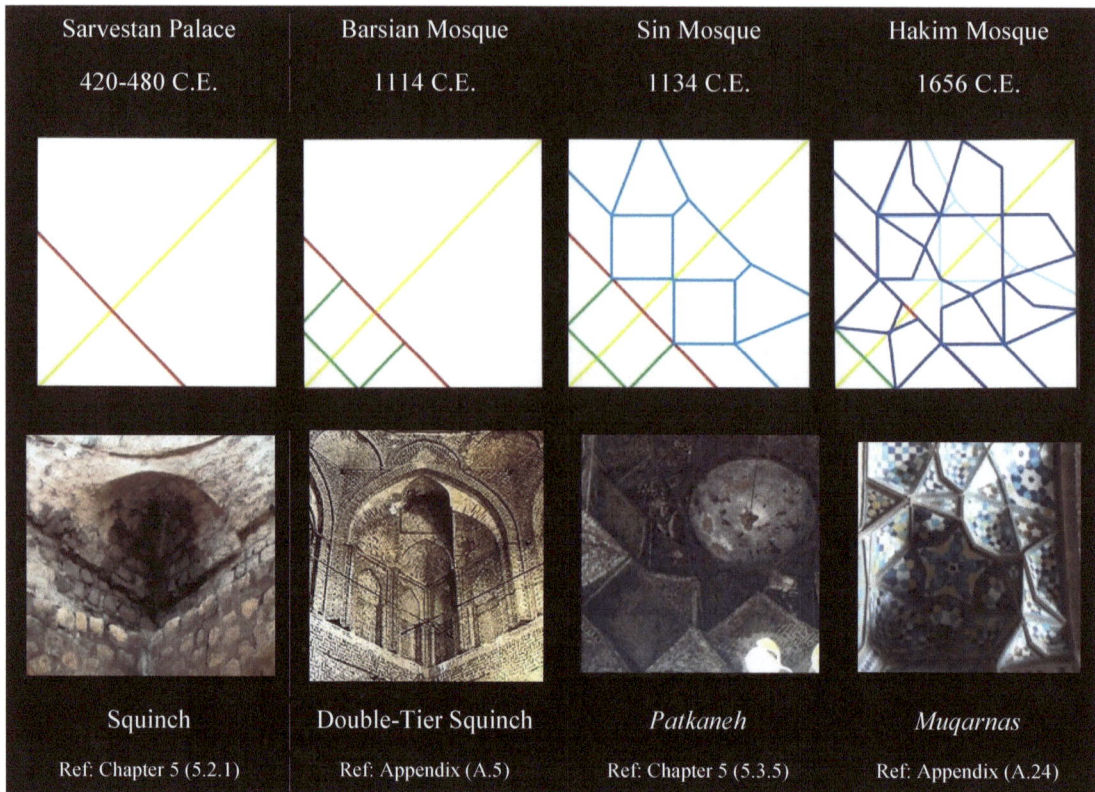

Figure 6.6   Demonstrating the evolution of *muqarnas*; from left to right, squinch (red), double-tier squinch (green), *patkaneh* (light blue), and *muqarnas* (dark blue) (Source: Author).

6.4   Further Study

The proposed methodology used in this study can be employed to investigate further about the evolution of *muqarnas* outside the borders of Iran, in order to trace the initial appearance of *Patkaneh*. In addition, the investigation about the development of forms can be further expanded over from 14th century to now, with the purpose of obtaining more details about the architectural characteristics of mature *muqarnas*.

Bibliography

Abolghasemi, L. & Omranipour, A., 2005. هنر و معماری اسلامی ایران. Trans.:Islamic Arts and Architecture of Iran. Tehran: Vezarat Maskan Shahrsazi.

Ahmad, K.b., 2004. ترتیب کتاب العین. Trans.: The Encyclopedia of Letter E.Qom: Osveh.

Al-Asad, M., 1995. The muqarnas: a geometric analysis. In Necipoglu, G. *The topkapi scroll– geometry and ornament in Islamic architecture*. Santa Monica: The Getty Center for the History of Art and Humanities. pp.349–59.

Al-Kashi, G.A.-D.M.J., 1427. مفتاح الحساب. Trans.: Key to Arithmetics. Samarkand: Hand-Written.

Al-Kashi, G.a.-d.M.J., 1954. رساله طاق و ازج. Trans.: Treatise on Arch and Vault. Hand-Written.

Allen, T., 1988. *Five Essays on Islamic Art*. California: Solipsist.

Al-Zabidi, M.b.M.b.A.A.-R.a.-M., 2008. تاج العروس من جواهر القاموس. Trans.: Taj al-Arus Arabic Encyclopedia. Kuwait: Tab'a Kuwait.

Amin, M.M. & Ibrahim, L., 1990. *Architectural Terms in Mamluk Documents*. Cairo: American University at Cairo Press.

Arenas, D.L.d., 1633. *Carpintería de lo Blanco*. Madrid: Cuarta Edicion.

Aslanapa, O., 1971. *Turkish Art and Architecture*. New York: Praeger.

Azarian, M.K., 1998. A Summary of Mathematical Works of Ghiyath ud-din Jamshid Kashani. *Journal of Recreational Mathematics*, 29, pp.32-42.

Babaie, S., 2008. *Isfahan and its Palaces: Statecraft and Shi'ism and the Architecture of Conviviality in Early Modern Iran*. Edinburgh: Edinburgh University Press.

Baker, P.L. & Smith, H., 2009. *Iran*. 3rd ed. Buckinghamshire: Bradt Travel Guides.

Behehsti, M. & Bidhendi, M.G., 2009. دانشنامۀ تاریخ معماری ایران‌شهر. Trans.: Encyclopedia of Iranian Architectural History. Tehran: Farhangestan Honar.

Behrens-Abuseif, D., 1993. Mukarnas. In P. Bearman et al., eds. *Encyclopaedia of Islam*. 2nd ed. Leiden: E. J. Brill. pp.501-06.

Besenval, R., 1984. *Technologie de la Voûte dans l'Orient Ancien*. Paris: Éditions Recherche sur les Civilisations.

Besenval, R., 2000. فن آوری تاق در خاور کهن. Trans.: Vaulting Techniques in Ancient East. Translated by M. Habibi. Tehran: Miras Farhangi.

Bier, L., 1986. *Sarvistan: A Study in Early Iranian Architecture (Monographs on the Fine Arts)*. Pennsylvania: Pennsylvania State University.

Bina-Motlagh, M., 1969. *Scheich Safi von Ardabil*. Göttingen: Georg-August-Universität zu Göttingen.

Blair, S.S., 1986. *The Ilkhanid Shrine Complex at Natanz, Iran*. Cambridge: Center for Middle Eastern Studies, Harvard University.

Blair, S.S., 1986a. *The Ilkhanid Shrine Complex at Natanz, Iran*. Cambridge: Center for Middle Eastern Studies, Harvard University.

Blair, S.S., 1986b. The Octagonal Pavilion at Natanz: a Reexamination of Early Islamic Architecture in Iran. In *Muqarnas*. Leiden: Brill. pp.69-94.

Blair, S.S., 1986. 'The Octagonal Pavilion at Natanz: a Reexamination of Early Islamic Architecture in Iran. In *Muqarnas*. Leiden: Brill. pp.69-94.

Blair, S.S. & Bloom, J.M., 1994. *The Art and Architecture of Islam*. New Haven: Yale University Press.

Blake, S., 1999. *Half the World: The Social Architecture of Safavid Isfahan*. Costa Mesa: Mazda Publishers.

Bloom, J.M., 1988. The introduction of the muqarnas into Egypt. *Muqarnas: an Annual on Islamic Art and Architecture*, 5, pp.21-28.

Blunt, W., 1966. *Isfahan: Pearl of Persia*. London: Elek Books.

Blunt, W., 1966. *Ispahan: Perle de la Perse*. London: Stein and Day.

Blunt, W., 2009. *Isfahan: Pearl of Persia*. London: Pallas Athene.

Bokharayi, S.A.-D.A., 2010. دیوان اشعار شهاب الدین عمعق بخارایی. Trans.: Poetry Compilations Bukharai.Tehran: Azma.

Bozorgmehri, Z., 1996. هندسه در معماری. Trans.: Geometry in Architecture. Tehran: Sobhan Noor.

Bozorgmehri, Z., 2002. معماران ایرانی. Trans.: Persian Builders. Tehran: Miras Farhangi.

Bruke, A., Elliot, M. & Mohammadi, K., 2004. *Lonely Planet Iran*. Hong Kong: Lonely Planet Publications Pty Ltd.

Bryson, N., 1983. *Vision and Painting: The Logic of the Gaze*. New Haven: Yale University Press.

Bulatov, M.S., 1988. *Geometric harmonization in the architecture of Central Asia from the ninth to the fifteenth century*. Moscow: Historical-theoretic research Nauka.

Burckhardt, T., 2009. *Art of Islam: Language and Meaning*. Bloomington: World Wisdom.

Burke, A., Elliott, M. & Mohammadi, K., 2004. *Iran*. London: Lonely Planet.

Byron, R., 1982. *The Road to Oxiana*. New York: Oxford University Press.

Byron, R., 2007. *The Road to Oxiana*. New York: Oxford University Press.

Castri, M.L. & Campisi, T., 2007. The Muqarnas Wooden Ceiling and the Nave Roofing in the Palatina Chapel of Palermo: Geometries, Failures and Restorations. In *ICOMOS IWC - XVI Symposium*. Florence, 2007.

Chorbachi, W., 1989. The Tower of Babel: Beyond Symmetry in Islamic Design. *Computers and Mathematics, with Applications*, 17, pp.751-89.

Christie, A.H., 2010. *Traditional methods of pattern designing; an introduction to the study of the decorative art*. New York: Cornell University Library.

Clarke, C.P., 1893. The Tracing Board in Modern Oriental and Medieval Operative Masonry. *Transactions of the Lodge Quatuor Coronati*, 20766, pp.99-110.

Corbin, J. & Strauss, A., 1990. Grounded Theory Research: Procedures, Canons, and Evaluative Criteria. *Qualitative Sociology*, 13(1), pp.3-20.

Creswell, J.W., 2013. *Research Design_ Qualitative, Quantitative, and Mixed Methods Approaches*. New York: SAGE Publications.

Dehkhoda, A.A., 1997. لغت نامه دهخدا. Trans.: Dehkhoda Persian Encyclopedia.Tehran: Tehran University Press.

Dieulafoy, J., 1887. *La Perse, la Chaldée et la Susiane*. Paris: Hachette et Cie.

Diez, E., 1917. *Die Kunst der islamischen Völker*. Berlin: Akademische Verlagsgesellchaft Athenaion.

Diez, E., 1993. Jarbas. In P. Bearman et al., eds. *Encyclopaedia of Islam*. 2nd ed. Leiden: E. J. Brill.

Dold-Samplonius, Y., 1992. Practical Arabic Mathematics: Measuring the muqarnas by Al-Kashi. *Centaurus*, 35, pp.193-242.

Dold-Samplonius, Y., 1996. How Al-Kashi measures the muqarnas: A second Look. In M. Folkerts, ed. *Mathematische Probleme im Mittelalter: Der lateinische und arabische sprachbereich*. Wiesbaden: Wolfenbütteler Mittelalter-Studien. pp.56-90.

Dold-Samplonius, Y. & Harmsen, S., 2004. Muqarnas, Construction and Reconstruction. In K. Williams & F.D. Cepeda, eds. *Nexus V: Architecture and Mathematics*. Fucecchio (Florence): Kim William Books. pp.69-77.

Dold-Samplonius, Y. & Harmsen, S.L., 2005. The Muqarnas Plate found at Takht-i Sulayman: A New Interpretation. *Muqarnas: an Annual on Islamic Art and Architecture*, 22, pp.85-94.

Dold-Samplonius, Y., Harmsen, S., Krömker, S. & Winckler, M., 2002. *Magic of Muqarnas, A Video About Muqarnas In The Islamic World*. Heidelberg: IWR.

Écochard, M., 1977. *Filiation de monuments grecs, byzantins et islamiques: une question de géométrie*. Paris: Paul Geuthner.

Edwards, C. & Edwards, D., 1999. The evolution of the shouldered arch in medieval architecture. *Architectural History*, 42, pp.68–95.

Elkhateeb, A.A., 2012. Domes in the Islamic Architecture of Cairo City: A Mathematical Approach. *Nexus Network Journal*, 14, pp.151–76.

Emisson, M. & Smith, P., 2000. *Researching the Visual*. London: SAGE Publication Ltd.

Fallahfar, S., 2010. فرهنگ واژه های معماری سنتی ایران Trans.: Encyclopedia of Traditional Architecture of Iran. Tehran: Kavosh Pardaz.

Farhangi, S.M., 2003. سیمای میراث فرهنگی قزوین. Trans.: Features of Cultural Heritage in Qazvin. Qazvin: Edareh Kol Amuzesh.

Fayz, A., 1977. گنجینه آثار قم. Trans.: Edifice Treasury of Qom. 1st ed. Qom: Mehr Ostevar Publication.

Fereshteh-Nejad, S.M., 2010. فرهنگ معماری و مرمت معماری. Trans.: Encyclopedia of Architecture and Restoration. Tehran: Arkan Danesh.

Fernández-Puertas, A., 1993. Mukarbas. In P. Bearman et al., eds. *Encyclopaedia of Islam*. 2nd ed. Leiden: E. J. Brill. pp.500-01.

Ferrante, M., 1968. Le Pavillon de Hast Bihist, ou les Huit Paradis, à Ispahan: Relevés et Problèmes s'y rattachant. In G. Zander, ed. *In Travaux de Restauration de Monuments Historiques en Iran*. Rome: IsMEO. pp.399-420.

Firuzabadi, Y., 1983. القاموس المحيط. Trans.: Concise Encyclopedia. Beirut: Dar Al-Fikr.

Fisher, W.B., Avery, P., Hambly, G.R.G. & Melville, C., 1991. *The Cambridge History of Iran*. Cambridge: Cambridge University Press.

Franz, J.M., 1994. A critical framework for methodological research in architecture. *Design Studies*, 15(4), pp.433-47.

Galdieri, E., 1979. *Esfahan, Ali Qapu: An Architectural Survey*. Rome: IsMEO.

Galdieri, E., 1984. *Esfahan: Masgid-i Gum'a*. Rome: IsMeo.

Garofalo, V., 2010. A Methodology for Studying Muqarnas: the Extant Examples in Palermo. *Muqarnas: an Annual on Islamic Art and Architecture*, 27, pp.357-406.

Getlein, M., 2005. *Living With Art*. New York: McGraw-Hill.

Ghazarian, A. & Ousterhout, R., 2001. Muqarnas drawing from thirteenth-century Armenia and the use of architectural drawings during the middle ages. *Muqarnas: an Annual on Islamic Art and Architecture*, 18, pp.141-54.

Godard, A., 1954. *Le Tombeau de Mawlânâ Ḥasan Kâshî à Sulṭânîyè*. Paris: PUF.

Godard, A., 1965. *The art of Iran*. Translated by M. Heron. California: Praeger.

Godard, A., 1977. The Mausoleum of Öljeitü at Sultaniya. In A.U. Pope & P. Ackerman, eds. *In A Survey of Persian Art from Prehistoric Times to the Present*. Tehran: Soroush Press. pp.1103-08.

Godard, A., 1990. طاقهای ایرانی. Trans.: Iranian Vaults. Translated by K.A. Afsar. Tehran: Yasavoli.

Golabchi, M., 2007. Art of Architectural Technology. In *First Conference on Structure and Architecture*. Tehran, 2007.

Golombek, L., 2003. The Safavid Ceramic Industry at Kirman. *Iran*, 41, pp.253-70.

Golombek, L. & Wilber, D.M., 1988. *The Timurid Architecture of Iran and Turan*. Princeton: Princeton University Press.

Gombrich, E.H., 1994. *The sense of order: a study in the psychology of decorative art*. 2nd ed. New York: Phaidon Press.

Goury, J. & Jones, O., 1842. Plans, Elevations, Sections and Details of the Alhambra. In *Drawings Taken on the Spot in 1834 by the Late M. Jules Gouty and in 1834 and 1837 by Owen Jones*. London: Jones.

Grabar, O., 1966. The Earliest Islamic Commemorative Structures. *Ars Orientalis*, 6, pp.7-46.

Grabar, O., 1983. Reflections on the Study of Islamic Art. *Muqarnas: an Annual on Islamic Art and Architecture*, 1, pp.25-32.

Grabar, O., 1988. Geometry and Ideology: The Festival of Islam and the Study of Islamic Art. In F. Kazemi & R.D. McChesney, eds. *In A Way Prepared: Essays on Islamic Culture in Honor of Richard Bayly Winder*. New York: New York University Press.

Grabar, O., 1990. *The Great Mosque of Isfahan*. New York: New York University Press.

Grabar, O., 1992. Art and Architecture. In Hayes, J.R. *The Genius of Arab Civilization: Source of Renaissance*. New York: New York University Press.

Groat, L.N. & Wang, D., 2013. *Architectural Research Methods*. 2nd ed. New Jersey: Wiley.

Grube, E., 1974. Wall Paintings in the Seventeenth Century Monuments of Isfahan. *Iranian Studies*, 3/4(7), pp.511-42.

Grube, E.J., 1981. Il-Khanid Stucco Decoration, Notes on the Stucco Decoration of Pir-i Bakran. In *Isfahan: Contributions to the first and second Convegno internazionale sull'arte e sulla civiltà islamica*. Venezia: La tipografica. pp.88-96.

Grütter, J.K., 2004. زیبایی شناسی در معماری. Trans.: The Aesthetics of Architecture. Translated by J. Pakzad & A. Homyoun. Tehran: Shahid Beheshti University.

Haghnazarian, A., 2006. کلیساهای ارامنه‌ی جلفای نو اصفهان. Trans.: Armenian Churches at the New Jolfa of Isfahan. Tehran: Farhangestan Honar.

Haji Ghasemi, K., 1996. *Ganjname: Masajid-e Isfahan*. Tehran: Shahid Beheshti University.

Halaweh, N.A., 2010. *Comparative History In Islamic Context; Al Muqarnasat*. Jordan: University of Jordan.

Halimi, M.H., 2011. زیبایی‌شناسی خط در مسجد جامع اصفهان. Trans.: The Aesthetics of Calligraphy at the jame' of Isfahan. Tehran: Ghadiani.

Hamekasi, N., Samavati, F.F. & Nasri, A., 2011. Interactive Modeling of Muqarnas. In Cunningham, D. & Isenberg, T., eds. *Computational Aesthetics in Graphics, Visualization, and Imaging*. Vancouver, 2011.

Harb, U., 1978. *Ilkhanidische Stalaktiten gewölbe: Beiträgezu Entwurf und Bautechnik*. Berlin: Reimer.

Harmsen, S.L., 2006. *Algorithmic Computer Reconstructions of Stalactite Vaults - Muqarnas - in Islamic Architecture*. Heidelberg: Heidelberg University.

Harmsen, S., Jungblut, D. & Krömker, S., 2007. Seljuk Muqarnas along Silk Road. *Zentrum für Wissenschaftliches Rechnen der Universität Heidelberg*, pp.1-11.

Hatim, G.A., 2000. معماری اسلامی ایران در دوره سلجوقی. Trans.: The Islamic Architecture of Iran in the Seljuk Period. Tehran: Jihad Daneshgahi (Majid).

Hautecoeur, L., 1931. De la trompe aux Mukarnas. *Gazette des Beaux–Arts*, 6, pp.26–51.

Herdeg, K., 1990. Mosque al-Hakim, Isfahan. In Herdeg, K. *Formal Structure in Islamic Architecture of Iran and Turkistan*. New York: Rizzoli International Publications. pp.21-24.

Herzfeld, E.E., 1942. Damascus: Studies in Architecture: I. *Ars Islamica*, 9, pp.10–40.

Hill, D., 1964. *Islamic Architecture and Its Decoration*. Chicago: The University of Chicago Press.

Hillenbrand, R., 1987. Abbasid Mosques in Iran. *Revista degli Studi Orientali*, LIX, pp.175-212.

Hillenbrand, R., 2004. *Islamic Architecture: Form, Function, and Meaning*. Edinburgh: Edinburgh University Press.

Hoag, J.D., 2004. *Islamic Architecture (History of World Architecture)*. 3rd ed. New York: Phaidon Press.

Hoeven, S.V.D. & Veen, M.V.D., 2010. Muqarnas. *Mathematics in Islamic Arts*, pp.1-21.

Holod-Tretiak, R., 1973. *The Monuments of Yazd*. Cambridge: Harvard University.

Homayooni, J., 1996. *History of Sarvestan*. Tehran.

Honarfar, L., 1965. گنجینه آثار تاریخی اصفهان. Trans.: Treasury of Historical Monuments of Isfahan. Isfahan: Saghafi.

Honarfar, L., 1995. *Orientation with Historical City of Isfahan*. Tehran: Golha.

Hutt, A., 1978. Iran. In Michel, G. *Architecture of Islamic World*. New York: Thames and Hudson. p.252.

Hutt, A. & Harrow, L., 1978. *Islamic Architecture: Iran*. London: Scorpion Publishers.

Ibn-Jabayr, A.A.M.b.A., 1907. رحلة. Trans.:Travel. Leiden: E.J. Brill.

Isfahani, J.A.-D.M.A.A.-R., 2000. دیوان استاد جمال الدین محمد بن عبدالرزاق اصفهانی. Trans.: Poetry Compilations of Jamal. Tehran: Negah.

Javadi, A., 1984. معماری ایرانی. In *84 پژوهشگر ایرانی 33 قلم به مقاله*. Trans.: 84 Papers from 33 Iranian Scholars. Tehran: Mojarrad. pp.671-74.

Jazbi, A. & Al-Kashi, G.A.-D.M.J., 1987. رساله طاق و ازج. Trans.: Treatise on Arch and Vault. Tehran: Soroush.

Jenab, M.S.A., 2007. آثار و ابنیهی تاریخی اصفهان. Trans.: Historical Buildings and monuments of Isfahan. Isfahan: Shahrdari Isfahan.

Karimi, K., 2005. بی تا، گزارش مرمتی تعمیرات بناهای تاریخی نایین از سال 1354 تا 1384. Trans.: Restoration Report on the Historical Buildings of Nayin From 1975 to 2005. Tehran: Miras Farhangi.

Kazempourfard, H., 2006. *"Measurement of Historical Buildings" and "Islamic Architecture"*. Undergraduate Course Reports. Tehran: Please refer to Appendix B IUST and SRTTU.

Khaghani, S., 1996. دیوان خاقانی. Trans.: Poetry Compilations of Khaghani. Tehran: Markaz.

Kochak zadeh, M.R., 1998. *History of Qom and its Historical Mosques*. Tehran: Center of Islamic Information and Documentation.

Komaroff, L., 1991. Review of the Timurid Architecture of Iran and Turan, by Lisa Golombek and Donald M Wilber. *Journal of the American Oriental Society*, 111(3), pp.609–11.

Krömker, S., 2007. *Muqarnas Visualization in the Numerical Geometry Group*. Heidelberg: Heidelberg University.

Lorzadeh, H., 1981. احیای هنرهای از یاد رفته، فنون مقرنس. Trans.: Revival of Forgotten Arts: Muqarnas Techniques. Tehran: Author.

Loveday, H., 1993. *Iran*. Geneva: Editions Olizane SA.

Mahgoub, Y., 2008. *Architectural Resaerch Methods*. Qatar: Qatar University.

Mainstone, R.J., 1998. *Developments in structural form*. 2nd ed. London: Architectural Press.

Manzur, I., 1984. لسان العرب. Trans.: Arabic Language. Ghom: Nashr Adab.

Martin, H., 1964. *L'Art Musulman*. Paris: Flammarion.

Mehrabadi, R., 1973. آثار ملی اصفهان. Trans.: National Monuments of Isfahan. Tehran: Anjoman Asar Melli.

Memarian, G.H., 2008. معماری ایرانی. Trans.: Iranian Architecture. Tehran: Soroush Danesh.

Memarian, G.H., 2012. معماری ایران: نیارش. Trans.: Iranian Architecture: Structure. Tehran: Naghmeh No Andish.

Michell, G., 1978. *Architecture of the Islamic World*. London: Thames and Hudson.

Michell, G. & Grube, E.J., 1995. *Architecture of the Islamic World*. London: Thames and Hudson.

Morton, A.H., 1974. The Ardabil Shrine in the Reign of Shah Tahmasp. *Iran*, 12, pp.31-64.

Mostafavi, S.M.T., 1996. *Tehran Monuments*. Tehran: Garous Publication.

Mostafavi, S.M.T., 2003. اقلیم پارس. Trans.: The Land of Pars. Tehran: Anjoman Asar Mafakher Farhangi.

Mousavi, M., 2002. Excavations in the Western Part of the Monumental Complex of Shaykh Safi, Ardabil. In S.R. Canby, ed. *Safavid Art and Architecture*. London: The British Museum Press. pp.16-19.

Mulayim, S., 1982. *Anadolu Türk Mimarisinde Geometrik Suslemeler: Selcuklu Caği*. Ankara: Kültür ye Turizm Bakanligi.

Nasr, S.H., 1987. *Islamic art and spirituality*. New York: State University of New York Press.

Necipoglu, G., 1995. *The topkapi scroll–geometry and ornament in Islamic architecture*. Santa Monica: The Getty Center for the History of Art and Humanities.

Notkin, I.I., 1995. Decoding Muqarnas Drawings by a Sixteenth–Century Bukharan Master Builder Muqarnas. *Muqarnas: an Annual on Islamic Art and Architecture*, 12, pp.148–71.

Oghabi, M.M., ed., 1999. دائره المعارف بناهای تاریخی، مساجد تاریخی. Trans.: Encyclopedia of Historical Monuments and Mosques. Tehran: Pajuheshgah Farhang Honar Eslami.

O'Kane, B., 1996. *Studies in Persian Art and Architecture*. Columbia: Columbia University Press.

Paccard, A., 1980. *Traditional Islamic craft in Moroccan architecture*. Paris: Saint-Jorioz.

Patton, M.Q., 2001. *Qualitative Research and Evaluation Methods*. New York: SAGE Publications.

Pirnia, A.A.-K., 1991. گنبد در معماری ایران. Trans.: Dome in the Architecture of Iran. *Asar*, 20, pp.5-139.

Pirnia, M.K., 2003. سبک شناسی معماری ایرانی. Trans.: Typology of Iranian Architecture. 2nd ed. Tehran: Pajuhandeh.

Pope, A.U., 1930. *An Introduction to Persian Art since the Seventh Century*. London: Peter Davies.

Pope, A.U., 1934. The Historic Significance of Stucco Decoration in Persian Architecture. *The Art Bulletin*, 4(16), pp.321-32.

Pope, A.U., 1934. The Photographic survey of Persian Islamic Architecture. In Archaeology, A.I.f.I.A.a. *Bulletin of the American Institute for Iranian Art and Archaeology*. New York: The Institute. pp.21-38.

Pope, A.U., 1965. *Persian Architecture*. New York: G. Braziller.

Pope, A.U., 1977b. Architectural Ornament. In A.U. Pope & P. Ackerman, eds. *A survey of Persian Art*. Tehran: Soroush Press. pp.1258-364.

Pope, A.U., 1977. The Fourteenth Century. In A.U. Pope & P. Ackerman, eds. *In a Survey of Persian Art from Prehistoric Times to the Present*. Tehran: Soroush Press. pp.1052-102.

Pope, A.U., 1997a. *A Survey of Persian Art*. New ed. Tehran: Soroush Press.

Pope, A.U. & Ackerman, P., 1977. *A Survey of Persian Art from Prehistoric Times to the Present*. Tehran: Soroush Press.

Pope, A.U. & Ackerman, P., 2005. *A Survey of Persian Art from Prehistoric Times to the Present*. 3rd ed. Santa Ana: Mazda Publishers.

Post, B., Jeene, K. & Veen, M.v.d., 2009. Muqarnas - Wiskunde in de Islamitische kunst. In *Concrete Publishing*. Utrecht, 2009.

Pourjavady, N. & Booth-Clibborn, E., 2001. *The Splendour of Iran*. London: Booth-Clibborn Editions.

Pugachenkova, G.A. & Rempel, L.I., 1965. *History of the art of Uzbekistan from ancient times to mid–nineteenth century*. Moscow: Iskusstvo.

Ra'iszadeh, M. & Mofid, H., 2010. مبانی معماری سنتی در ایران. Trans.: Principles of Traditional Architecture in Iran. 4th ed. Tehran: Mola.

Rizvi, K., 2000. *Transformations in Early Safavid Architecture: The Shrine of Shaykh Safi al-din Ishaq Ardabili in Iran (1501-1629)*. Boston: Massachusetts Institute of Technology.

Rizvi, K., 2002. The Imperial Setting: Shah Abbas at the Safavid Shrine of Shaykh Safi in Ardabil. In S.R. Canby, ed. *Safavid Art and Architecture*. London: The Beritish Museum Press. pp.9-15.

Safaeipour, H., 2009. بررسی توسعه شکل سازه ای با تاکید بر پوشش پتکانه در معماری ایران. Trans.: Studying of Development of Structural Form Through Analyzing Patkaneh Vaulting in Iran's Architecture. Tehran: Tarbiat Modares University.

Sarre, F.P.T., Herzfeld, E. & Berchem, M.V., 1911. *Archaeologische Reise im Euphrat–und Tigris–Gebiet*. Berlin: D. Reimer.

Schroeder, E., 1977. Standing Monuments of the First Period. In A.U. Pope & P. Ackerman, eds. *A Survey of Persian Art*. Tehran: Soroush Press. pp.931-66.

Schroeder, E., 1977. The Seljuk Period. In A.U. Pope & P. Ackerman, eds. *In A Survey of Persian Art*. Tehran: Soroush Press. pp.981-1045.

Shahidi Mazandarani, H., 2004. *Sagozasht- e Tehran*. Tehran: Donya Publication.

Sha'rbaf, A.A., 1996. گره و کاربندی. Trans.: Gereh and Kaarbandi. Tehran: Sobhan Noor.

Sha'rbaf, A.A., 2006. گزیده آثار استاد علی اصغر شعرباف: گره و کاربندی. Trans.: Gereh and Kaarbandi. Tehran: Farhangestan Honar.

Shirazi, F. & Ja'far, M.N., 1998. آثار عجم. Trans.: The Monuments of Ajam. Tehran: Amir Kabir.

Siroux, M., 1947. La Masjid-e-Djum'a de Yezd. *Bulletin de l'Institut Français d'Archéologie Orientale*, 44, pp.76-119.

Smith, M.B., 1947. *The vault in Persian architecture : a provisional classification, with notes on construction*. Baltimore: Johns Hopkins University.

Snelling, F.J., 1995. *Upon culture, logic and aesthetics*. Istanbul: Archnet.org.

Sourdel-Thomine, J. & Wilber, D.N., 1974. *Monuments seljoukides de Qazwin en Iran*. Paris: Librarie orientaliste P. Geunthner.

Stierlin, H., 1976. *Ispahan : image du paradis*. Paris: La Bibliothèque des arts.

Strand, K.J. & Weiss, G.L., 2005. *Experiencing Social Research: A Reader*. Boston: Pearson Education Inc.

Sultanzadeh, H., 1985. *History of Iranian School*. Tehran: Agah Publication.

Tabba, Y., 1980. *"Visions of Iraq: Topography and Historical Monuments.* [Online] Available at: http://archnet.org/library/images/one-image.jsp?location_id=9358&image_id=208859.

Tabba, Y., 2009. Muqarnas. In J.M. Bloom & S.S. Blair, eds. *The Grove Encyclopedia of Islamic Art and Architecture.* New York: Oxford University Press.

Taboroff, J.H., 1981. Bistam. In P.D. Diss., ed. *Iran: The Architecture, Setting and Patronage of an Islamic Shrine.* Ann Arbor: University Microfilms. pp.115-16.

Tabrizi, M.H.b.K., 2003. فرهنگ برهان قاطع. Trans.: Borhan Encyclopedia. Tehran: Amir Kabir.

Takahashi, S., 1973. *Muqarnas: A Three-Dimentional Decoration of Islam Architecture.* [Online] Available at: http://www.tamabi.ac.jp/idd/shiro/muqarnas/.

Tavassoli, ., 2004. هنر هندسه: پویایی اشکال، احجام کروی ابوالوفای بوزجانی. Trans.: The Art of Geometry: Fluidity of Forms, Spherical Forms of Bouzjani. Tehran: Payam Peyvand No.

Teimouri, E., 2010. چهارراه سرچشمه تهران. Trans.: Sarcheshmeh Junction of Tehran. *Bukhara*, (76), pp.135-37.

Thackeston, W.M., 1989. *A century of Princes: Sources on Timurid History and Art.* Cambridge: The Aga Khan Program for Islamic Architecture at Harvard University.

Varjavand, P., 2009. سبک‌شناسی هنر معماری در سرزمینهای اسلامی. Trans.: Typology of Architectural art in Islamic Lands. 6th ed. Tehran: Elmi Farhangi.

Wilber, D.N., 1955. *The Architecture of Islamic Iran: The Il-Khanid Period.* New York: Greenwood Press.

Wilber, D.N., 1969. *The Architecture of Islamic Iran.* New York: Greenwood Press.

Wilber, D.N., 1972. *The Masjid-i 'Atiq of Shiraz.* Shiraz: Shiraz: Asia Institute, Pahlavi University.

Yaghan, M.A., 1998. Structural Genuine Muqarnas Dome: Definition, Unit Analysis, and a Computer Generation System. *Journal of King Saud University*, 10, pp.17-52.

Yaghan, M.A., 2000. Decoding the Two-Dimensional Pattern Found at Takht-i Sulayman into Three-Dimensional Muqarnas Forms. *Iran: Journal of the British Institute of Persian Studies*, 38, pp.77-95.

Yaghan, M.A., 2001a. The Muqarnas Pre-Designed Erecting Units: Analysis, Definition of the Generic Set of Units, and a System of Unit-Creation as a New Evolutionary Step. *Architectural Science Review*, 44(3), pp.296-318.

Yaghan, M.A., 2001b. *The Islamic Architectural Element "Muqarnas": Definition, Geometrical Analysis, and a Computer Generation System.* Vienna: Phoibos Verlag.

Yaghan, M.A., 2003a. Gadrooned-Dome's Muqarnas-Corbel: Analysis and. *Architectural Science Review*, 46(1), pp.69-88.

Yaghan, M.A., 2003b. Teaching Architectural-Visual-Experience Through Virtual Reality Using VRML: Muqarnas Example. *Journal of King Abdulaziz University*, 1, pp.27-42.

Yaghan, M.A., 2005. Self-Supporting "Genuine" Muqarnas Units. *Architectural Science Review*, 48(3), pp.245-55.

Yaghan, M.A., 2010. The evolution of architectural forms through computer visualisation: muqarnas example. In *Electronic Visualisation and the Arts*. London, 2010.

Yaghan, M.A. & Hideki, M., 1995. Muqarnas Typology: A Tool for Definition and a Step in Creating a Computer Algorithm for Muqarnas Generation System. In *42nd Annual Conference of the Japanese Society for the Science of Design*. Tsukuba, 1995.

Yaghoubi, H. & Beheshti, A., 2004. *Travel Guide for the Province Isfahan.* Tehran: Rouzane.

Yeomans, R., 1999. *The Story of Islamic architecture.* Reading: Garnet.

Zamani, A., 1972. مقرنس تزئینی در آثار تاریخی اسلامی ایران. هنر و مردم, Trans.: Art and the Public. Decorative Muqarnas in theIslamic Monuments of Iran. 103(9), pp.8-25.

Zendehdel, H., Norouzi, M. & Salimi, Z., 1998. *Tourist Guide of Iran: The Qazvin Province.* Tehran: Iran Gardan Publication.

Zomarshidi, H., 2009. *The mosque in architecture of Iran.* Tehran: Zaman.

Appendix A

Introduction

In order to achieve a complete picture for seeking the historical and form progress of the studied decorative phenomenon, this investigation required as much evidence as possible (Groat & Wang, 2013). For that, the author has tried to gather a comprehensive database of existing *muqarnas* structures in Iran. The data collection was conducted during two years and 100 samples were recorded from 51 different buildings, located in 25 different cities. After detailed evaluation of the samples and their two–dimensional plans and three–dimensional forms, 20 samples were chosen to be investigated more specifically in the manuscript and the rest are presented here in this appendix.

Hence, in the following pages, 80 different structures are arranged in a chronological order and named as Sample 1 to Sample 80 with a table providing some details of the building information, accordingly. Furthermore, there are 80 pertaining figures named as Figure S-1 to S-80 that shows the image, the two–dimensional pattern plan and the three–dimensional view of these samples. The mentioned structures belong to 39 different buildings, which are introduced as A.1 to A.39, accordingly, where A is the initial of Appendix.

The copyright of the majority of photos, as well as all two–dimensional plans, and three–dimensional forms belong to the author and anyone interested in using them is encouraged to contact him in advance for proper citation.

A.1. Ardeshir Palace, 224 CE

The Palace of Ardashir Pāpakan (Dezh–e Ardashir Pāpakān), also known as the Atash–kadeh, i.e. fire–temple, is a castle located on the slopes of the mountain on which Dezh Dokhtar is situated on. Built in AD 224 by King Ardashir I of the Sassanian Empire, it is located two kilometers (1.2 miles) north of the ancient city of Gor, i.e. the old city of Piruz–Apad in Pars, in ancient Persia (Iran).

The structure is 104 m by 55 m. The iwan is 18 m high, although it has partially collapsed. The structure was built of local rocks and mortar with plasterwork on the insides. The style of the interior design is comparable to that of Tachar palace at Persepolis. It contains three domes, among other features, making it slightly larger and more magnificent than its predecessor, the nearby castle of Dezh Dokhtar. However, it seems that the compound was designed to display the royalty image of Ardashir I, rather than being a fortified structure for defense purposes. That is why perhaps it would be best to refer to the structure as a "palace" rather than a "castle", even though it has huge walls on the perimeters (twice as thick as Ghal'eh Dokhtar), and is a contained structure. From the architectural design, it seems the palace was more of a place of social gathering where guests would be introduced to the imperial throne.

What is particularly interesting about this palace is that its architectural design does not exactly fall into that of the Parthians or even Sassanian category; the design is a unique design particular to architects of Fars. The palace was built next to a picturesque pond that was fed by a natural spring, perhaps in connection with the Persian goddess of water and growth, Anahita. The spring is thought to have fed a royal garden, in the same way that Cyrus had his garden (bustan) built at Pasargadae. (Byron, 2007; Dieulafoy, 1887).

Sample 1   Ardeshir Plalace building information.

Name	**Ardeshir Plalace**
Variant Names	Ardeshir Papakan Plalace; Firuzabad Palace; Piruz Apad; Ardashir–Khwarrah; Palace of Ardashir Pāpakan; Dezh–e Ardashir Pāpakān
Location	Firuzabad, Fars, Iran
Position	Cornering
Date	224
Period	Sassanid
Century	3rd
Building Type	Palatial
Building Usage	Palace
Copyright	Hamidreza Kazempour (3D and plan)

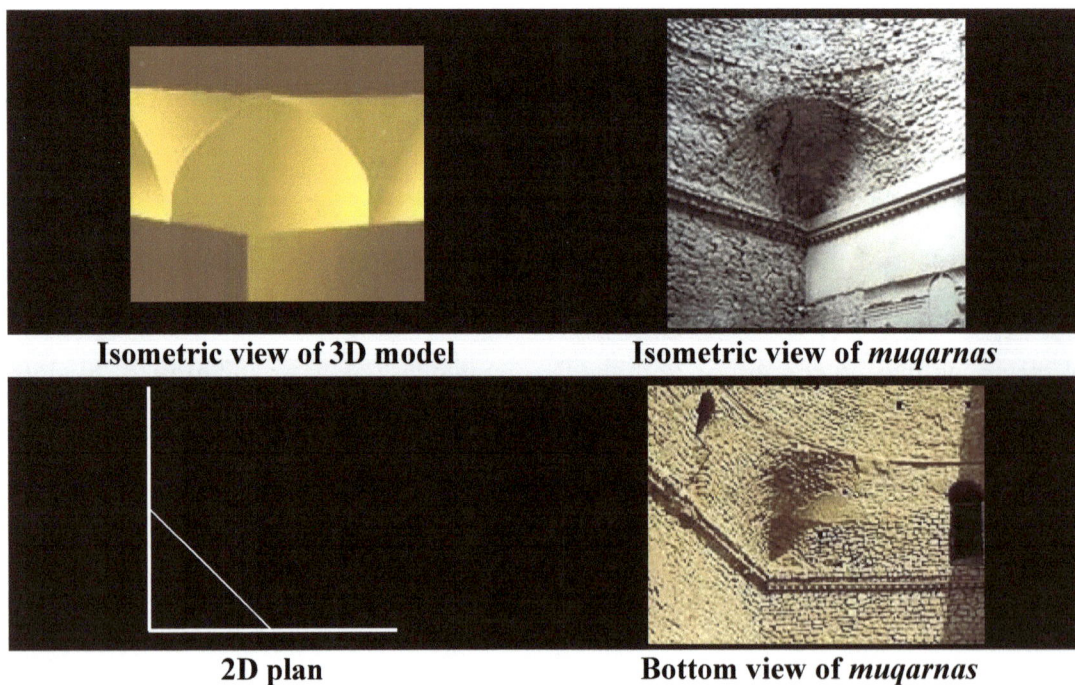

Isometric view of 3D model   **Isometric view of *muqarnas***

2D plan   **Bottom view of *muqarnas***

Figure S.1   View of cornering in Ardeshir Plalace.

A.2. Gonbad Qabus, 1006 CE

Gonbad–e Qabus tower is a monument in Gonbad–e Qabus, Iran, and a UNESCO World Heritage Site since 2012. The Tower in the central part of the city reaches 72 metres (including the height of the platform). The baked–brick–built tower is an enormous decagon building with a conic roof, which forms the golden ratio Phi, that equals 1.618. The interiors contain the earliest examples of *Muqarnas* decorative styles. The decagon with its 3 meter–thick wall, divided into 10 sides, has a diameter of 17 m. The Tower was built on such a scientific and architectural design that at the front of the Tower, at an external circle, one can hear one's echo.

The tower was built in 1006 AD on the orders of the Ziyarid Amir Shams ol–Ma'āli Qabus ibn Wushmgir. It is located 3 km north of the ancient city of Jorjan, from where the Ziyarid dynasty ruled. The tower is over 1000 years old. A Kufic inscription at the bottom of the tower reads in Arabic: "This tall palace for the prince Shams ul–Ma'ali, Amir Qabus ibn Wushmgir ordered to build during his life, in the year 397 the lunar Hegira, and the year 375 the solar Hegira". Even though the inscription does not explicitly refer to the rumor that the tower was built for the tomb for the prince Ziyarid ruler, it is believed that the Sultan's body was put in a glass coffin and was suspended from the ceiling of the tower.

Pope describes the tower as a flanged, cylindrical, slightly tapering tower with a conical roof on a small hill, built as the tomb of Shams al–Ma'ali Qabus. Qabus was an astrologer, poet, calligrapher, and patron of numerous scholars and writers, including Ibn Sina. He reigned in Gurgan until his assassination in 1012, five years after initiation of the tomb's construction. There is no access to the roof, which has a small opening in the eastern side, and no underground chamber. Reports suggest that Qabus' glass coffin was suspended within the dome, the morning sun striking his body through the eastern opening. The interior is undecorated and without fenestration. On the exterior face, between the ten flanges, are two rows of inscriptions in brick–formed Kufic (Pope & Ackerman, 1977).

Sample 2   Gonbad Qabus building information.

Name	Gonbad Qabus Tower
Variant Names	Gonbad–e Qabus; Gonbad–e Qābus; Gonbad–e Kāvus; Gonbad–e Ghābus; Gonbad–i Ghāboos
Location	Gonbad Qabus, Gorgan, , Iran
Position	Entrance Corner
Date	1006
Period	Ziyarid
Century	11th
Building Type	Monumental; Funerary
Building Usage	Monumental, Shrine
Copyright	Hamidreza Kazempour (3D and plan)

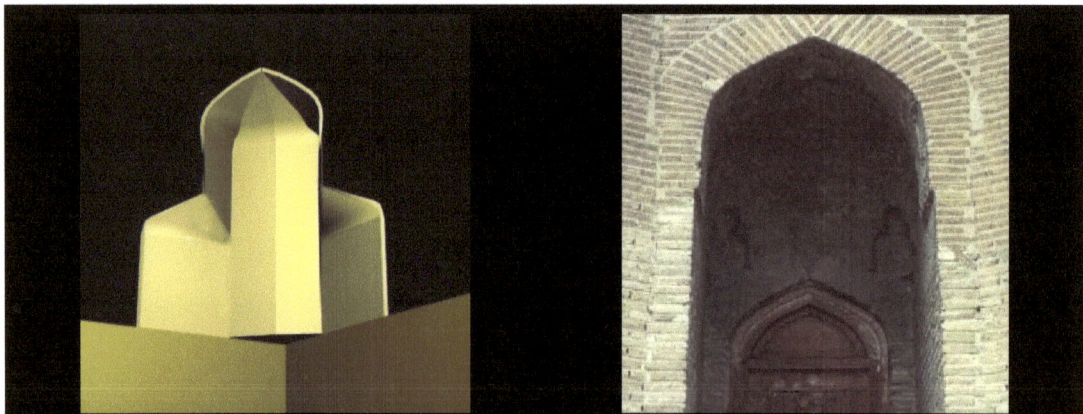

Isometric view of 3D model   **Isometric view of *muqarnas***

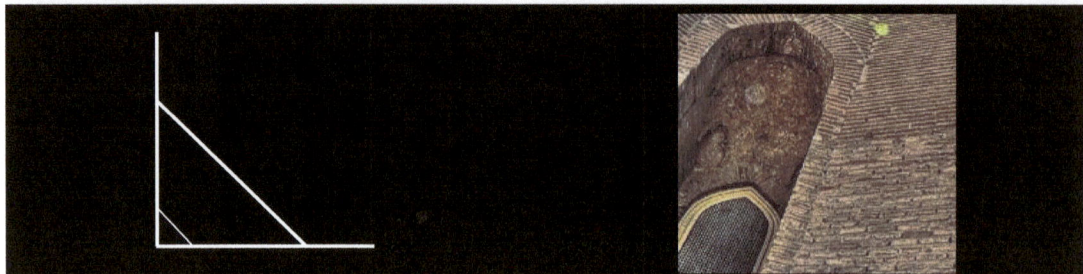

2D plan   **Bottom view of *muqarnas***

Figure S.2   View of cornering in Gonbad Qabus.

A.3. Na'in Jame' Mosque, 1062 CE

Typical of pre–Seljuk mosques, the Masjid–e Jame at Na'in exhibits a simple hypostyle plan, which has remained uncomplicated despite the additions and alterations of subsequent years. A courtyard is accessed through the arcades that are built from bays of irregular spacing and number. The courtyard façade probably dates to reconstruction work of the Seljuk period, although the most unusual feature – the angled piers flanking the central nave on the southwestern (qibla) side – is dated to the original period of construction.

The qibla axis is emphasized by these angled piers, and also by the slightly increased width and height of the central nave, forming a lip which projects above the arcade roofline. This structure represents an emergent form of the monumental portal that would later pervade Iranian mosque design.

The minaret represents an important transition from the early square form to the Iranian minarets of the 11th and 12th centuries. Maintaining the early square plan at the base, a tall tapering octagonal mid–section rises to a short cylindrical shaft that terminates in a cornice decorated with carved stucco. The cornice holds a brick railing, forming a balcony upon which stands a thin cylindrical cap, pierced with apertures and resembling a dovecote. Later building additions enclose the minaret which was originally freestanding. The transitional form of this minaret and its relatively unadorned state suggest that it is pre–Seljuk, contemporary with the oldest areas of the mosque.

The Masjid–e Jame at Na'in is renowned for the extensive and masterful carved stucco of the mihrab and adjacent bays, including the oldest extant epigraphic friezes in Iran. Stylistically it bridges the stucco decoration of the Sasanian and Abbasid periods with that of the Seljuks; effusive vegetal forms released from earlier geometric constraints (Michell, 1978; O'Kane, 1996; Pope, 1977b; Schroeder, 1977; Hillenbrand, 1987).

Sample 3   Na'in Jame' Mosque building information, Entrance.

Name	Masjid Jame' Na'in
Variant Names	Friday Mosque of Na'in, Masjid Jami' of Na'in, Masjid–e Jame, Congregational Mosque, Masjid–i Jami' and Minaret of Na'in, Masjid Jami' and Minaret of Na'in, Masjid Jami and Minaret of Nain
Location	Na'in, Iran
Position	Entrance
Date	960–1062/348 AH
Period	Buyid
Century	10th
Building Type	Religious
Building Usage	Mosque
Copyright	Hamidreza Kazempour

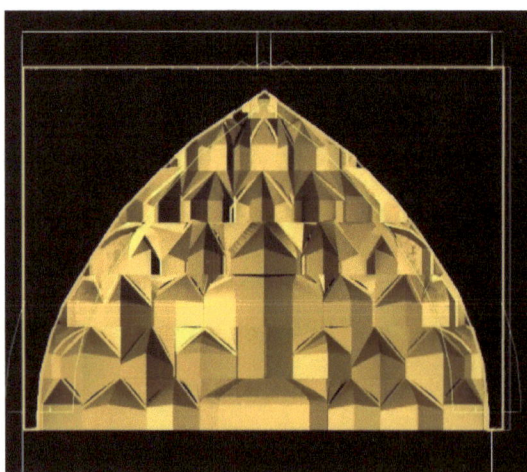

Isometric view of 3D model   Isometric view of *muqarnas*

2D plan   Bottom view of *muqarnas*

Figure S.3   View of *muqarnas* in Na'in Jame' Mosque, Entrance.

A.4. Varzaneh Jame' Mosque, 1100 CE

The city of Varzaneh, also written as Varsaneh, is located 105 km in Southeast of Isfahan. The city has an about 910 year old mosque, Varzaneh's Jame' mosque, which was built in ca. 1100, on the remains of a fire temple dating back to the Sassanid era. However, the present structure of the mosque has been renovated in 1430 during the Timurid era.

The city is most famous because its people are very much attached to their old tradition and language. The women of this township attend the midday prayer enveloped in white head–to–toe vails or Chadors, creating a nice image.

The mosque has a 20 m high minaret and many tileworks that date back to its renovation time. The iwan leading to the prayer chamber incorporates the name of Shah Rukh (the son of Timur Lang who took Isfahan in 1417). It is on the mihrab tilework that the date 1444 is recorded, after the Quranic inscription (3:38–9). The different appearance of the north iwan results from 17th–century Safavid repairs. The guardian will suggest you walk behind the mosque leaving by the left–hand door, facing the prayer chamber, so that you can see the exterior profile of the dome which still retains just a little of two bands of ceramic decoration around the zone of transition.

Varzaneh has another treat for those interested in language because most of the residents, formerly a long–surviving Zoroastrian community, still speak a form of Pahlavi in the home (Baker & Smith, 2009).

Sample 4   Varzaneh Mosque building information, Dome chamber.

Name	Masjid Varzaneh
Variant Names	Masjid Varzaneh; Varzaneh Grand Mosque; Varzaneh Jame' Mosque
Location	Varzaneh, Isfahan, Iran
Position	Dome chamber
Date	Ca. 1100– Ca. 1400 (renovation)
Period	Seljuk, Timurid
Century	15th
Building Type	religious
Building Usage	mosque
Copyright	Hamidreza Kazempour

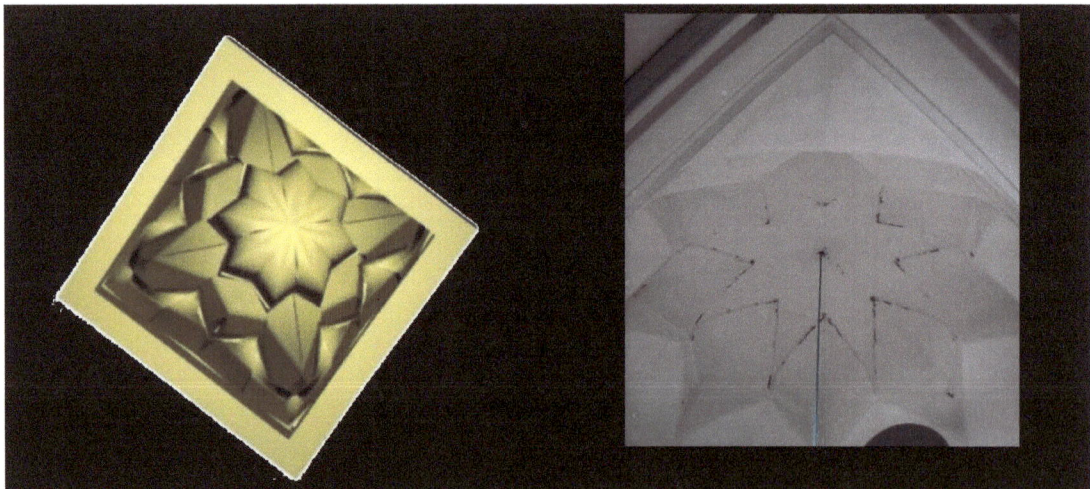

Isometric view of 3D model   Isometric view of *muqarnas*

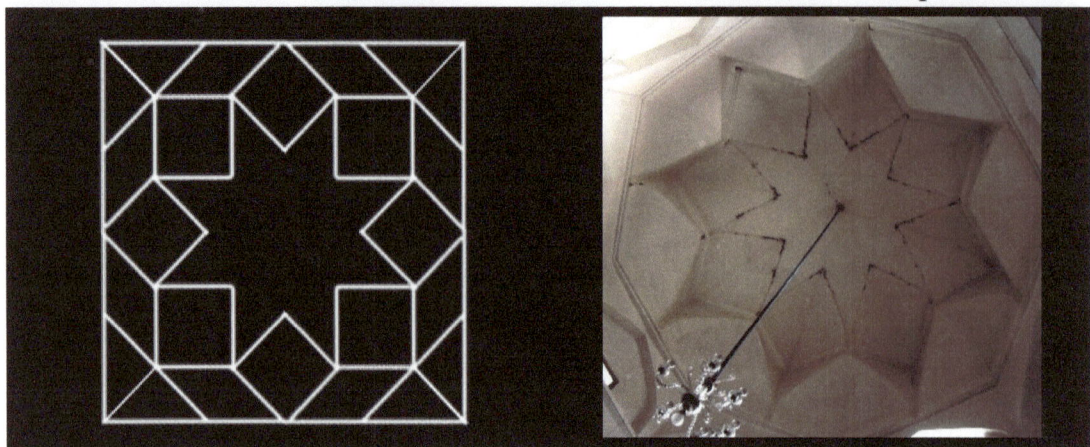

2D plan   Bottom view of *muqarnas*

Figure S.4   View of *muqarnas* in Varzaneh Mosque, Dome chamber.

A.5. Barsian Jame' Mosque, 1114 CE

Barsian located 42 kilometres on North–East of Isfahan is a village in which there is a nice historical site consisting of a minaret, a mosque and a Caravanserai. The locals pronounce the name of the village as "Bersioon" that is probably an equivalent of "Parsian", which in turn literary means "Persian", indicating that the village was of significant reputation in earlier times. Caravanserais were normally attached to mosques and this mosque is specifically interesting because of its old minaret and its very fine brickworks, as well as its exceptional mihrab inside the prayer hall, having unusual brick carving decoration (Pope, 1977b).

The old minaret was built first in 1097 C.E but the mosque was added later to the complex in 1134. Therefore, the mosque does not follow the more typical "four iwan style" of Iranian architecture. The prayer hall of the mosque has four double–tier squinches to provide a smooth transition from the octagon base to the circular dome (Hillenbrand, 2004).

Although the building went through a long historical development and changes, the Seljuk identity of its architecture can still be felt and the squinches are similar to the one in the Shrine of the davazdah Imams, Yazd (Grabar, 1992; Pope & Ackerman, 2005).

Sample 5   Barsian Jame' Mosque building information.

Name	**Masjid Jame' Barsian**
Variant Names	Friday Mosque of Barsian, Masjid–i–Jami' Bersian, Masjid–i Jami, Masjid–i Jomeh, Masjid–i–Jomeh, Bersian Mosque
Location	Barsian, Iran
Position	Dome chamber squinch
Date	1114/514 AH
Period	Seljuk
Century	12th
Building Type	Religious
Building Usage	Mosque
Copyright	Hamidreza Kazempour (3D and plan)

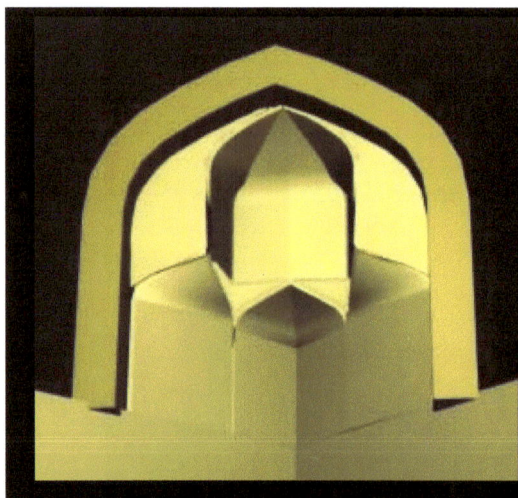

Isometric view of 3D model   Isometric view of *muqarnas*

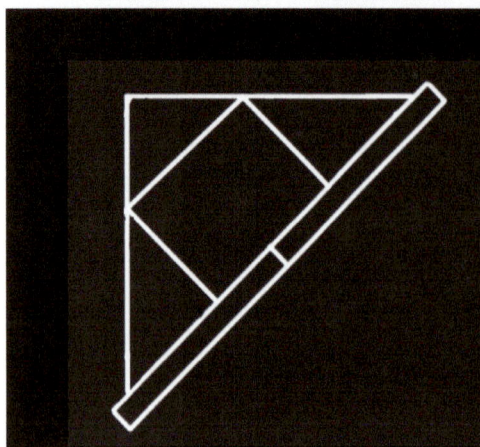

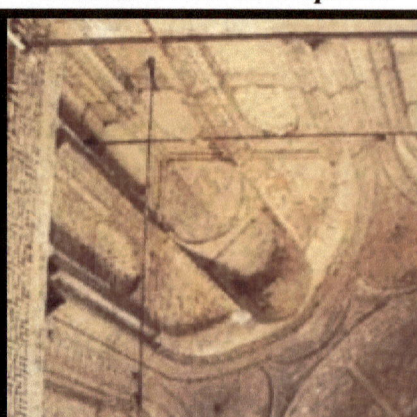

2D plan   Bottom view of *muqarnas*

Figure S.5   View of *muqarnas* in Barsian Jame' Mosque.

A.6. Isfahan Jame' Mosque, 1118–1667 CE

Located in Isfahan, the Friday mosque of Isfahan is a prominent architectural expression of the Seljuk rule in Persia (1038–1118). The mosque's plan evolved from a hypostyle plan with a rectangular inner court surrounded by prayer halls comprised of round columns carrying a wooden roof, to a four–iwan plan established/augmented in the twelfth century after the additions of the four iwans, the southern (southwest) domed chamber, the two minarets flanking it, and the northern domed chamber The exact date of current entrance gate on the southeastern area, which was restored in 1804 as part of the restoration projects is obscure. On the opposite side, on the southwest part, another gate, still in use, dates from 1590–1, the period of Shah Abbas's rule. A large monumental gate, no longer in use today, is located on the north, adjoining the northeast wall of the northern dome. It dates from 1366.

In 1086 Nizam al–Mulk, Abu al–Fath Malik Shah's vizier, ordered the building of a domed chamber (15 meters per side, approximately 30 meters high) on the southwest. This chamber was designed by the architect Abul Fath, who is sometimes credited with the construction of both domes. Two preserved inscriptions on the dome's drum mention the names of Abu Malik Shah and Nizam al–Mulk. The ribbed dome rests on a *muqarnas* transitional zone. Commissioned by Taj al–Mulk (the successor of Nizam), the northeast dome was built in 1088–9 for Terkan Khatun (Malik Shah's wife). Smaller in size and placed on the same lateral longitudinal axis as the southwest dome, the northeast dome rests on a square base of square, massive piers (with three slim round engaged columns), with an octagonal transitional zone formed by four squinches, on top of which rests another zone of sixteen arches with a drum comprising an inscription band with religious inscription. Ten double–ribs emerge from the dome's drum and ascend to inscribe a pentagon.

The southwest iwan, preceding the domed chamber with the mihrab, is the most prominent among the other iwans. Visually, it is flanked by two towers and referred to in the vernacular as sofe–e saheb or "the high [dignified] space of the master". The three other iwans in the middle of each court elevation repeat this motif. Inscription bands decorate the mihrab date mainly from the time of Shah Tahmasp (reg. 1531–32) and Shah Abbas II (reg. 1642–67). The iwan's ceiling dates from the 15th century; most of its walls are covered with Safavid statements. Under the iwan's pavement, columns and bases of an earlier mosque were found.The southeast and northwest iwans both have later Safavid architectural elements. The names of both iwans indicate their pedagogical roles/associations, called the sofe of the master (Ustadh). It comprises a multitude of small brick *muqarnas* units, whose edges are delineated by glazed dark blue lines. Each *muqarnas* cluster, as it ascends, ends with a star–shaped form in which is inscribes geometric arabesques in dark blue.The southeast iwan, called the sofe of the student (shagird), displays Safavid motifs of tilework. It is composed of larger *muqarnas* units than east iwan. Each face of the *muqarnas* units is decorated with very small square pieces of glazed tiles in dark blue points and lines forming a larger geometric arabesque inscribing an epigraphic element in lighter blue. (Byron, 1982; Galdieri, 1984; Golombek & Wilber, 1988; Hoag, 2004; Michell, 1978; Pourjavady & Booth-Clibborn, 2001; Yeomans, 1999; Blair & Bloom, 1994; Grabar, 1990)

Sample 6   Isfahan Jame' Mosque building information, Shagerd Iwan, Mihrab.

Name	Masjid Jame' Isfahan
Variant Names	Friday Mosque of Isfahan, Masjid–i Jami' Isfahan, Masjid–i Jami of Isfahan, Masjid–e Jomeh, Great Mosque, Masjid–e Jame, Masjid–e Jameh, al–Jami' al–Kabir, Masjid–i Jami' Isfahan, Masdjid–i Djum'a, Congregational Mosque
Location	Isfahan, Iran
Position	DOWN – Shagerd Iwan/ Mihrab
Date	1118–1157
Period	Seljuk
Century	12th
Building Type	Religious
Building Usage	Mosque
Copyright	Hamidreza Kazempour

Sample 7   Isfahan Jame' Mosque building information, Darvish Iwan.

Masjid Jame' Isfahan	Position: Darvish Iwan	Date: 1118–1157

Sample 8   Isfahan Jame' Mosque building information, Jam Minaret.

Masjid Jame' Isfahan	Position: Jam Minaret	Date: 1480

Sample 9   Isfahan Jame' Mosque building information, Majlesi Entrance.

Masjid Jame' Isfahan	**Position:** Alameh Majlesi Entrance	**Date:** 1501–1667

Sample 10   Isfahan Jame' Mosque building information, Eastern Entrance.

Masjid Jame' Isfahan	**Position:** Eastern Entrance	**Date:** 1501–1667

Sample 11   Isfahan Jame' Mosque building information, Eastern Entrance.

Masjid Jame' Isfahan	**Position:** Eastern Entrance	**Date:** 1501–1667

Sample 12   Isfahan Jame' Mosque building information, Shagerd Iwan.

Masjid Jame' Isfahan	**Position:** Shagerd Entrance	**Date:** 1501–1667

Sample 13   Isfahan Jame' Mosque building information, Hakim Iwan.

Masjid Jame' Isfahan	**Position:** Hakim Iwan	**Date:** 1501–1667

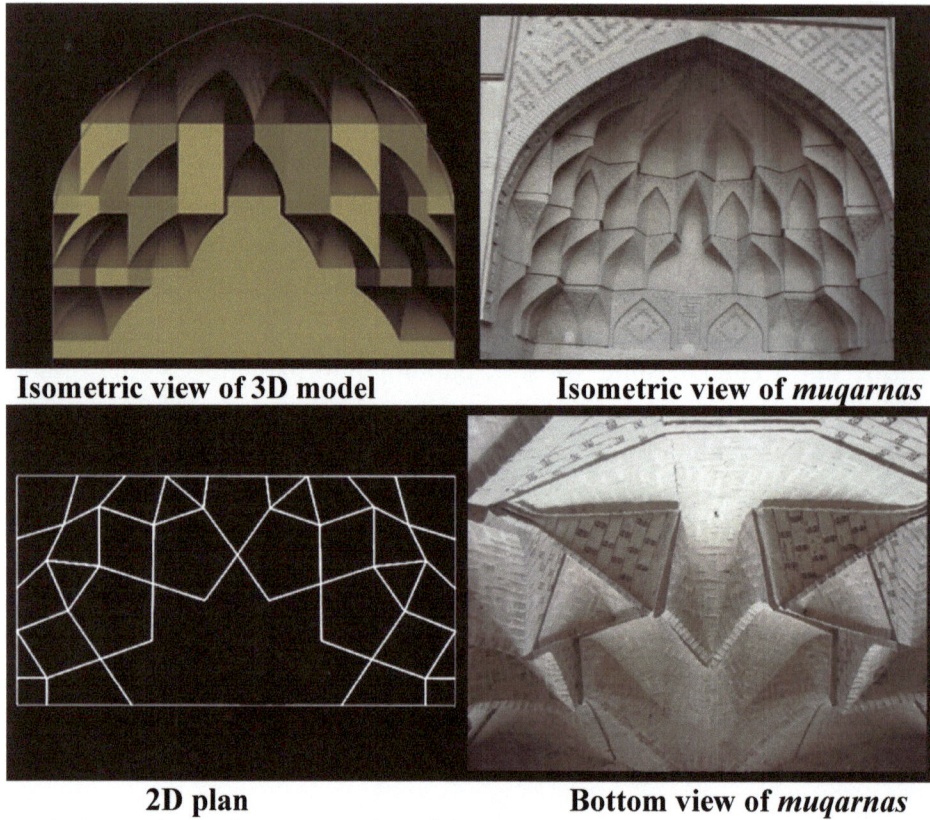

Isometric view of 3D model   **Isometric view of *muqarnas***

2D plan   **Bottom view of *muqarnas***

Figure S.6   View of *muqarnas* in Isfahan Jame' Mosque, Shagerd Iwan, Mihrab.

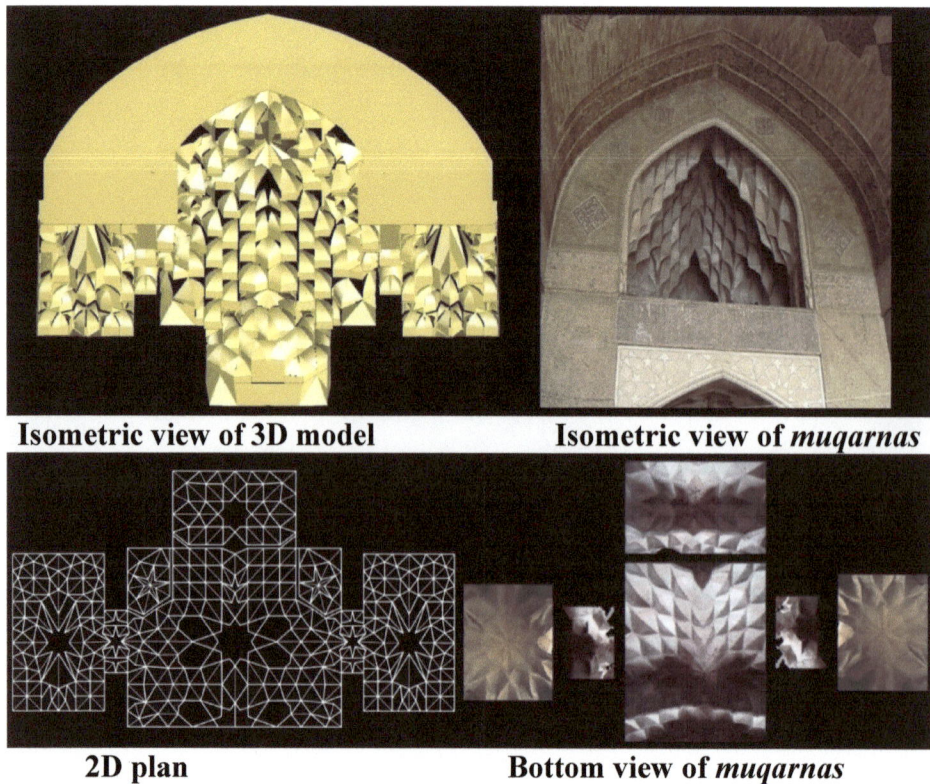

Isometric view of 3D model   **Isometric view of *muqarnas***

2D plan   **Bottom view of *muqarnas***

Figure S.7   View of *muqarnas* in Isfahan Jame' Mosque, Darvish Iwan.

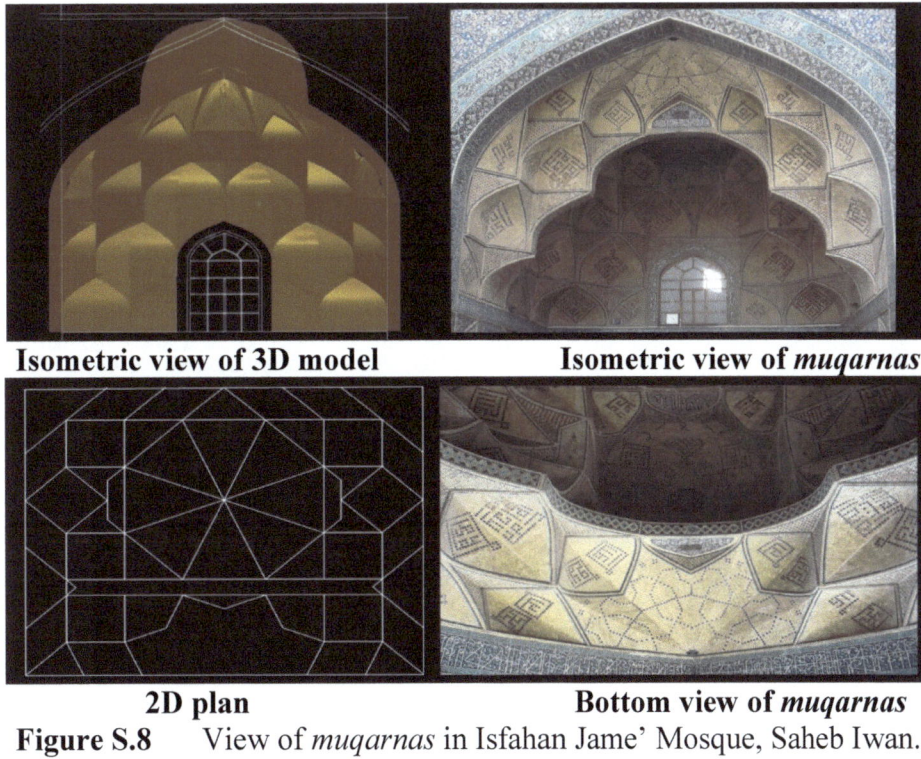

Isometric view of 3D model   **Isometric view of *muqarnas***

2D plan   **Bottom view of *muqarnas***

Figure S.8   View of *muqarnas* in Isfahan Jame' Mosque, Saheb Iwan.

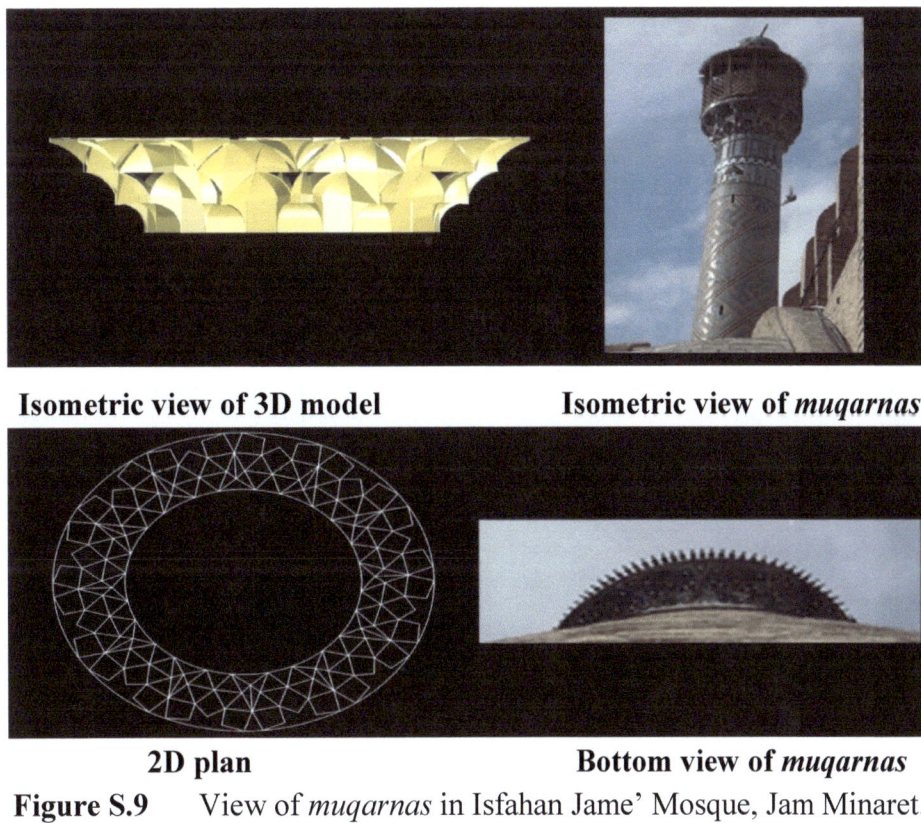

Isometric view of 3D model   **Isometric view of *muqarnas***

2D plan   **Bottom view of *muqarnas***

Figure S.9   View of *muqarnas* in Isfahan Jame' Mosque, Jam Minaret.

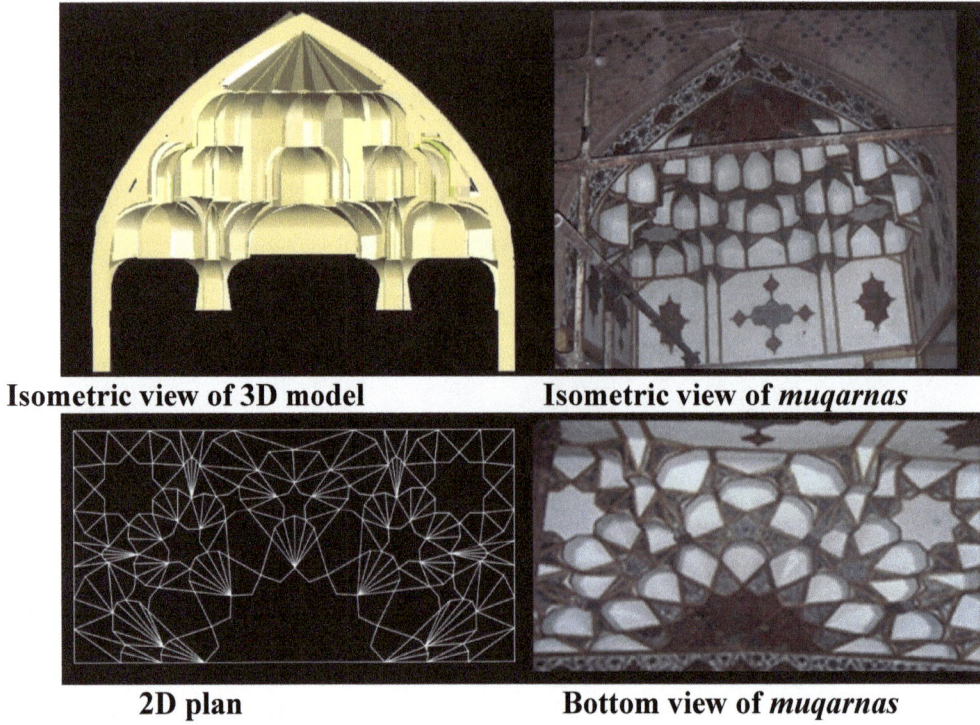

Isometric view of 3D model   **Isometric view of *muqarnas***

2D plan   **Bottom view of *muqarnas***

Figure S.10   View of *muqarnas* in Isfahan Jame' Mosque, Majlesi Entrance.

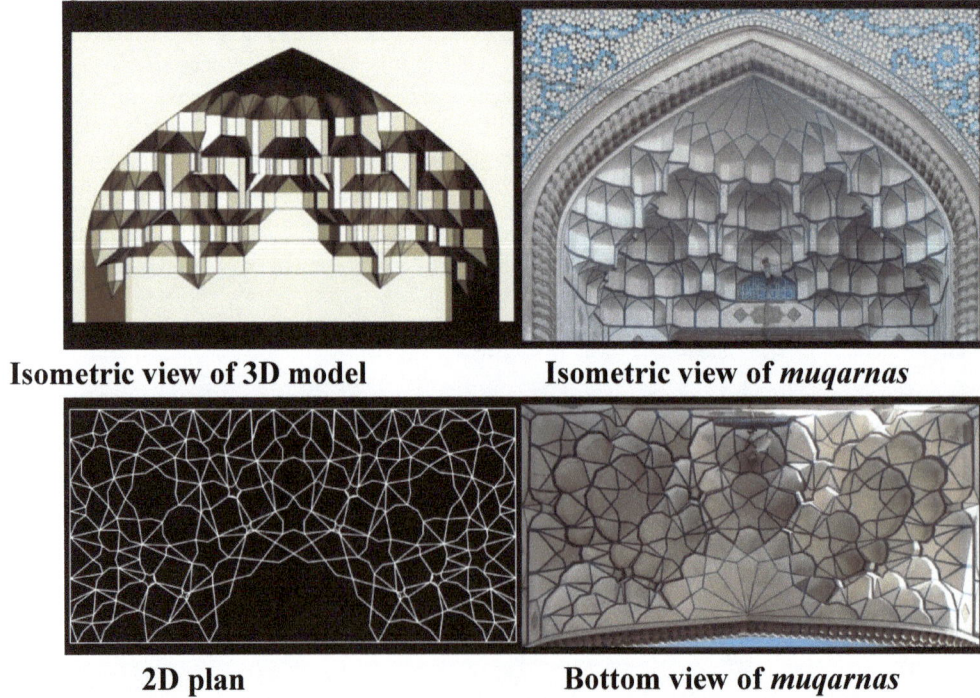

Isometric view of 3D model   **Isometric view of *muqarnas***

2D plan   **Bottom view of *muqarnas***

Figure S.11   View of *muqarnas* in Isfahan Jame' Mosque, Eastern Entrance.

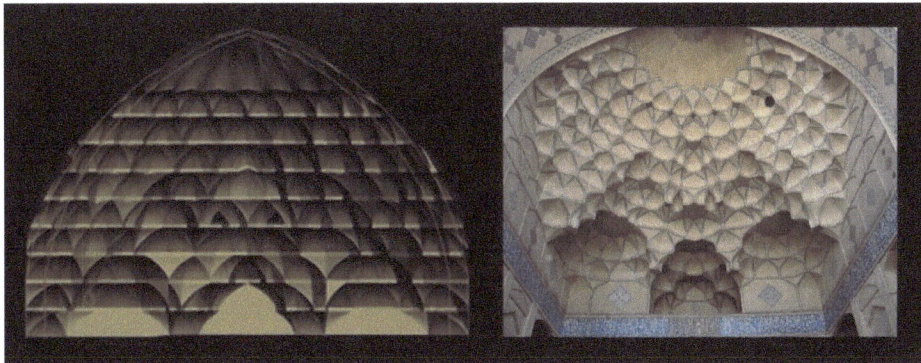

Isometric view of 3D model **Isometric view of *muqarnas***

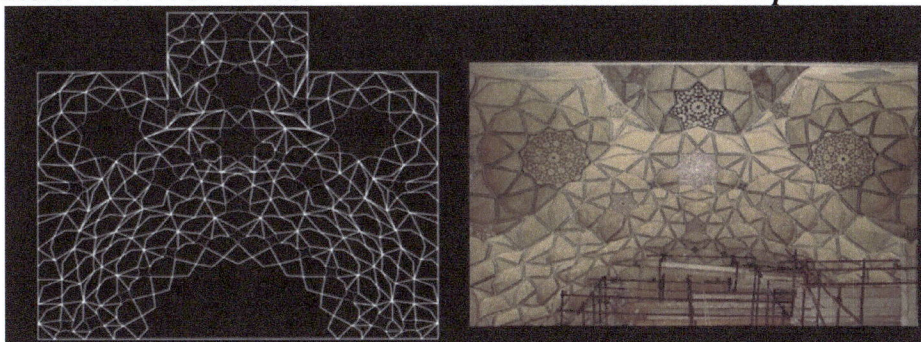

2D plan **Bottom view of *muqarnas***

Figure S.12 View of *muqarnas* in Isfahan Jame' Mosque, Shagerd Iwan.

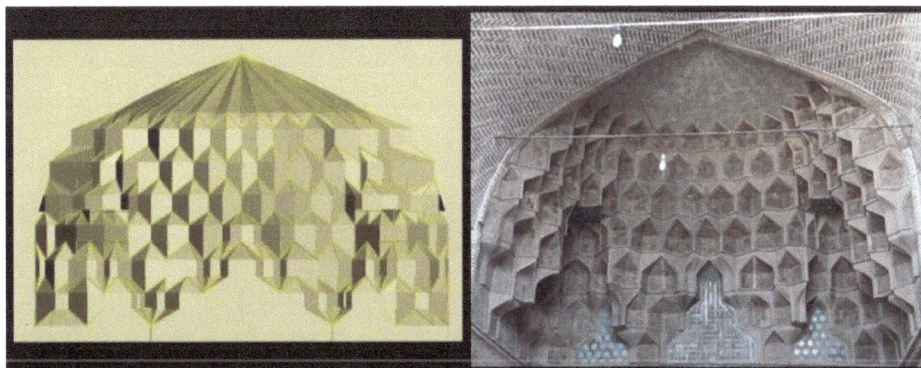

Isometric view of 3D model **Isometric view of *muqarnas***

2D plan **Bottom view of *muqarnas***

Figure S.13 View of *muqarnas* in Isfahan Jame' Mosque, Hakim Iwan.

A.7. Abd Al–Samad Mausoleum, 1304 CE

Abd al–Samad, a shaykh of the Suharwardi Sufi order, died in Natanz in 1299. During the decade that followed, the site of his grave was developed by the vizier Zayn al–Din Mastari into what has survived to be one of the best preserved of Ilkhanid shrine complexes. The much–admired façade of colorful glazed tile, terracotta, and stucco bends slightly to bring together the four constituent structures that lie behind; a four–iwan mosque, an octagonal sanctuary, a minaret, and a mosque from the 1930's fronted by a khanaqah portal. The organization of these structures at varying angles and on multiple floor levels reflects the difficulty with which they were inserted into the existing built context. The square courtyard mosque is faced by two stories of rooms that link four iwans of varying depths. Construction is of baked brick, with a coat of white plaster. *Muqarnas* vaults are found in the north and south iwans. Two bays at the rear of the south iwan flank the mihrab, leading to the domed octagonal sanctuary, which abuts the main façade.

Restoration during the 1970's revealed that this sanctuary predates the courtyard mosque and was originally a freestanding Buyid pavilion from 389/999. In its original state the pavilion was open on all sides to a vaulted ambulatory supported by columns. This pavilion constitutes the earliest dated example of an octagonal form in Iran and although the form is not unusual in tomb towers in Iran, the later extant examples are closed. (Hutt, 1978; Pope, 1977; Blair, 1986b; Blair, 1986a).

Sample 14   Sheikh Abd Al–Samad Mausoleum building information, Minaret.

Name	Khanghah Sheykh Abd al–Samad
Variant Names	Shaykh 'Abd al–Samad Shrine Complex, Shrine of Shaykh 'Abd al–Samad, Shrine Complex of Shaykh 'Abd al–Samad: Octagonal Pavilion, Masjid–i Shaykh 'Abd al–Samad, Mosque of Shaykh 'Abd al–Samad, Majmu'ah–i Shaykh 'Abd al–Samad, Shrine Complex of Shaykh Abd al Samad, Shaykh Abd al–Samad Mosque,
Location	Natanz, Iran
Position	Minaret
Date	1304–25/703–25 AH
Period	Ilkhanid
Century	13th
Building Type	Funerary, Religious
Building Usage	Tomb, Shrine
Copyright	Hamidreza Kazempour

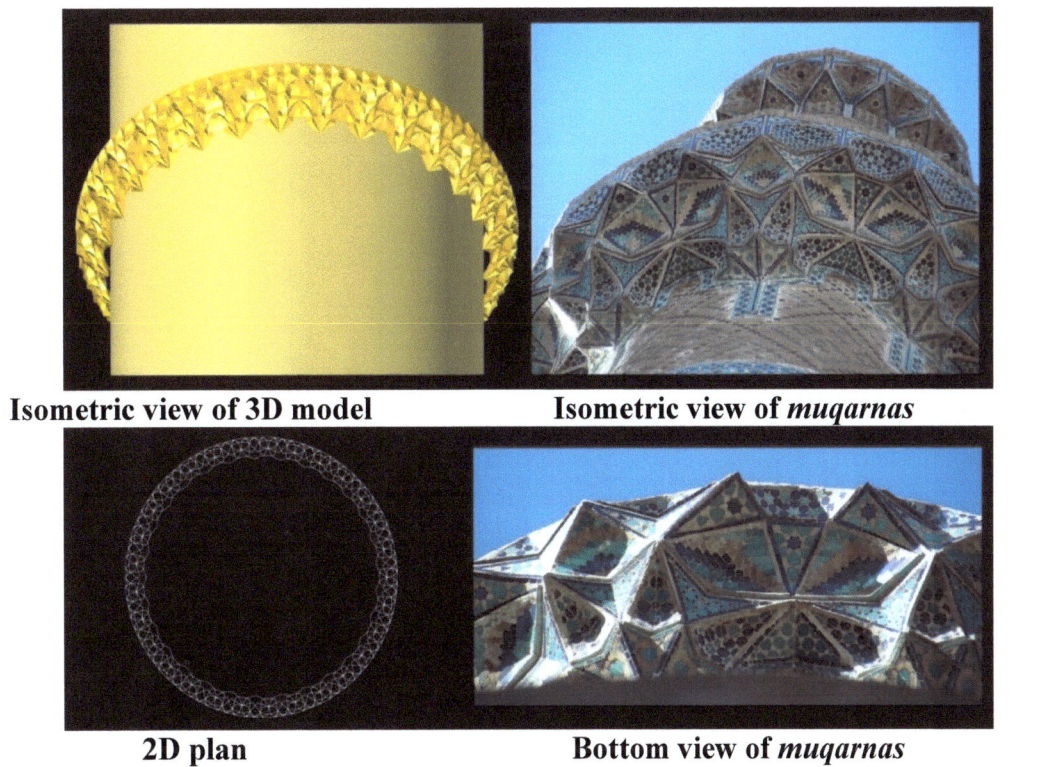

Isometric view of 3D model **Isometric view of *muqarnas***

2D plan **Bottom view of *muqarnas***

Figure S.14 View of *muqarnas* in Sheikh Abd Al–Samad Mausoleum, Minaret.

Sample 15 Na'in Jame' Mosque building information, Natanz iwan, lower.

Abd al–Samad	**Position:** Iwan-Down	**Date:** 1304–25

Sample 16 Natanz Jame' Mosque building information, North iwan, upper.

Abd al–Samad	**Position:** Iwan-Up	**Date:** 1304–25

Sample 17 Natanz Jame' Mosque building information, North iwan, lower.

Abd al–Samad	**Position:** Portal	**Date:** 1304–25

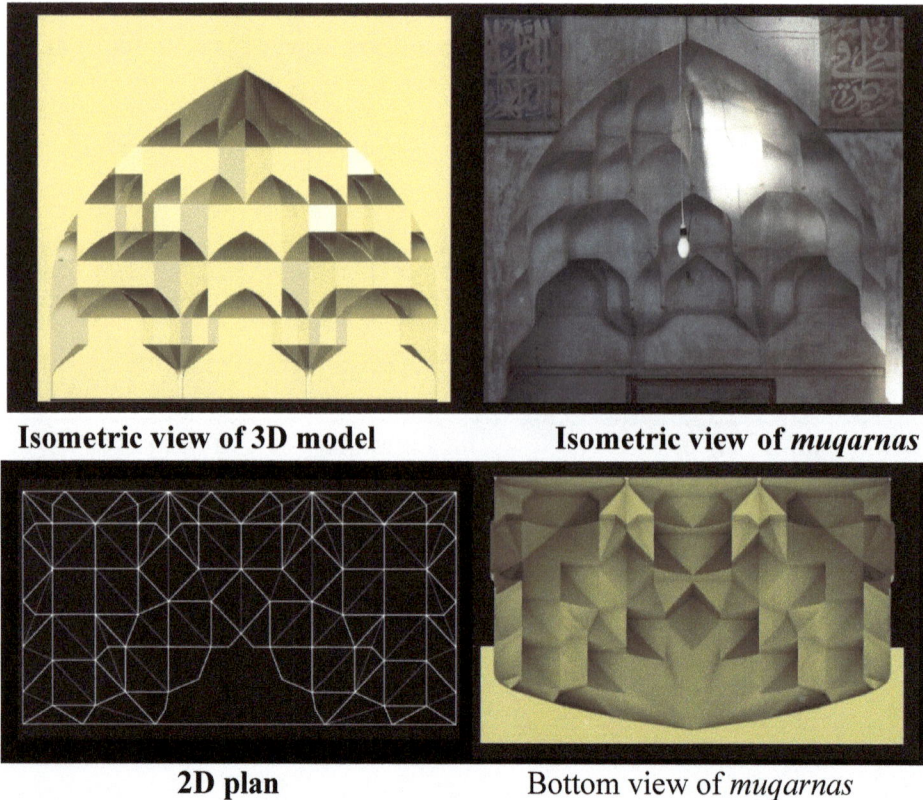

Isometric view of 3D model **Isometric view of *muqarnas***

2D plan Bottom view of *muqarnas*

Figure S.15 View of *muqarnas* in Natanz Jame' Mosque, North iwan, lower.

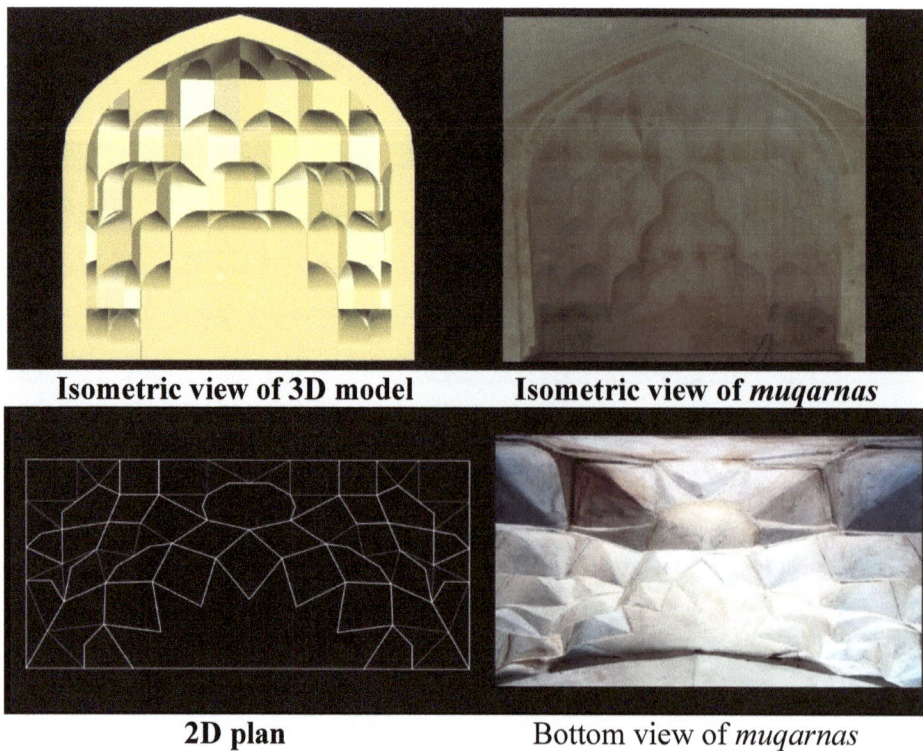

Isometric view of 3D model **Isometric view of *muqarnas***

2D plan Bottom view of *muqarnas*

Figure S.16 View of *muqarnas* in Natanz Jame' Mosque, North iwan, upper.

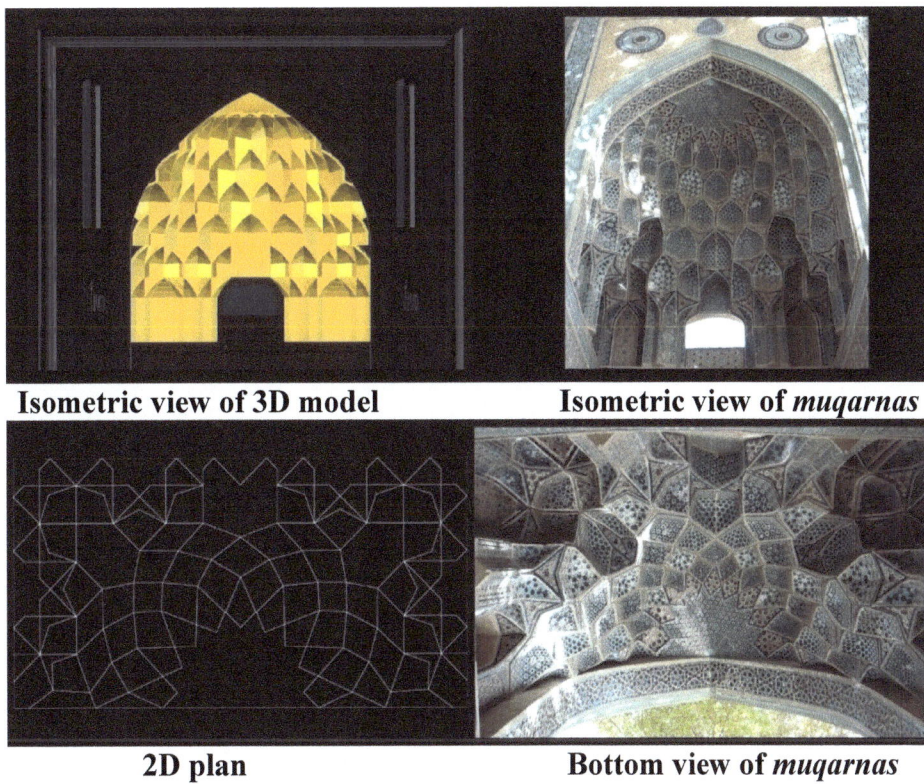

Isometric view of 3D model **Isometric view of *muqarnas***

2D plan **Bottom view of *muqarnas***

Figure S.17 View of *muqarnas* in Natanz Jame' Mosque, North iwan, lower.

A.8. Ashtarjan Jame' Mosque, 1315 CE

Commissioned by an Ilkhanid administrator from Ashtarjan and dated by two separate inscriptions to 1315–16, this mosque has survived in excellent condition. Construction appears to have been swift, executed in phases over a brief period. The decoration is of greater import than the primary structure, which consists of a narrow rectangular courtyard lined on three sides by arcaded prayer halls; on the fourth side, a vaulted iwan fronts a dome chamber. The main entrance portal with flanking cylindrical minarets is located on the northern exterior façade, off axis with the sanctuary – possibly due to constrictions of pre–existing structures. The mosque is built with a mud brick core faced with fired brick. The dome chamber appears to have been erected slightly earlier, utilizing thick plaster over the mud brick core. The use of mud brick is unusual in the vicinity of Isfahan, where fired brick prevails.

The diversity of patterns, materials, and combination of materials employed in the decoration represent a compendium of decorative techniques from around the country. Exposed brick is found only in the arcades, although possibly this area originally had a plaster coat as is employed extensively in the rest of the structure. The plaster is carved in a variety of decorative designs, including geometric patterns; floral motifs in high relief; simulated brick bond, and brick end–plugs in some areas arranged to configure rectangular Kufic inscriptions. On the sanctuary walls and dome, traces indicate that plaster was painted in a vivid array of colors. The lofty stucco mihrab describes an uncommon proportion, reaching the zone of transition. Across the interior of the dome, decorative terracotta elements form eight radial ribs; diverse painted plaster patterns fill the interstices. The northern entrance portal is lavishly decorated with mosaic faience, glazed and unglazed terracotta.

An inscription on an eastern pier of the courtyard notes repairs undertaken in 1476 by a pious Ashtarjani, in the name of the Aq Qoyunlu ruler, Uzun Hasan (Golombek & Wilber, 1988; Pope, 1977; Wilber, 1969).

Sample 18   Ashtarjan Jame' Mosque building information.

Name	Masjid Jame' Ashtarjan
Variant Names	Friday Mosque of Ashtarjan, Masjid–i Jami' Ashtarjan, Masjid–i Jami' of Ashtarjan, Masjid–e Jame, Masjid–i Jami, Masjed–e Jomeh, Masjed Jomeh–e Oshtorjan, Masjid–i Jami' at Ashtarjan
Location	Ashtarjan, Iran
Position	Dome chamber, Corner
Date	1315–16/715 AH
Period	Aq Qoyunlu, Ilkhanid
Century	14th
Building Type	Religious
Building Usage	Mosque
Copyright	Hamidreza Kazempour

Isometric view of 3D model   **Isometric view of _muqarnas_**

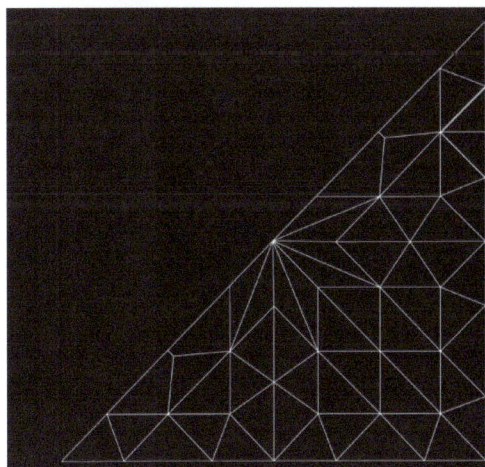

2D plan   **Bottom view of _muqarnas_**

Figure S.18   View of _muqarnas_ in Ashtarjan Mosque.

A.9. Varamin Jame' Mosque, 1322 CE

Varamin Jame' mosque was built in 1322 in varamin, 42 km South of Tehran, the capital of Iran. The mosque is famous as being the earliest surviving Ilkhanid mosque which has a four-iwan plan type. Furthermore, the building is famous for having relatively unusual aspect ratios. For instance, it has a small square courtyard, compared to the whole structure dimensions. The western part of Varamin's Jame' Mosque had been destroyed almost completely but the remaining parts of it, including the mosque's mihrab, main portal, Eastern portal and iwans, were restored and conserved.

The exterior of the building is decorated by glazed and unglazed tiles, terracotta and plaster. The iwan of the domed chamber is decorated with richly ornamented geometric patterns in terracotta, above which an inscription band runs horizontally and marks the beginning of the iwan's vault, which comprises a cluster of *muqarnas* units. These units are in turn constructed out of smaller bricks. The dome of the square sanctuary chamber was recently renovated and rises on top of a sixteen–sided drum with alternating windows, resting on an octagonal transitional area formed by four squinches. It is decorated with a sunburst medallion at its center from which descends an arabesque of a diamond geometric motif that expands with the curve of the dome. The spandrels of the arches of the four squinches are ornamented with glazed terracotta and incorporate in their vaults three tiers of *muqarnas*es.

For most contemporary architectural historians, the Friday mosque of Varamin continues to epitomize the first crystallization of a four–iwan mosque by the Ilkhanids, but also a monumentality achieved by the delicacy of ornament, the display of rich materials, and the play of light and shadow of the sculptural architectural elements (Blair & Bloom, 1994; Michell, 1978; Pope, 1977b; Pourjavady & Booth-Clibborn, 2001; Wilber, 1955).

Sample 19   Varamin Jame' Mosque building information.

Masjid Jam'e Varamin (35.320000°N, 51.650000°E)	
Variant Names	Friday Mosque of Varamin, Masjid–i Jami' Varamin, Masjid–e Jame, Congregational Mosque, Masjed–e Jameh
Location	Varamin, Iran
Position	Dome chamber squinches
Date	1322/721–22 AH
Period	Ilkhanid
Century	14th
Building Type	Religious
Building Usage	Mosque
Copyright	Hamidreza Kazempour

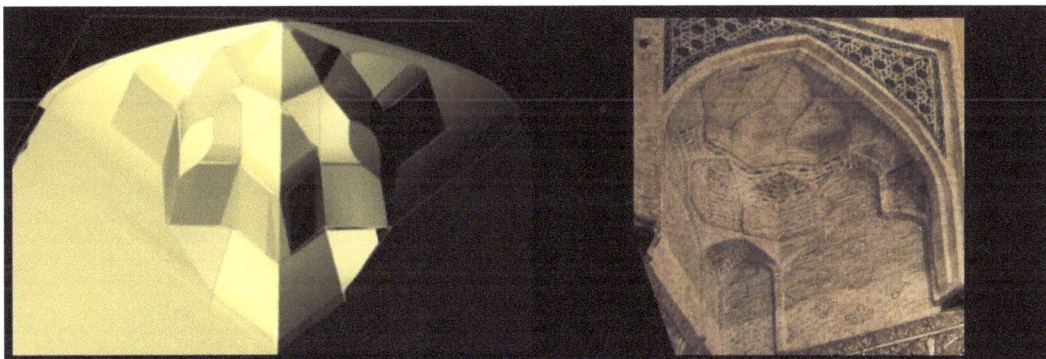

Isometric view of 3D model   **Isometric view of *muqarnas***

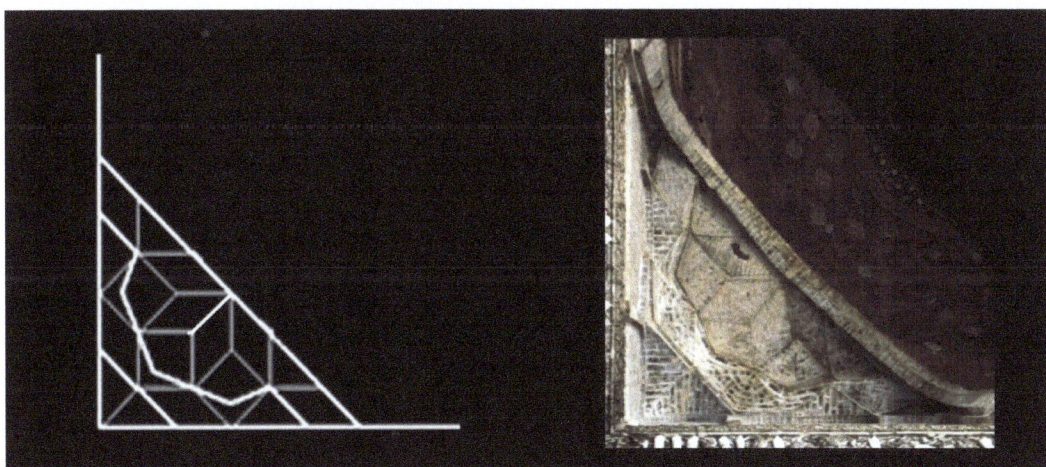

2D plan   **Bottom view of *muqarnas***

Figure S.19   View of *muqarnas* in Imamieh School, Iwan I.

A.10. Imamieh School, 1325 CE

The Madrasa–i Imami is a theological college in Isfahan and is significant as one of the earliest Iranian madrasas still in existence. Dated to 1325, this baked brick structure is built around a courtyard in typical Seljuq style, with four iwans or vaulted halls in the centre of each elevation. It adjoins the more famous tomb of a respected theologian named Baba Qasim, erected by Abu al–Hasan al Talut al Damghani in 1340–41. This tomb attributed with miraculous powers was lost in the nineteenth century and now has been rebuilt to be associated again with the madrasa.

The two–storied madrasa measures 92 by 72 meters and each side of the internal court consists of cells for student accommodation, flanking a central common space for prayers and study. This functional arrangement is expressed in the four nearly identical courtyard facades as series of continuous niches that engulf a central iwan. The iwans, interrupt the flat roofline but are not the dominant visual feature, instead the mihrabs (or niches) form the elevation's principal accent.

Decorative mosaic tile–work marks the earliest phase of the transformation from purely geometrical design to intricate floral patterns, which was later perfected under the Timurid in the fifteenth century. The central mihrab or niche marking the direction of prayer in the madrasa is renowned worldwide as a ceramic mosaic masterpiece. Assembled through a painstaking process of cutting each piece of tile, this turquoise blue artwork is made of glazed and painted ceramic on plaster, with Kufic inscriptions around the frame of the niche.

The Madrasa–i Imami is a functioning school and the structure marks an important stage in the evolution of architecture of residential theological colleges as contrasted to institutions visited by students and scholars (Hillenbrand, 2004; Blunt, 2009).

Sample 20   Imamieh School building information, Iwan I.

Name	Madrasa Imami
Variant Names	Imami Madrasa, Madrasah–i Imami, Madrasa–yi Imami, Baba ghasem School, Emamieh school, Emamiah, Imamiah, Baba Qasim, Baba Qasem school, Madreseh
Location	Isfahan, Iran
Position	Iwan I
Date	1325/726 AH
Period	Muzaffarid
Century	14th
Building Type	Educational
Building Usage	Madrasa
Copyright	Hamidreza Kazempour

Isometric view of 3D model   **Isometric view of *muqarnas***

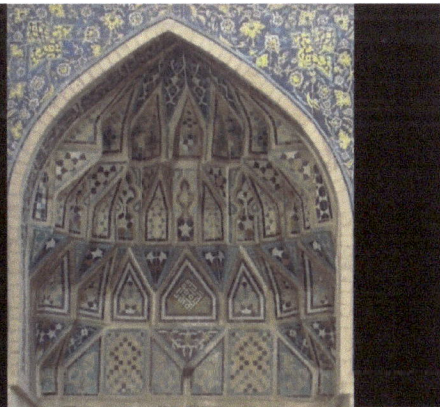

2D plan   **Bottom view of *muqarnas***
Figure S.20   View of *muqarnas* in Imamieh School, Iwan I.

Sample 21   Imamieh School building information Iwan II.

Madrasa Imami	Position: IwanII	Date: 1325

Isometric view of 3D model　　Isometric view of *muqarnas*

2D plan　　Bottom view of *muqarnas*

Figure S.21   View of *muqarnas* in Imamieh School, Iwan II.

Sample 22   Imamieh School building information, Iwan III.

Madrasa Imami	Position: IwanIII	Date: 1325

Isometric view of 3D model　　Isometric view of *muqarnas*

2D plan　　Bottom view of *muqarnas*

Figure S.22   View of *muqarnas* in Imamieh School, Iwan III.

A.11. Sheikh Safi Mausoleum, 1334 CE

Shaykh Abu'l Fath Ishaq, known as Safi al–Din Ardabili (b. 1252/3), is the eponymous founder of the Safavid order of Sufiism and is hence considered the founder of the Safavid Dynasty. Upon his death in 1334, he was buried in a tomb tower adjoining his khanqah outside the city walls of Ardabil. His burial site became a center of pilgrimage soon after, one richly endowed by Safavid rulers, many of whom were also buried there. Caravanserais, hospices, khans, baths and soup kitchens were built in Ardabil to serve the pilgrims, supported by wealthy waqfs. Following the Afghan invasion of Iran in 1722 and subsequent divestment in the shrine and the city, pilgrimage traffic dwindled. The shrine fell into disrepair; neglect and war damage took their toll until repairs were undertaken in the second half of the nineteenth century. The Iran Archaeological Service undertook extensive restoration of the shrine in the 1940s; the site continues to draw over a hundred thousand pilgrims and tourists every year.

The overall plan of the complex was developed, modified and renovated over many centuries. It consists of tombs and shrine structures clustered around a single rectangular courtyard centered on a fountain. The exterior of the complex is irregular and features plain brick and tiled walls. The main entrance to the shrine courtyard is through a small forecourt, which is located at the end of a long and narrow walled garden established in the fifteenth century. The garden is aligned northwest–southeast, whereas the shrine buildings (with the exception of the Chini–khana) are aligned with qibla slightly east of south (Bina-Motlagh, 1969; Golombek & Wilber, 1988; Hillenbrand, 2004; Hutt, 1978; Morton, 1974; Mousavi, 2002; Rizvi, 2002; Rizvi, 2000).

Sample 23   Sheikh Safi building information.

Name	**Sheikh Safi Ardebili**
Variant Names	Imamzadah Shaykh Ṣafi al–Din Ardabili, Aramgah–I Shaykh Ṣafi al–Din, Aramgah–I va Khanaqah Shaykh Ṣafi al–Din, Shaykh Ṣafial–Dīn Shrine, Shrine of Shaikh Safi, Sheikh Safi Mausoleum, Shaykh Safi al–Din Tomb, Shrine of Sheikh Safi, Mausoleum of Shah Safi
Location	Ardabil, Iran
Position	Chini Khaneh
Date	c.1350–94/751–96AH, c.1448–60/852–64AH
Period	Safavid, Timurid
Century	14th, 15th, 16th
Building Type	Funerary, religious
Building Usage	mausoleum, shrine
Copyright	Hamidreza Kazempour

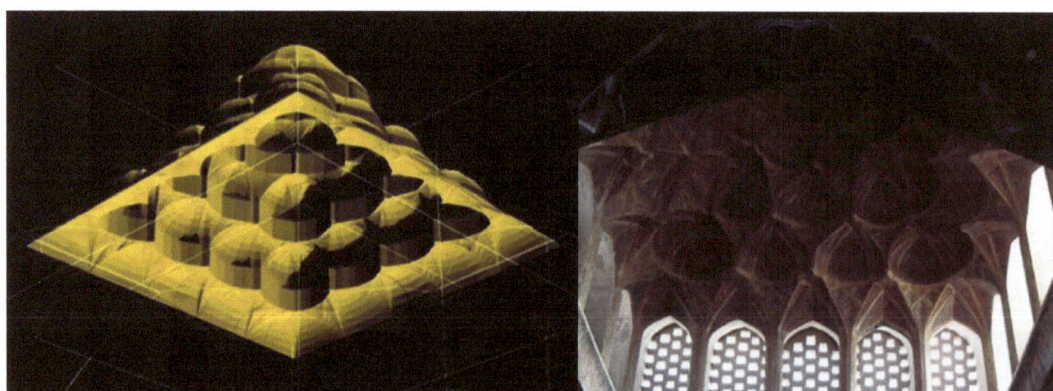

Isometric view of 3D model   Isometric view of *muqarnas*

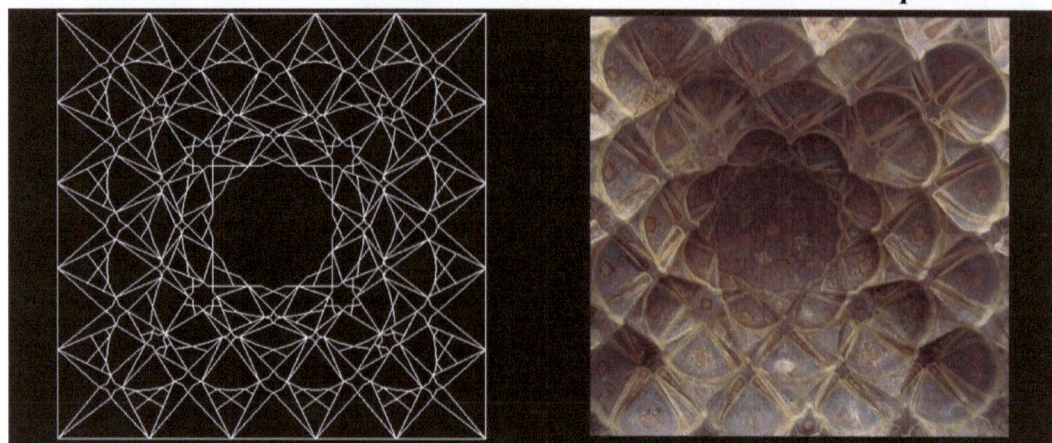

2D plan   Bottom view of *muqarnas*

Figure S.23   View of *muqarnas* in Sheikh Safi.

A.12. Eziran Jame' Mosque, 1335 CE

This building is located on the Eziran village, 30 kms of the east of Isfahan and it was built during Ilkhanid era. The main building of the mosque was consisted of yard, porch, iwan and dome but nowadays only dome, remaining corridors and yard porches has been left. The building dome has a square plan which has been changed to octagon by eight sharp arches at the top and them has been turned to a circle by sixteen wall arches.

The inner side of the bigger corner arches has a simple *muqarnas*. At the smaller arches side, some plans with bricklayers have been developed. This dome has three facades in different sides. Exterior building is simple and brick and there is a belt cornice of brick only at the neck of the dome which repeating the phrase of "Allah u Akbar".

This building has no historical cornice, but based on the style of the mosque building and its similarity with the Dashti and Kaj Mosques at the same region, is considered as the Fourteen century (Wilber, 1955).

Sample 24   Eziran Mosque building information, Corner I.

Name	**Masjid Jame' Eziran**
Variant Names	Masjid–i–Jami, Masjid–i Jami, Masjid–i Jomeh, Masjid–i–Jomeh
Location	Eziran, Iran
Position	Under Dome – Corner
Date	1335 / 735 AH
Period	Ilkhanid
Century	14th
Building Type	Religious
Building Usage	Mosque
Copyright	Hamidreza Kazempour

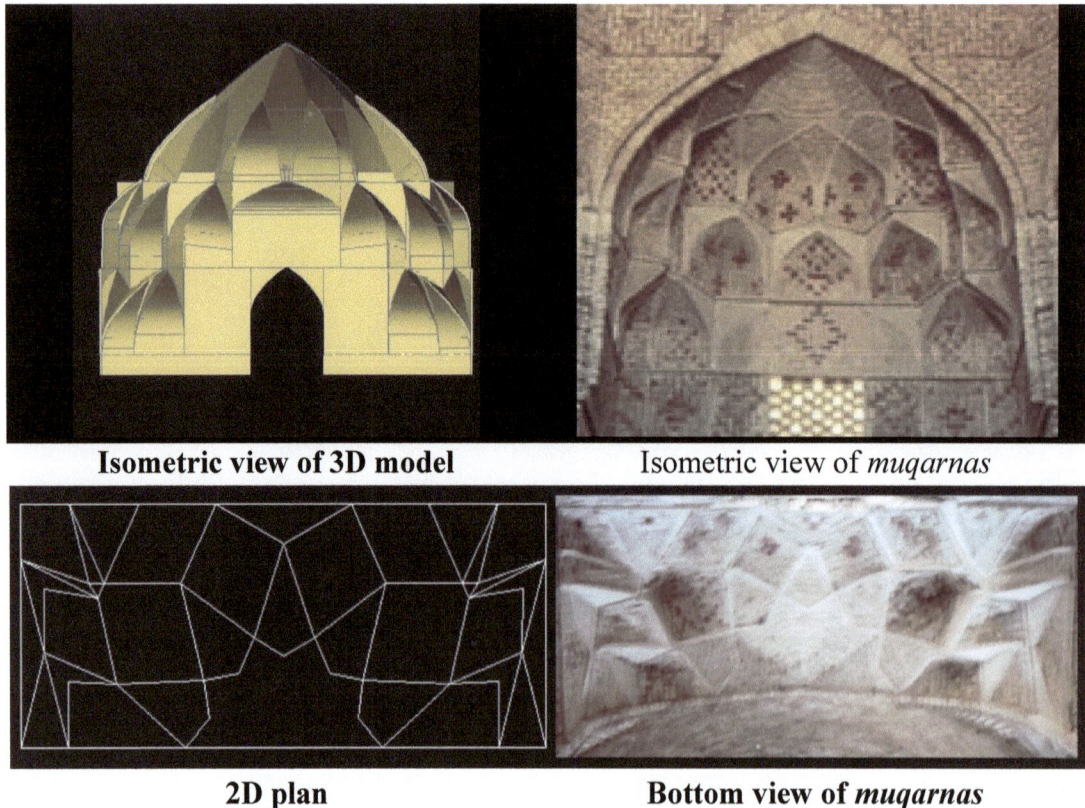

Isometric view of 3D model   **Isometric view of *muqarnas***

2D plan   **Bottom view of *muqarnas***

Figure S.24   View of *muqarnas* in Eziran Mosque, Corner I.

A.13. Qom Jame' Mosque, 1350 CE

This mosque has been constructed on a rectangular plan. The main skeleton of the dome is related to the early mid–14th century and ranks second in antiquity to the old Jame' mosque of Qom. According to valid historical records the current dome dates back to 529 A.H. Its tall south facing porch and colored encrustation of plasterwork of the dome is related to the Safavid period. The construction of the northern porch and nocturnal areas in an east–west direction are of the Qajar era.

This Mosque has two entrances which one of them is located from by–lane to northern nave and the other, which is the main Entrance of the building, is located in the west side. This building has a tall portal with plaster *muqarnas* arch. The main portal of the mosque is exactly located opposite of the Jahangir khan School which represents the relationship between religious and educational places (Kochak zadeh, 1998; Zomarshidi, 2009; Fayz, 1977).

Sample 25 Qom Jame' Mosque building information, Iwan I

Name	Masjid Jame' Ghom
Variant Names	Masjid–e Jame, Masjid–i Jami, Masjed–e Jomeh, Congregational Mosque
Location	Qum, Iran
Position	Entrance
Date	1350
Period	Ilkhanid, Qajar
Century	14th, renovated 19th
Building Type	Religious
Building Usage	Mosque
Copyright	Hamidreza Kazempour

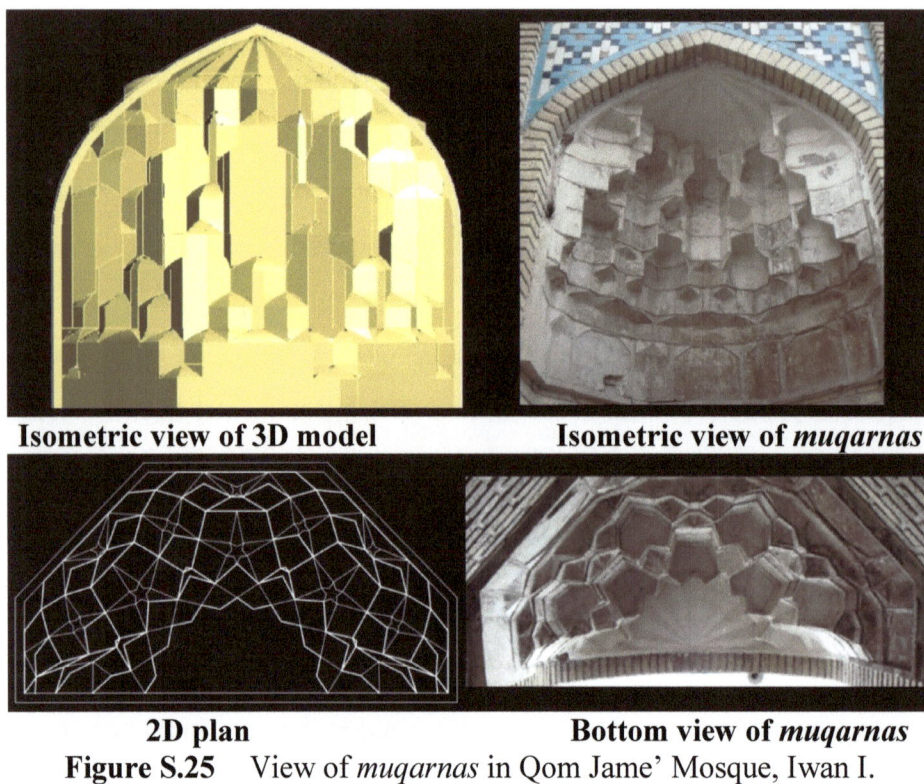

Isometric view of 3D model   **Isometric view of *muqarnas***

2D plan   **Bottom view of *muqarnas***

Figure S.25   View of *muqarnas* in Qom Jame' Mosque, Iwan I.

Sample 26   Qom Jame' Mosque building information, Iwan II

Masjid Jame' Ghom	**Position:** Entrance II	**Date:** 1350

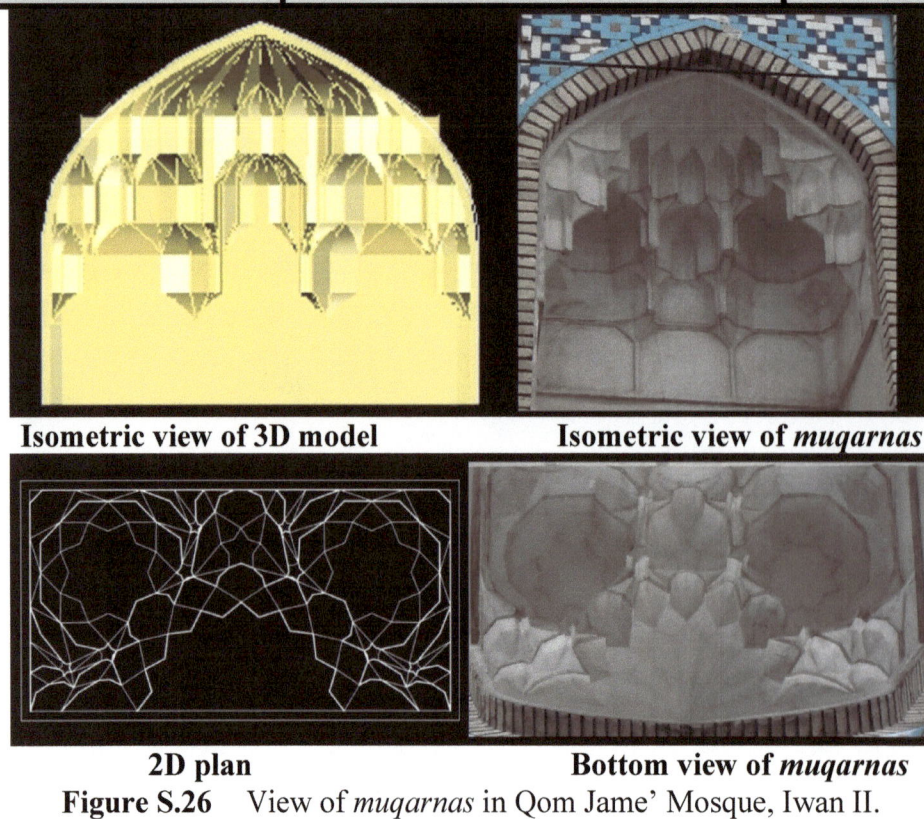

Isometric view of 3D model   **Isometric view of *muqarnas***

2D plan   **Bottom view of *muqarnas***

Figure S.26   View of *muqarnas* in Qom Jame' Mosque, Iwan II.

Sample 27   Qom Jame' Mosque information, Entrance Portal I.

Masjid Jame' Ghom	**Position:** Entrance III	**Date:** 1350

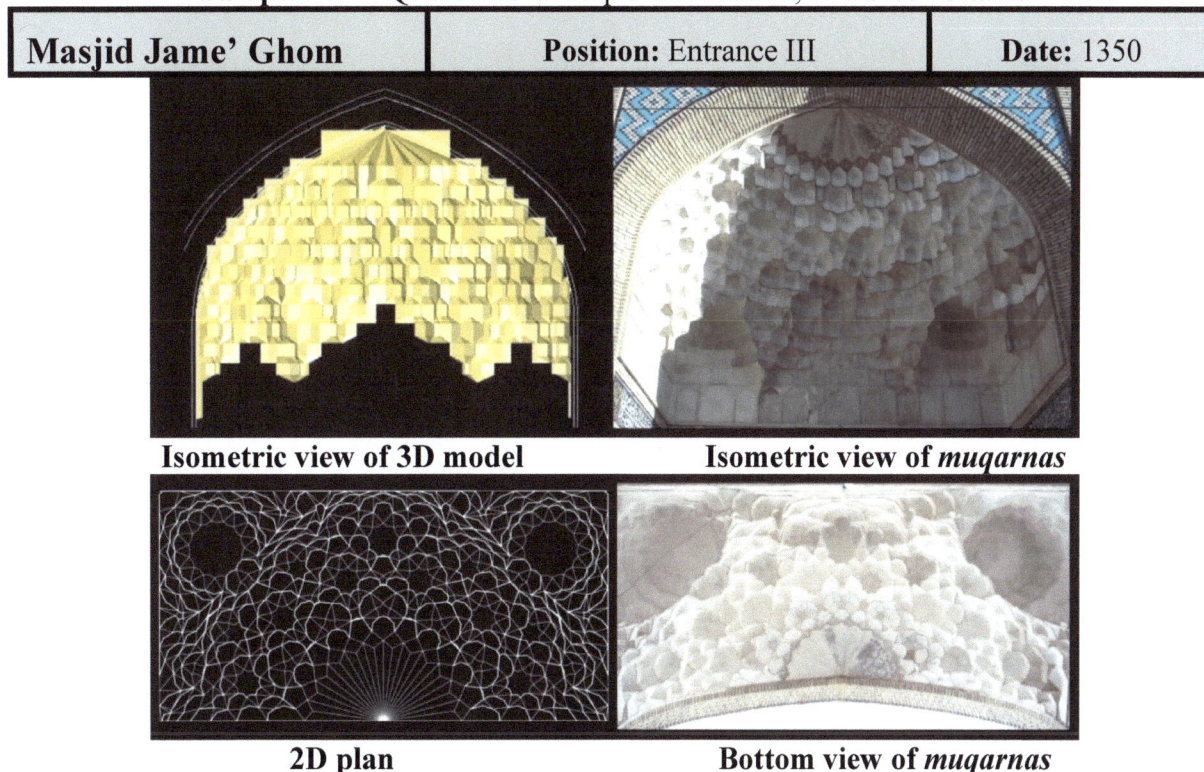

Isometric view of 3D model   **Isometric view of *muqarnas***

2D plan   **Bottom view of *muqarnas***

Figure S.27   View of *muqarnas* in Qom Jame' Mosque, Entrance Portal I.

Sample 28   Qom Jame' Mosque building information, Porta,II.

Masjid Jame' Ghom	**Position:** Iwan	**Date:** 1350

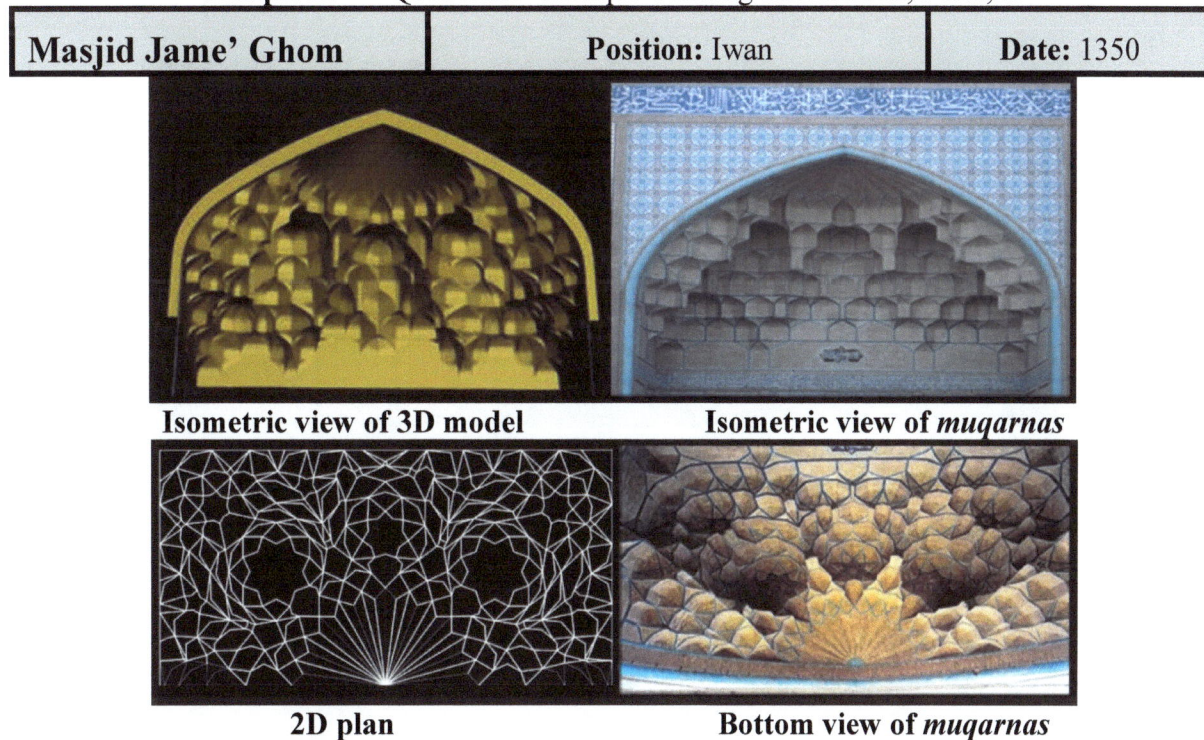

Isometric view of 3D model   **Isometric view of *muqarnas***

2D plan   **Bottom view of *muqarnas***

Figure S.28   View of *muqarnas* in Qom Jame' Mosque, PortalII

Sample 29   Qom Jame' Mosque building information, Under Dome – Corner.

Masjid Jame' Ghom	**Position:** Corner, dome chamber	**Date:** 1350

Isometric view of 3D model   **Isometric view of *muqarnas***

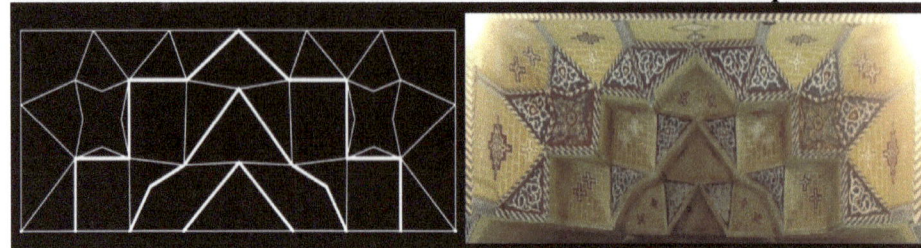

2D plan   **Bottom view of *muqarnas***

Figure S.29   View of *muqarnas* in Qom Jame' Mosque, Under Dome – Corner

Sample 30   Qom Jame' Mosque building information, Under Dome – Middle.

Masjid Jame' Ghom	**Position:** Centre, dome chamber	**Date:** 1350

Isometric view of 3D model   **Isometric view of *muqarnas***

2D plan   **Bottom view of *muqarnas***

Figure S.30   View of *muqarnas* in Qom Jame' Mosque, Under Dome – Middle.

A.14. Yazd Jame' Mosque, 1365 CE

The Friday Mosque of Yazd Situated adjacent to the center of the town of Yazd, was founded in the twelfth century; however, what stands on the site today is the new mosque built in 1324 under the Ilkhanids, and later augmented in 1365 under the Muzaffarids. Three important aspects distinguish the Friday mosque of Yazd: its structural innovation, its remarkable decoration, and its being the "earliest" mosque upon which later fifteenth–century mosques in the Yazd region were modeled. The mosque encloses a rectangular open court. The complex's main portal iwan is located on the east side of the court; the domed chamber with the main iwan preceding it occupies the center of the southern side of the court. A distinctive feature of the complex is the tall eastern portal iwan surmounted by two soaring minarets on each side. The dome of the vestibule rests on eight squinches, from which ascend eight ribs that intersect at the top of the dome and form a smaller octagonal perforated nucleus.

Another important structural feature of the mosque is the dome of the prayer chamber. It consists of two domes: a dome inside a dome, with a space of a half meter separating them at their base. The prayer halls were built ca. 1360 of mud brick. The mihrab is a half–octagonal niche in the qibla wall, fully decorated with a glazed tile mosaic revetment of floral motifs, predominantly in blue and white and surmounted by a *muqarnas* vault. The mihrab's pointed arch is framed by a calligraphic inscription band in white thuluth script on a blue background.

The vault of the soaring portal iwan is surmounted by a *muqarnas* units in blue, white, and ochre, ascending to a smooth semi–dome medallion decorated in concentric rings of floral arabesque. These *muqarnas* units inscribe, as they spatially cluster, star–shaped surfaces of various sizes containing calligraphic medallions or floral patterns. Script bands run horizontally and vertically, bordering the pointed arch of the iwan and the surfaces of its recesses. (Golombek & Wilber, 1988; Holod-Tretiak, 1973; Pope & Ackerman, 1977; Siroux, 1947; Wilber, 1969).

Sample 31   Yazd Jame' Mosque building information.

Name	**Masjid Jame' Yazd**
Variant Names	Friday Mosque of Yazd, Masjid–i Jami Yazd, Masjid–e Jame, Masjed–e Jameh, Congregational Mosque
Location	Yazd, Iran
Position	Entrance
Date	1470/874–75 AH
Period	Ilkhanid, Muzaffarid
Century	14th
Building Type	Religious
Building Usage	Mosque
Copyright	Hamidreza Kazempour

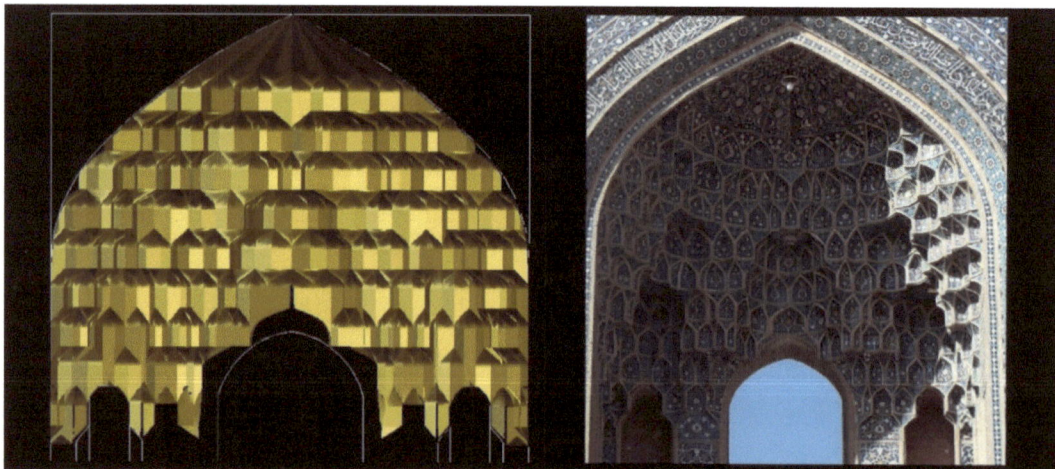

Isometric view of 3D model   **Isometric view of *muqarnas***

2D plan   **Bottom view of *muqarnas***

Figure S.31   View of *muqarnas* in Yazd Jame' Mosque.

Sample 32 Yazd Jame' Mosque building information.

Masjid Jame' Yazd	**Position:** Mihrab	**Date:** 1365

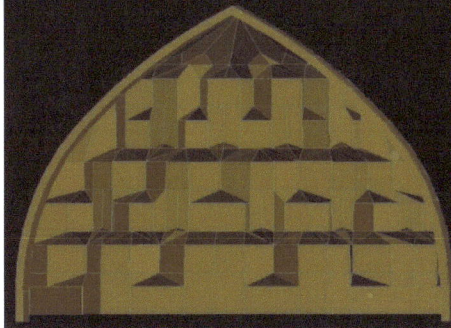

Isometric view of 3D model

Isometric view of *muqarnas*

2D plan

Bottom view of *muqarnas*

Figure S.32 View of *muqarnas* in Yazd Jame' Mosque.

Sample 33 Yazd Jame' Mosque building information.

Masjid Jame' Yazd	**Position:** Corner, dome chamber	**Date:** 1365

Isometric view of 3D model

Isometric view of *muqarnas*

2D plan

Bottom view of *muqarnas*

Figure S.33 View of *muqarnas* in Yazd Jame' Mosque.

A.15. Mozaffarieh School, 1367 CE

The Muzaffarid madrasa, locally known as the iwan of Umar (Suffeh–i Umar), was erected on the southeast side of the Friday mosque in the fourteenth century and is particularly remarkable for its superb mosaic faience decoration in floral and geometric designs. An inscription on the soffit of the iwan of the madrasa gives the name of the Muzaffarid Sultan Mahmud (reg. 1358–1374) as the patron of the addition to the Friday Mosque. The original extant section of the building consists of an iwan leading to a rectangular hall with transverse vaulting. The central bay of the qibla hall is covered with a lantern and encompasses a tile mosaic mihrab with *muqarnas* hood. While hazarbaf tiles in geometric patterns enliven the soffit of the iwan, the *muqarnas* above the mihrab is revetted with light blue, dark blue, black and white tiles as well as unglazed tiles.

Historians of architecture consider the Friday Mosque of Isfahan to be a masterpiece of brick architecture. While similar in magnitude to mosques found in Syria and Cordoba, it also presents new elements, highly esteemed for their structural ingenuity and complexity. The amalgam of decoration compositions produced by the variety of brick patterns, the meticulous work in carved stucco, colored panels of floral, geometric and epigraphic motifs, all render the Friday mosque of Isfahan a highlight of Seljuk architecture (Byron, 1982; Galdieri, 1984; Golombek & Wilber, 1988; Hoag, 2004; Michell, 1978; Pourjavady & Booth-Clibborn, 2001; Yeomans, 1999; Blair & Bloom, 1994; Grabar, 1990).

Sample 34   Mozafariyeh School building information, Mozafariyeh Entrance.

Name	Mozafariyeh School
Variant Names	Friday Mosque of Isfahan, Masjid–i Jami' Isfahan, Masjid–i Jami of Isfahan, Masjid–e Jomeh, Great Mosque, Masjid–e Jame, Masjid–e Jameh, al–Jami' al–Kabir, Masjid–i Jami' Isfahan, Masdjid–i Djum'a, Congregational Mosque
Location	Isfahan, Iran
Position	Mozafariyeh Entrance
Date	1366–67/768 AH (madrasa)
Period	Mozaffarid
Century	14th
Building Type	Educational, Religious
Building Usage	Madrasa, School
Copyright	Hamidreza Kazempour

Isometric view of 3D model   **Isometric view of *muqarnas***

2D plan   **Bottom view of *muqarnas***

Figure S.34   View of *muqarnas* in Mozafariyeh School, Mozafariyeh Entrance.

Sample 35   Mozafariyeh School building information, Iwan I.

Mozafariyeh School	Position: Iwan I	Date: 1366

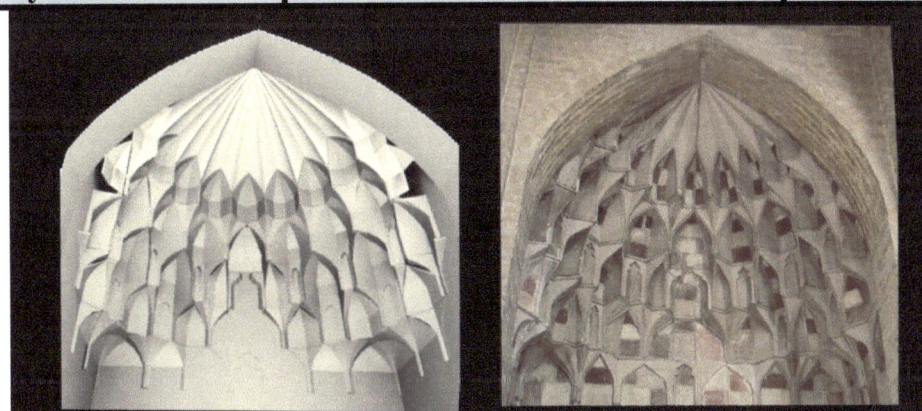

Isometric view of 3D model   **Isometric view of *muqarnas***

2D plan   **Bottom view of *muqarnas***

Figure S.35   View of *muqarnas* in Mozafariyeh School, Iwan I.

Sample 36 Mozafariyeh School building information, Iwan II.

Mozafariyeh School	Position: Iwan II	Date: 1366

Isometric view of 3D model **Isometric view of *muqarnas***

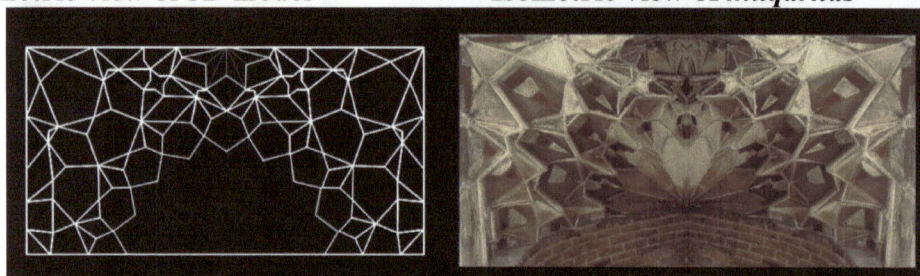

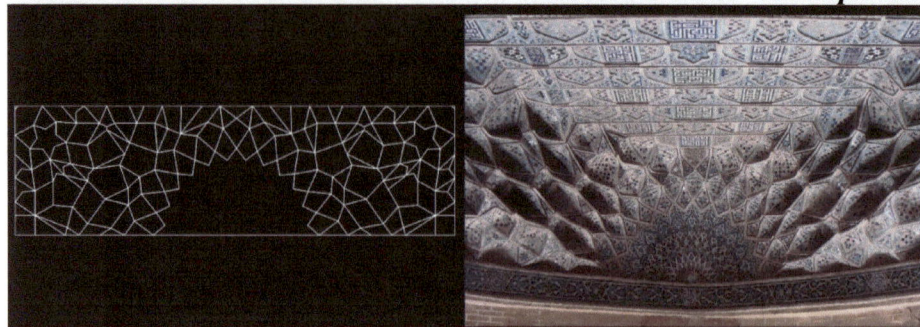

2D plan **Bottom view of *muqarnas***

Figure S.36 View of *muqarnas* in Mozafariyeh School, Iwan II.

Sample 37 Mozafariyeh School building information, Mihrab.

Mozafariyeh School	Position: Mihrab	Date: 1366

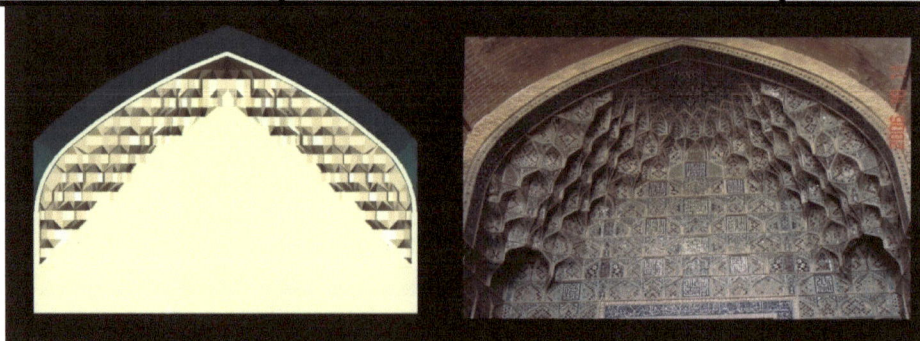

Isometric view of 3D model **Isometric view of *muqarnas***

2D plan **Bottom view of *muqarnas***

Figure S.37 View of *muqarnas* in Mozafariyeh School, Mihrab.

A.16. Darb Imam Mosque, 1453 CE

The shrine of Darb–i–Imam, which means the shrine of the Imams, is a funerary complex consisting of an ancient cemetery, shrine structures, and courtyards representing different construction periods and styles within the present day Dardasht quarter of Isfahan. In the Seljuk period, the cemetery existed beyond city walls in the city's northwest quadrant, but in the tenth century, it was incorporated within the city in the Sunbulistan or Chumulan quarter. The complex is unusual in having two closely spaced domes, but is renowned for the excellence of the mosaic tile work on its iwan (a vaulted hall, walled in three sides with one end open), portals and galleries.

The first structures were built by Jalal al–Din Safarshah, a vassal of Sultan Muhammadi the governor of Isfahan during the Qara Qoyunlu reign in 1453. A richly embellished pishtaq (a projecting entrance porch), a square vestibule, and a mausoleum to Jahan Shah Aq Qoyunlu's mother, alone remain from the first stage of construction. The brilliance in execution of the shrine's celebrated mosaic faience pishtaq is well matched by the elegance of the console of the shrine's dome. The shrine's portal was copied in the Qutbiyah mosque of 1543, which was later removed to the Chihil Sutun garden. (Blunt, 1966; Hutt, 1978; Golombek & Wilber, 1988; Blake, 1999).

Sample 38   Darb Imam Mosque building information, Entrance Portal.

Name	Masjid Darb Imam
Variant Names	Darb–i Imam Shrine, Imamzadah Darb–i Imam, Darb–i Imam Imamzada, Darb–i Imam Imamzade, Shrine of Darb–e Imam, Imamzadeh Darb–i Imam, Imamzada Darb i Imam
Location	Isfahan, Iran
Position	Entrance
Date	1453/857 AH
Period	Qara Qoyunlu, Safavid
Century	15th
Building Type	funerary, religious
Building Usage	mausoleum, shrine
Copyright	Hamidreza Kazempour

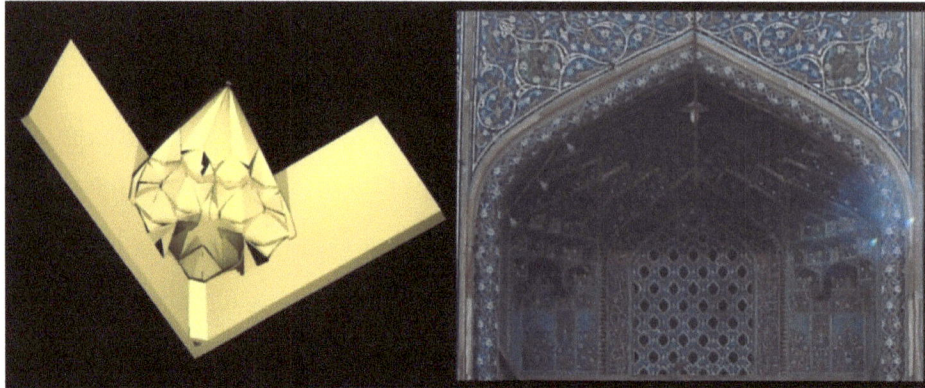

Isometric view of 3D model　　　　**Isometric view of *muqarnas***

2D plan　　　　**Bottom view of *muqarnas***

Figure S.38　　View of *muqarnas* in Darb Imam Mosque, Entrance Portal.

Sample 39　　Darb Imam Mosque building information, Iwan.

Masjid Darb Imam	**Position:** Iwan	**Date:** 1453

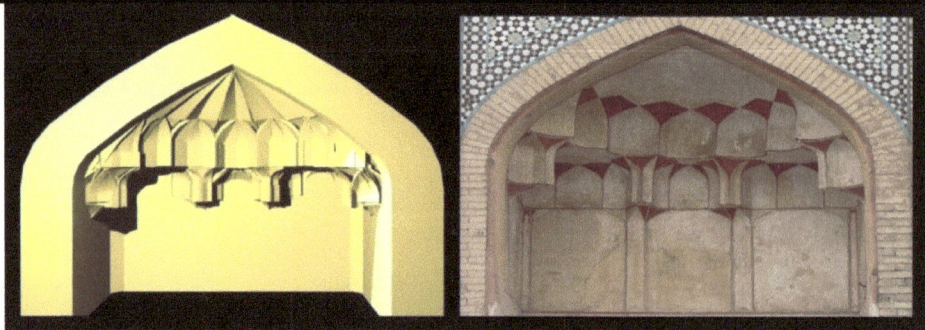

Isometric view of 3D model　　　　**Isometric view of *muqarnas***

2D plan　　　　**Bottom view of *muqarnas***

Figure S.39　　View of *muqarnas* in Darb Imam Mosque, Iwan.

A.17. Qazvin Jame' Mosque, 1501–1732 CE

The Friday mosque of Qazvin is one of the oldest mosques in Iran. Its construction was first ordered in 807 (192 AH) by the Abbasid Caliph Harun Al–Rashid. Under the Seljuk leaders (1038–1194), two iwans were added to its north. After a renovation in the eleventh century, the twelfth century saw the construction of the main prayer hall, a dome, a courtyard, and a religious school. Under the Safavids (1501–1732), the southern and western iwans and arcades were added, and the Qajar period (1779–1924) witnessed a major renovation and expansion. The mosque follows the four–iwan typology; each iwan is centered on a large courtyard with a central fountain.

The main entrance to the mosque is on the east side through a portal, linked with a narrow corridor to the mosque courtyard with its central fountain. This court also contains a ten–step staircase, located between the fountain and the east iwan that leads down to a canal. Two smaller courtyards are located at the northwest and northeast corners of the mosque. The mosque is constructed of brick, which is clad with tiles and inscriptions in some areas. The main prayer hall is the most ornamented part of the mosque. The upper part of the walls is ornamented in different floral patterns and small polychrome tiles. Carved Kufic and Sols inscriptions decorate the walls. This prayer hall is roofed with a large double–shelled Seljuk dome that rests on the hall's thick columns. On its exterior, the dome is decorated with tiles and floral patterns.

A half–dome roofing the south iwan also has a double–shell structure. The two northern minarets are clad with colorful tiles in floral patterns. The interior of the iwans are decorated with *muqarnas*; in the north iwan, these *muqarnas* are made by stucco, while those in the southern iwan are of exposed brick (Godard, 1965; Hatim, 2000; Pope, 1965; Pope, 1997a; Sourdel-Thomine & Wilber, 1974; Farhangi, 2003).

Sample 40 Qazvin Jame' Mosque building information.

Name	**Masjid Jame' Qazvin**
Variant Names	Friday Mosque of Qazvin, Masjid–i Jami' Qazvin, Masjid–e Jame, Congregational Mosque, Jame Kabir, Masjid–i Jami' of Qazvin, Ghazvin Masjid Jame', Masjid–i Jomeh of Qazvin, Qazvin Kabir Mosque, Gahzvin Ateegh Mosque
Location	Qazvin, Iran
Position	Iwan
Date	1501–1732 / 900–1130 AH
Period	Safavid
Century	16th – 18th
Building Type	Religious
Building Usage	Mosque
Copyright	Hamidreza Kazempour

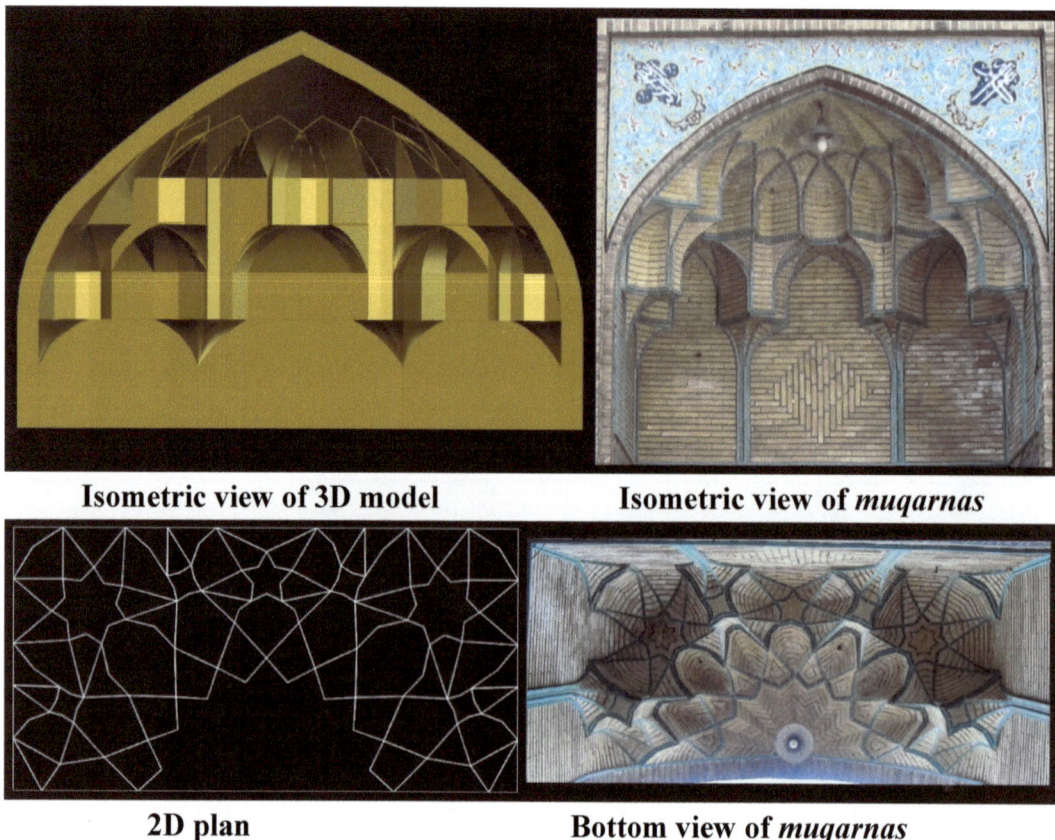

Isometric view of 3D model   **Isometric view of *muqarnas***

2D plan   **Bottom view of *muqarnas***

Figure S. 40 View of *muqarnas* in Qazvin Jame' Mosque.

285

A.18. Shazdeh Hosein Shrine, 1568 CE

The holy shrine of Hazrat Imamzadeh Hussain, popularly known as Shahzadeh Hussain, is situated at the southwest of Qazvin city and is one of the famous and magnificent places of pilgrimage in Iran.

The building complex of the shrine is a grand masterpiece of Islamic architecture and its tile–works and mirror–works are a source of great attraction to both, domestic as well as foreign tourists and pilgrims. The complex comprises a large courtyard, an octagonal hall, and a small domed–room that houses the main grave of the shrine.

The entrance gate is very tall, above which there is a blue tile table on which 12 couplets of poetry are written in white, in the Nasta'liq script. On entering the gate, one would find himself in a large rectangular courtyard, at the middle of which is situated, an octagonal drinking water fountain. The walls of the courtyard have arch–like designs, behind which there are small rooms that were used in earlier times for prayers and supplications on Thursday nights and on other important days and festivals.

Directly opposite the main entrance gate, is situated the main building of the shrine, that opens up into an octagonal hall. There are two minarets on the two sides of the building. The main shrine of Shahzadeh Hussain is placed in the middle of a small domed–room, surrounded by the octagonal hall, with four different entrances. The hall as well as the domed–room is adorned with mirror works and blue tile–works that daze and impress every visitor (Zendehdel et al., 1998).

Sample 41   Shahzadeh Hossein Shrine building information.

Name	**Masjid Shahzadeh Hussein**
Variant Names	Shahzada Husayn Shrine, Imamzade Shahzadeh Hussain, Imamzadeh Shahzadeh Hussein Shahzade Hossein Tomb Complex, Mausoleum of Husseyn, Sanctuary of Shahzade Hussain
Location	Qazvin, Iran
Position	Entrance
Date	1568–1600/975–1008 AH
Period	Safavid
Century	17th
Building Type	Funerary
Building Usage	Tomb
Copyright	Hamidreza Kazempour

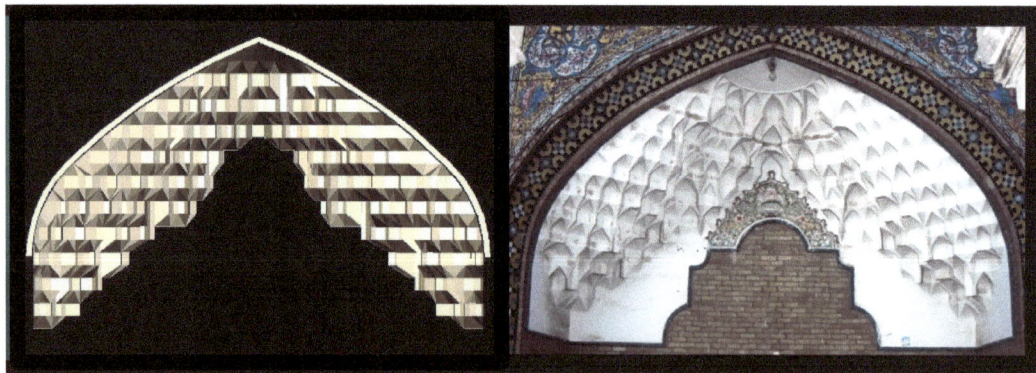

Isometric view of 3D model   **Isometric view of *muqarnas***

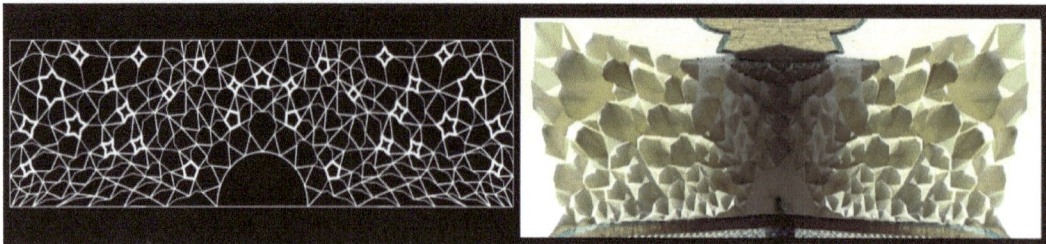

2D plan   **Bottom view of *muqarnas***

Figure S. 41   View of *muqarnas* in Shahzadeh Hossein Shrine.

A.19. Ali Qapu Palace, 1590 CE

The Ali Qapu is located on the west facade of the Shah Square (now *Maydan–i Imam*) facing the Mosque of Shaykh Lutfallah. Originally much smaller in scale, designed as a portal between the palace gardens and the square, the Ali Qapu grew with a series of additions over a sixty year period to accommodate court functions.

The additions include a double height hall (a columnar porch form usually located at ground level) which sits upon a two–story base. The hall provided a post from which royalty and courtiers could view any spectacle in the square below. Three more stories rise behind the hall, each with a different floor plan composed of reception halls and rooms of varying sizes and uncertain function.

Of note are the stucco *muqarnas* niches in the fifth floor 'music room'. Cutouts in the shape of ceramics and glassware create a delicate surface and intense chiaroscuro effect.

Along with the Chihil Sutun and Hasht Behesht, the Ali Qapu was restored by IsMEO – Istituto Italiano per il Medio ed Estremo Oriente for NOCHMI –National Organization for Conservation of Historic Monuments of Iran. The project, completed in 1977, received an Aga Khan Award for Architecture in 1980 (Babaie, 2008; Blair & Bloom, 1994; Galdieri, 1979; Hillenbrand, 2004; Michell, 1978).

Sample 42   Ali Qapu building information.

Name	Ali Qapu
Variant Names	'Ali Kapu, Ali Qapu, Ali Qapu Palace, Ali Kapu palace
Location	Isfahan, Iran
Position	Porch
Date	1590–91/998–99 AH, 1602/1010 AH, 1622/1031 AH
Period	Safavid
Century	17th
Building Type	Palatial
Building Usage	Palace
Copyright	Hamidreza Kazempour

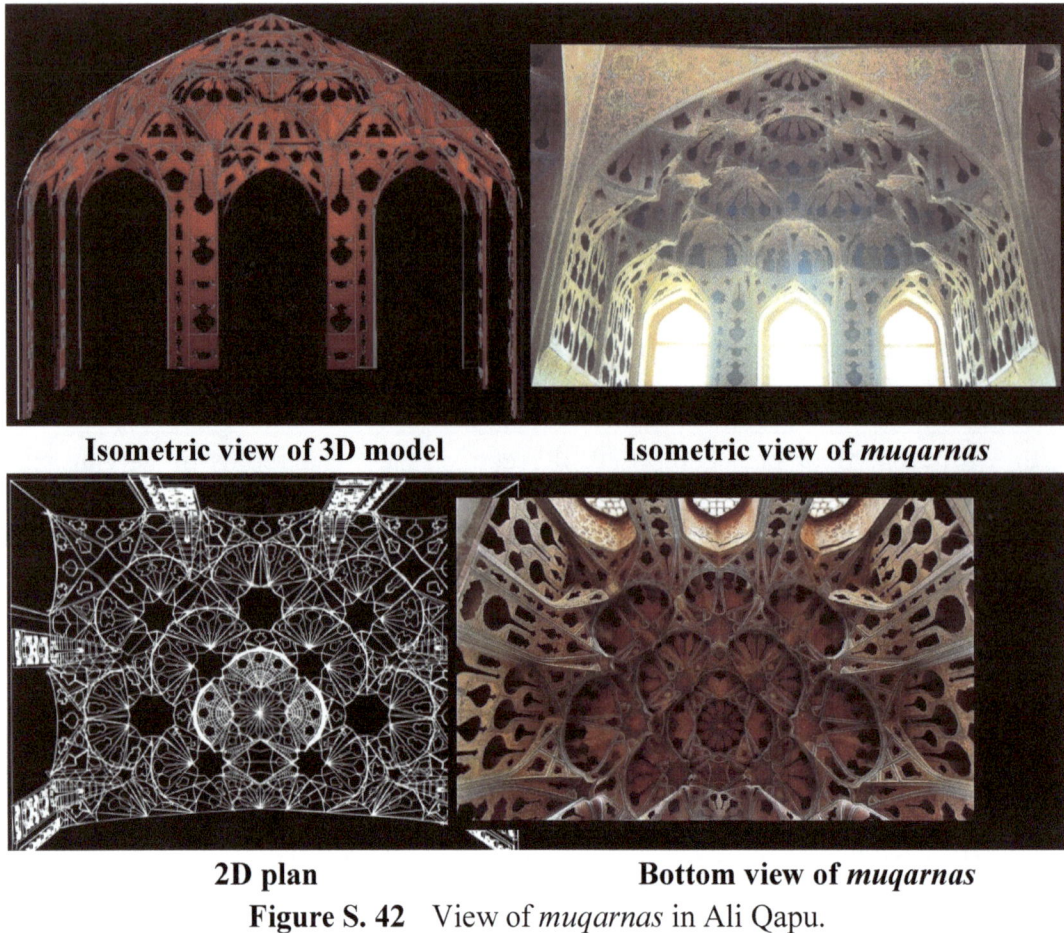

Isometric view of 3D model	**Isometric view of *muqarnas***
2D plan	**Bottom view of *muqarnas***

Figure S. 42   View of *muqarnas* in Ali Qapu.

A.20. George Church, 1611 CE

The Saint George Church, also known as the "Geork" church is located Northeast of the intersection between Hakim Nezami Street and Nazar Street, on the left hand side of the former Jolfa quarter. It stands behind a high wall and the view from the street is shown above. It is the third oldest church in Jolfa, constructed in 1611 C.E. The Church is principally famous for the blocks of stone from the original church in Armenia, which have been placed as an annex. These are regarded with great devotion by the Armenian community who light candles close to them and pray beside them. The interior of the church is plain, but there is a fine piece of tile work depicting the Adoration of the Magi above the entrance which was placed there in 1719 C.E. (Haghnazarian, 2006).

Sample 43   George Church Information.

Name	George Church
Variant Names	George Church, St. George Church, Gevork Church
Location	Isfahan, Iran
Position	Entrance
Date	1611/ 1131 AH
Period	Safavid
Century	17th
Building Type	Religious
Building Usage	Church
Copyright	Hamidreza Kazempour

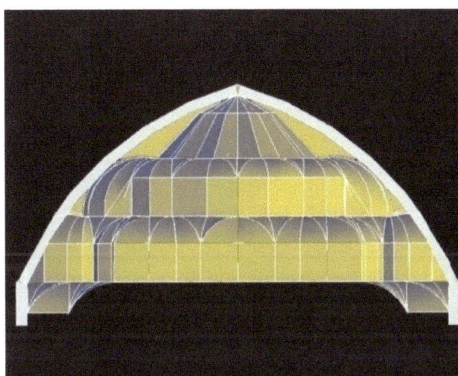

Isometric view of 3D model   Isometric view of *muqarnas*

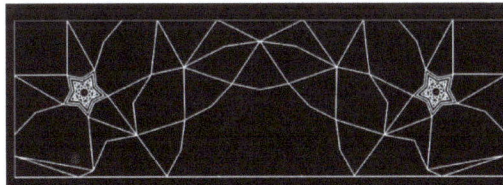

2D plan   Bottom view of *muqarnas*

Figure S. 43   View of *muqarnas* in George Church.

A.21. Sheikh Lotf Allah Mosque, 1617 CE

Shaykh Lutfallah Mosque, built in 1617–18, is located at the centre of the east side of Isfahan's grand royal square (512 by 159 meters), built by Shah 'Abbas I between 1590 and 1602. The royal square was an expression of Isfahan's emergence as the new political and economic capital of the Safavid dynasty. It is a multifunctional space with intensive mercantile activity, which also served as a house for royal ceremonial rituals, royal caravanserai, baths, a royal mint, and a hospital— all centred in the space of the square and the two story arcade surrounding it on four sides. The mosque was named in 1622 after Shaykh Lutfallah Maysi al–'Amili (d.1622).The mosque is more similar in type to a mausoleum than a four–iwan mosque.

The main entrance to the mosque is located on the east side of this small court. The structure itself is not aligned perpendicularly to the royal square's eastern wall, but lies at an angle (almost 45 degrees) against the royal square's wall. Its asymmetrical layout was initially introduced to reconcile the (southwest) direction of Mecca with the placement of the mihrab on the qibla wall, and adds visual complexity to the structure.

The portal iwan is elaborately ornamented in colourful mosaics. It is built as a recessed area on the eastern wall of the court, an elevated platform raised by four steps from the court level. An inscription band in white on a dark blue background runs horizontally on the three sides of the portal niche, above which begins the iwan's vault, comprising four clusters of *muqarnas* structures made of small glazed–tiles units. These four clusters ascend to inscribe a concentric floral medallion. The pointed arch doorway is located below the inscription band and is flanked by two panels of mosaics of floral arabesques with motifs in yellow, white, and blue on a dark blue background. These panels rest on top of a continuous marble dado. (Hoag, 2004; Michell, 1978; Blunt, 2009; Blair & Bloom, 1994).

Sample 44   Sheikh Lotf–Allah Mosque building information.

Name	Masjid Sheikh Lotf–Allah
Variant Names	Shaykh Lutfallah Mosque, Masjid–e Shaykh Lutf Allah, Sheikh Lutfallah Mosque, Masjid–i Shaikh Lutfullah, Sheikh Lotfollah Mosque
Location	Isfahan, Iran
Position	Entrance
Date	1617/1025–26 AH
Period	Safavid
Century	17th
Building Type	Religious
Building Usage	Mosque
Copyright	HamidrezaKazempour

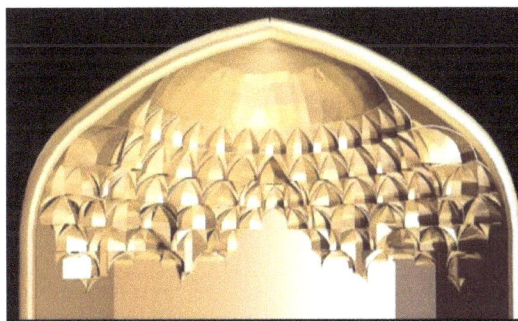

Isometric view of 3D model   Isometric view of *muqarnas*

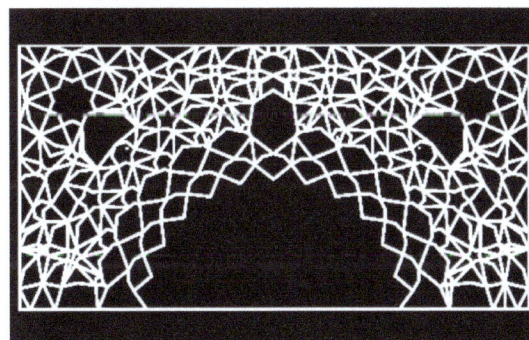

2D plan   Bottom view of *muqarnas*

Figure S. 44   View of *muqarnas* in Sheikh Lotf–Allah Mosque

A.22. Khan School, 1636 CE

This school is a building related to the Safavid era that has been made based on Shah Abbas's order, by Allah Verdi Khan, Fars governor, and has been continued later by his son Imam Qoli Khan and its building is ended in the year 1024 AH. Also he invited Ṣadr–al–Din Šhirāzi (Mollā Ṣadrā) to teach there.

In 1615, Imam Gholi Khan founded the serene Madrasa–e–Khan theological college for about 100 students. Then impressive portal at the entrance has beautiful and unusual type of stalactite moulding inside the outer arch and some intricate mosaic tiling with much use of red, in contrast to the tiles used in Yazd and Esfahan. The college (still in use) has a fine stone–walled inner courtyard and garden.

This school which is one of the greatest and most famous schools in Shiraz is located in the Isaac Baig neighborhood. The Khan School has a long 51 and the width 45 meters. Chambers have been established two–storey in every 4 sides of the school that its number counts to 70 Chambers. The beautiful entrance portal and mosaics tiling of vestibule entrance are the excellent decorations of this school. In the hallway, the date of building 1024 and the name of its architect Hossein Shamaei have been listed (Shirazi & Ja'far, 1998; Mostafavi, 2003).

Sample 45   Khan School information.

Name	Madrasa–i Khan
Variant Names	Khan Madrasa, Masjid–i Khan, Khan Mosque, Masjid–i Khan va Madrasa–i, Khan Mosque and Madrasa, Imam Gholi Khan Madrasa
Location	Shiraz, Iran
Position	Entrance
Date	1636/1036 AH
Period	Safavid
Century	17th
Building Type	educational, religious
Building Usage	madrasa, mosque
Copyright	Hamidreza Kazempour

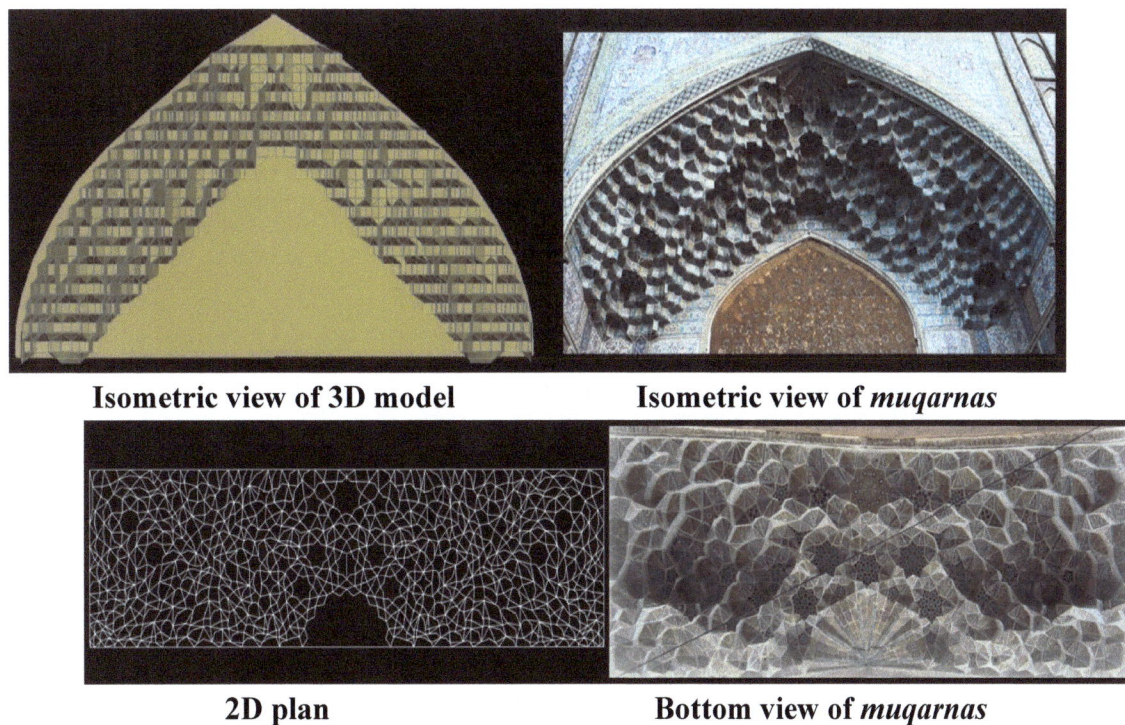

| Isometric view of 3D model | Isometric view of *muqarnas* |
| 2D plan | Bottom view of *muqarnas* |

Figure S. 45   View of *muqarnas* in Madrasa–i Khan.

A.23. Agha Noor Mosque, 1639 CE

Agha Nour mosque was founded during the closing years of the reign of the Shah Abbas I era and completed during the Shah Safi era in 1629/30 C.E. (1039 A.H.), therefore the names of the both kings were mentioned in the inscription above the portal of the mosque. This mosque was built under the supervision of Noureddin Mohammad Esfahani, who was one of the richest men in Isfahan. The nave of this mosque is one of the most beautiful naves in Isfahan. This nave has stone pillars with marble arches, which provide light during the day (Yaghoubi & Beheshti, 2004).

Sample 46   Agha Noor Mosque building Information.

Name	Masjid Jame' Agha Noor
Variant Names	Agha Nour Mosque, Masjid Jame' Agha Noor
Location	Isfahan, Iran
Position	Under Dome – Corner
Date	1639 /1039 AH
Period	Safavid
Century	16th
Building Type	Religious
Building Usage	Mosque
Copyright	Hamidreza Kazempour

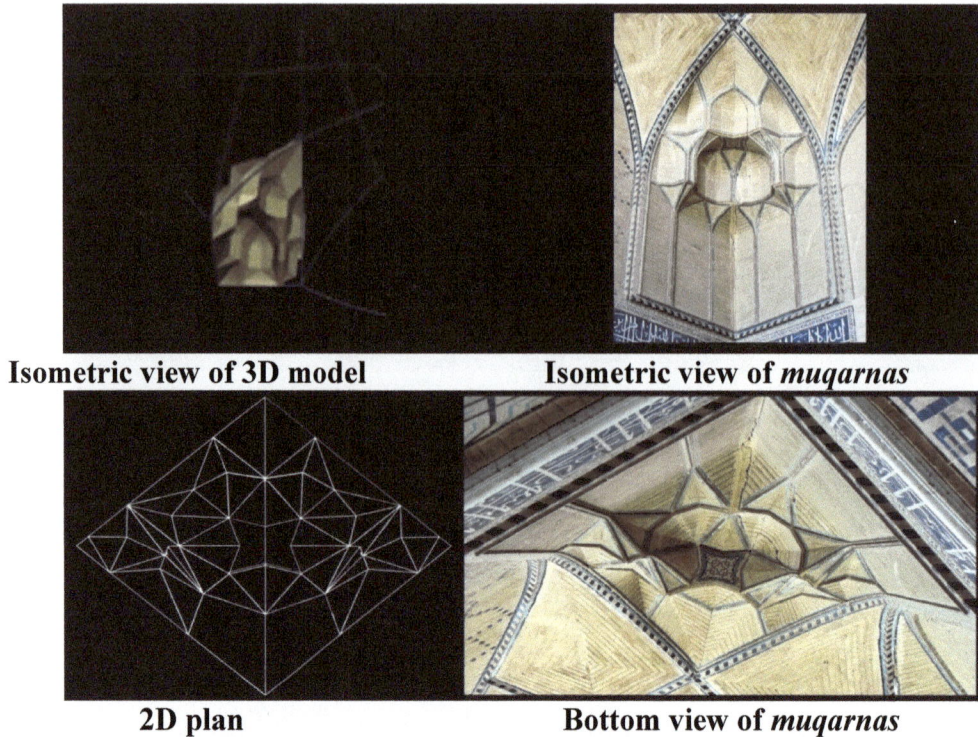

Isometric view of 3D model   **Isometric view of *muqarnas***

2D plan   **Bottom view of *muqarnas***

Figure S. 46 View of *muqarnas* in Agha Noor Mosque.

A.24. Hakim Mosque, 1656 CE

The Hakim Mosque is situated at the end of Painter and Dyer's Market (Bazaar–e–Rang Razan) part of Isfahan's Bazaar. It was built in mid *Jurjir (Jorjir) Portal from the 10th Century Mosque* seventeenth century, on the site of a tenth century mosque built by Ismail Ibn Abbad minister in the court of Fakhr–Al–Doleh or Moayed–Al–Doleh of the Dalaimite or Buyids dynasty, of which only the Jurjir (or Jorjir) Portal remains (discovered in 1955 hidden behind mud brick walls). The seventeenth century mosque was built by our ancestor Mohammad Davoud Hakim. Construction started in 1656/7 and finished in 1662/3. Mohammad Davoud was the king's doctor in the Safavid Courts of Shah Safi and Shah Abbas II. According to European travelers to Persia and India in the seventeenth century, Doctor Mohammad Davoud fell out of Shah's favour and went to India and was given the title Tagharrob Khan in the Court of the Emperor of India Aurengzib.

In between the above poems on a smaller plaque there are lines describing repairs in later years. One of which is written in between lines five and six of the poem in white on a turquoise background the following has been inscribed: "Repairs were carried out in the solar year 1323"; in between lines six and seven there is yet another inscription the following is written in Arabic One thousand and Eighty Five (1085 AH 1674/5 AD) and at the end of the poems on a small plaque attributing the above work to: "Master Ceramic Maker Mohammad 1085 AH" (Honarfar, 1995; Herdeg, 1990).

Sample 47   Hakim Mosque building information, East Entrance.

Name	Masjid Hakim
Variant Names	Masjed–e Hakim, Masjid–i–Hakim, Hakim Mosque, Mosque of al–Hakim
Location	Isfahan, Iran
Position	East Entrance
Date	1656–62/1067–73 AH
Period	Safavid
Century	17th
Building Type	Religious
Building Usage	Mosque
Copyright	Hamidreza Kazempour

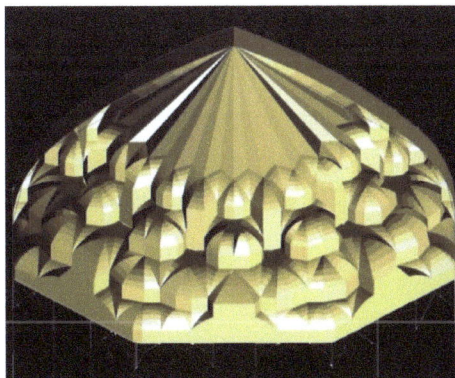

Isometric view of 3D model   **Isometric view of *muqarnas***

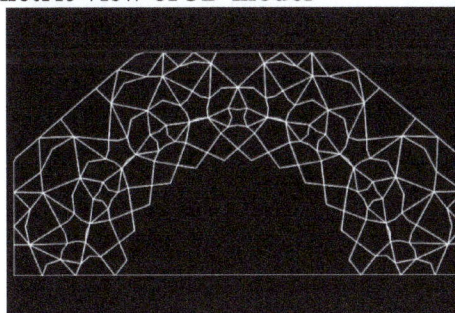

2D plan   **Bottom view of *muqarnas***

Figure S. 47   View of *Muqarnas* in Hakim Mosque, East Entrance.

Sample 48   Hakim Mosque building information, East Mihrab.

Masjid Hakim	**Position:** Eastern Mihrab	**Date:** 1656

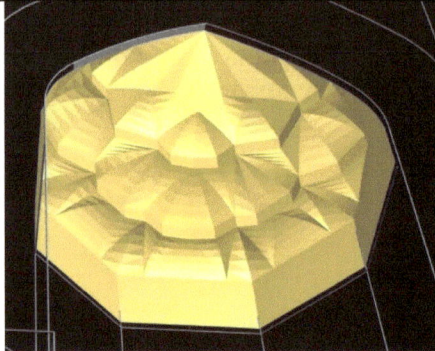

Isometric view of 3D model   **Isometric view of *muqarnas***

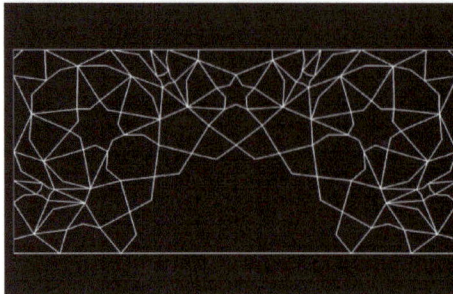

2D plan   **Bottom view of *muqarnas***

Figure S. 48   View of *Muqarnas* in Hakim Mosque, East Mihrab.

Sample 49   Hakim Mosque building information, Main Mihrab.

Masjid Hakim	**Position:** Main Mihrab	**Date:** 1656

Isometric view of 3D model   **Isometric view of *muqarnas***

2D plan   **Bottom view of *muqarnas***

Figure S. 49   View of *Muqarnas* in Hakim Mosque, Main Mihrab.

Sample 50   Hakim Mosque building information, North Entrance.

Masjid Hakim	**Position:** Entrance	**Date:** 1656

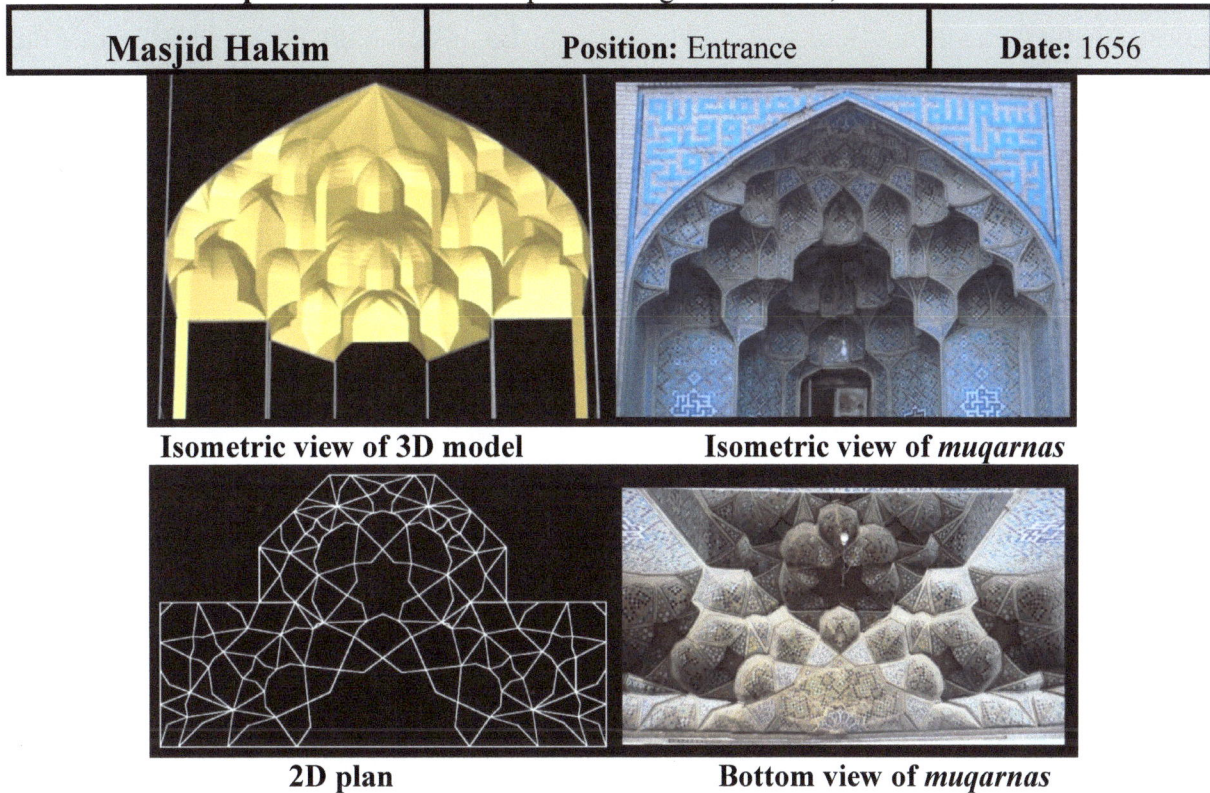

Isometric view of 3D model   **Isometric view of *muqarnas***

2D plan   **Bottom view of *muqarnas***

Figure S. 50   View of *Muqarnas* in Hakim Mosque, North Entrance.

Sample 51   Hakim Mosque building information, West Mihrab.

Masjid Hakim	**Position:** Western Mihrab	**Date:** 1656

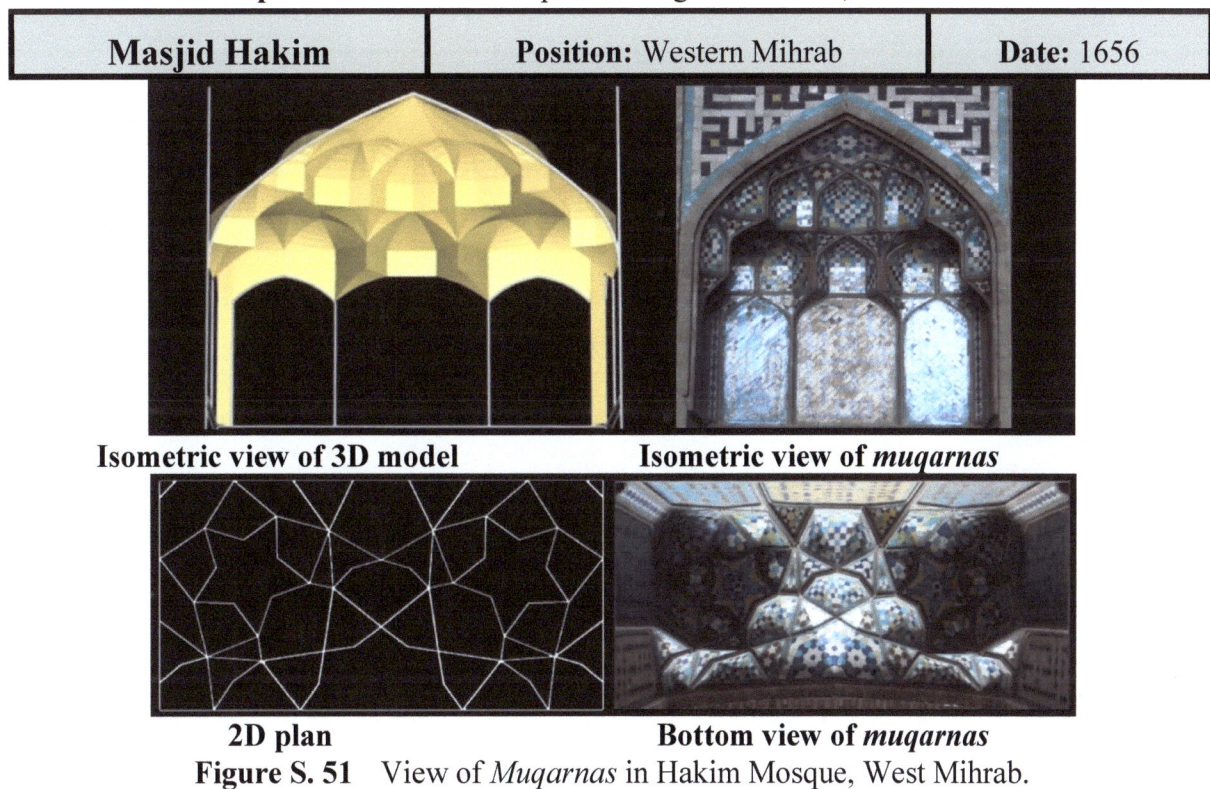

Isometric view of 3D model   **Isometric view of *muqarnas***

2D plan   **Bottom view of *muqarnas***

Figure S. 51   View of *Muqarnas* in Hakim Mosque, West Mihrab.

A.25. Hasht Behesht Pavilion, 1670 CE

Located in the center of the Garden of Nightingales (the Bagh–e Bulbul), the Hasht Behesht is one of Isfahan's two surviving Safavid pavilions. Built under Shah Sulaiman some twenty years after the Chihil Sutun, it is quite different in style from the earlier pavilion, although it exhibits the same concern for the interplay of interior and exterior spaces. "Hasht Behesht" translates as "Eight Paradises" and refers to a Timurid palace building type consisting of two stories of four corner rooms around a central domed space. In Isfahan, the corner rooms are octagonal, forming massive pillars that define four large openings leading to large porches in the south, east and west, and an iwan in the north.

The vault of the central space is detailed with polychrome *muqarnas* and is capped with a lantern. Nineteenth century engravings reveal that the interior was once covered in tiles and wall paintings that have since been removed. Some of the original mirror mosaic remains on the vault. Along with the Ali Qapu and Chihil Sutun, the Hasht Behesht was restored by IsMEO – Istituto Italiano per il Medio ed Estremo Oriente for NOCHMI – National Organization for Conservation of Historic Monuments of Iran. The project, completed in 1977, received an Aga Khan Award for Architecture in 1980 (Babaie, 2008; Pope, 1965; Pope, 1977b; Ferrante, 1968; Blair & Bloom, 1994).

Sample 52   Hasht Behesht Palace building information, Vault.

Name	Hasht Behesht
Variant Names	Hasht Bihisht, Hasht Bihisht Pavilion, Hasht Behesht Palace, Eight Paradises, Hasht Bihisht Palace, Hasht Behisht
Location	Isfahan, Iran
Position	Vault
Date	1670/1080 AH
Period	Safavid
Century	17th
Building Type	Palatial
Building Usage	Garden pavilion
Copyright	Hamidreza Kazempour

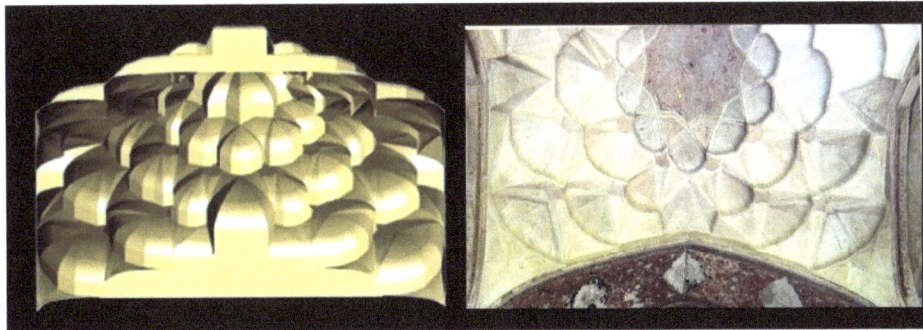

Isometric view of 3D model **Isometric view of *muqarnas***

2D plan **Bottom view of *muqarnas***

Figure S. 52 View of *muqarnas* in Hasht Behesht Palace, Vault.

Sample 53 Hasht Behesht Palace building information, Pavilion.

Hasht Behesht	**Position:** Eastern Pavilion	**Date:** 1670

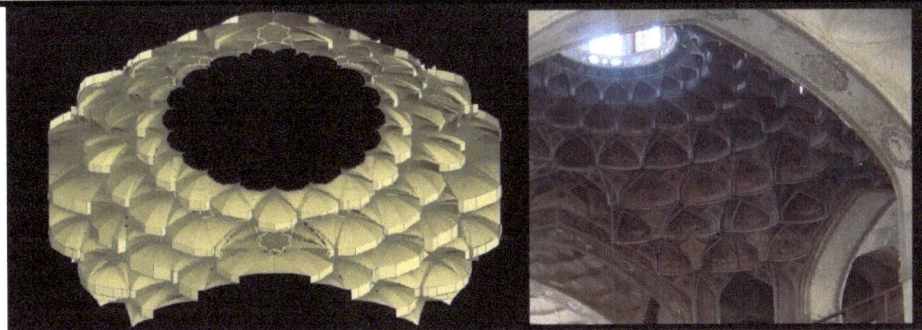

Isometric view of 3D model **Isometric view of *muqarnas***

2D plan **Bottom view of *muqarnas***

Figure S. 53 View of *muqarnas* in Hasht Behesht Palace, Pavilion.

A.26. Mulla Abd Allah School, 1678 CE

Molla Abdollah School is located in the bazaar of Isfahan, Iran, and was a premise for prayer and theological studies by reputed clergy, including Mullah Abdollah Shooshtari (one of the Safavid scientists), during the reign of Shah Abbas I of Persia. There isn't any specific date for this building, but based on its famous tiles, it can be attributed to Shah Abbas I.

The central school space is around a rectangle–shaped area and its main dimensions are 26 × 5.28 meters and its subsidiary sides are 3.80 meters. The school is built based on the two Iwan methods. On the wall of the south Iwan, there is an inscription of rock–themed about its dedication with the size of 35.1 × 20.1 meters, and the date of 1088 AH during the reign of King Soleiman is engraved on it.

Decoration on the southwest Iwan is executed by semi–dome using coffering and it has a Mihrab with *muqarnas* ornaments, gypsum plaster throughout the tiles and the dated and undated inscriptions. There is an inscription of tile with white thuluth and blue background around the centre part of Iwan which has some religious statement on it. Based on the two black and yellow narrow margins which have been continued over the inscription, it appears that the inscription has been added to Iwan during Qajar era.

A big Prayer room is located on the northwest of the school which lots of decorations is used for that and most of these decorations are focused on the centre. An inscription is also installed in the school entrance which is indicative of the actions of the Qajar era. The most important decorations of this inscription are its tiles that are made of seven–color traditional tile with plant motifs and arabesques and the blue and green background. The inscription text is a thuluth with engraved white color on a blue background. Some parts of the school have been repaired during Nader Shah Afshar, in the year 1158 AH (Godard, 1965).

Sample 54   Mulla Abdullah School building information.

Name	**Madrasa Mullah Abdullah**
Variant Names	Mullah Abdullah School, Molla Abdullah, Mulla Abd Allah
Location	Isfahan, Iran
Position	Entrance
Date	1678/1088 AH
Period	Safavid
Century	16th
Building Type	Educational, religious
Building Usage	Madrasa, Mosque
Copyright	HamidrezaKazempour

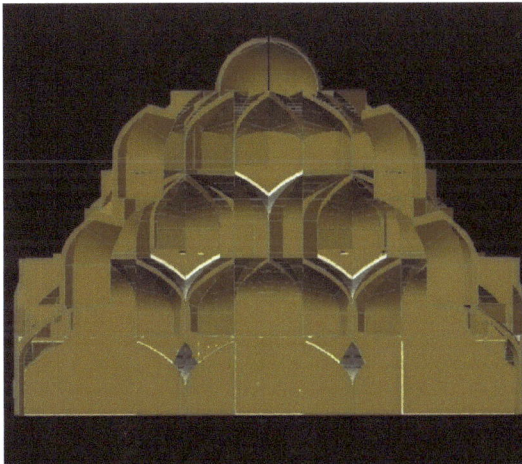

Isometric view of 3D model   **Isometric view of *muqarnas***

2D plan   **Bottom view of *muqarnas***

Figure S. 54   View of *muqarnas* in Mulla Abdullah School.

A.27. Nimavard School, 1691 CE

Nimavar School is a historical school in Isfahan, Iran. It's located in Nimavar Bazar and belongs to Safavid era. This school was built in 1691. Its uninspiring exterior, shown above, conceals one of the best examples of a "madrasa", or theological school, outside the Chahar Bagh Theological College.

Within there is a magnificent example of a traditional Persian garden planted within the traditional four–iwan courtyard. In the centre is a pool round which are ranged four sunken beds planted with trees. This design was called "chahar bagh" or "four gardens" and symbolised the Garden of Eden which was divided in this way. The traditional name was also adopted by Shah Abbas I for the name of the main street of Isfahan.

The Eastern Iwan, in front of you as you enter, is simple but elegant, the Northern one, on the left as you enter the courtyard, has a swastika–like pattern which I believe to be made up of the letters forming the name of God, 'Allah'. Others have suggested that this pattern also represents the circulatory motion around the 'Kabah'. A similar design is also found on one of the tile–work panels inside the Southern iwan of the Friday Mosque (Yaghoubi & Beheshti, 2004; Sultanzadeh, 1985).

Sample 55   Nimavard School building information, Entrance.

Name	Nimavar School
Variant Names	Madrasa Nimavar, Masjid Nimavar
Location	Isfahan, Iran
Position	Entrance
Date	1691/ 1117 AH
Period	Safavid
Century	17th
Building Type	Educational
Building Usage	Madrasa
Copyright	Hamidreza Kazempour

Isometric view of 3D model   **Isometric view of *muqarnas***

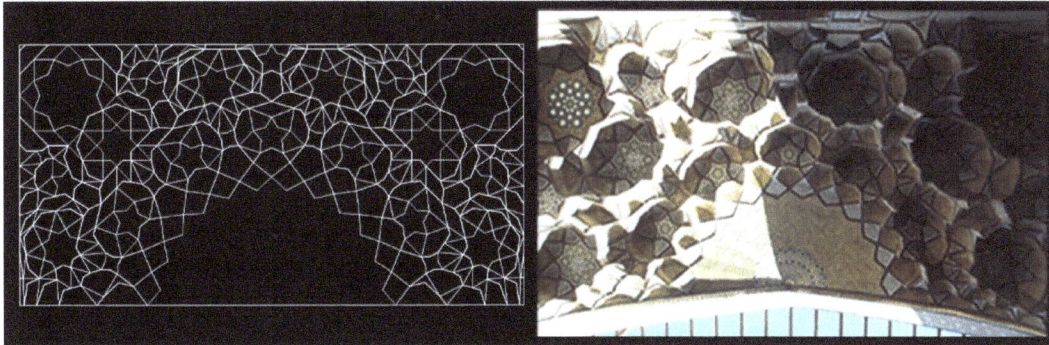

2D plan   **Bottom view of *muqarnas***

Figure S. 55   View of *muqarnas* in Nimavard School, Entrance.

Sample 56   Nimavard School building information, Northern Iwan.

Nimavar School	**Position:** North Iwan	**Date:** 1691

Isometric view of 3D model   **Isometric view of *muqarnas***

2D plan   **Bottom view of *muqarnas***

Figure S. 56   View of *muqarnas* in Nimavard School, Northern Iwan.

Sample 57   Nimavard School building information, Western Iwan.

Nimavar School	**Position:** Western Iwan	**Date:** 1691

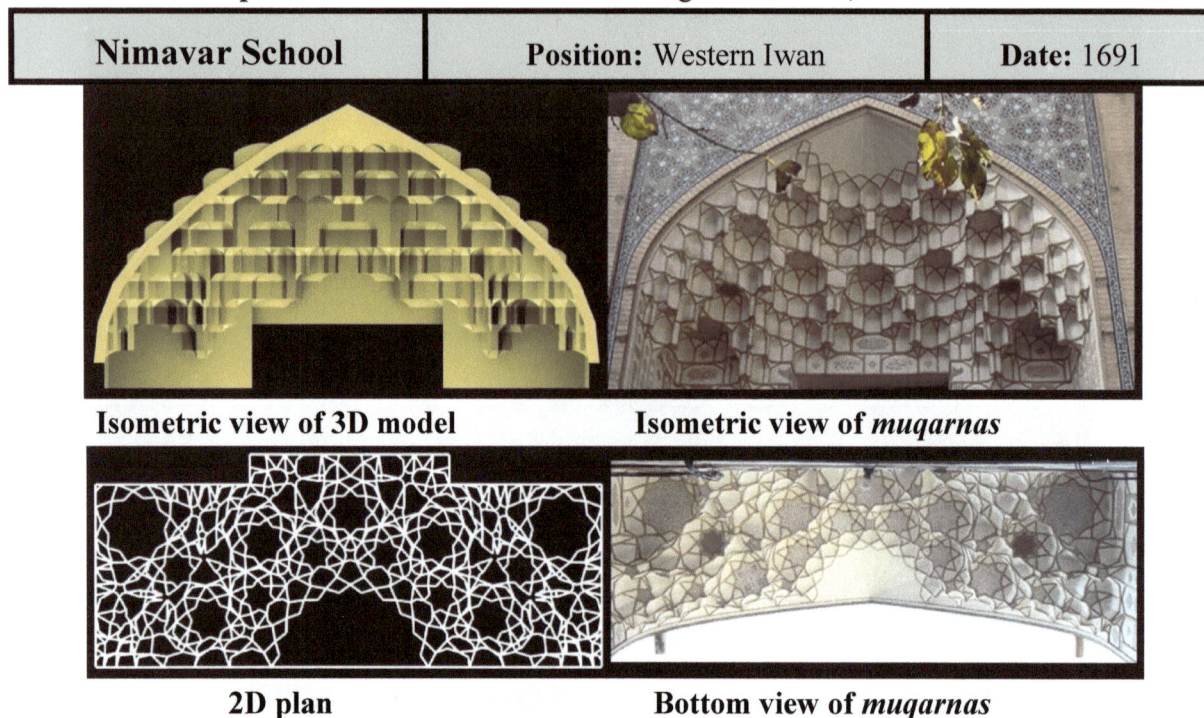

Isometric view of 3D model   **Isometric view of *muqarnas***

2D plan   **Bottom view of *muqarnas***

Figure S. 57   View of *muqarnas* in Nimavard School, Western Iwan.

A.28. Chahar Bagh School, 1706 CE

Better known as a part of the Sultani or Chahar Bagh Madrasa or the Madrasa Madar–i Shah, the Caravanserai and Bazaar of Shah Husain I is an integral constituent of one of the largest urban complexes in medieval Iran. A sophisticated architectural arrangement of diverse institutional functions, this complex consists of a madrasa, a mosque, a caravanserai, and a bazaar.

Construction of the complex began in 1704 under the patronage of Shah Hussein I, the last Safavid emperor, and continued until the end of the eighteenth century. The complex is attributed to the influence of the then–Mudarris (head religious scholar), Mir Muhammad Baqir Khatunabadi, on the reigning Shah Hussain I. The relationship between this scholar and ruler has been likened to that between the scholar Sheikh Lutfallah and the Safavid ruler Shah Abbas, which gave rise to Isfahan's famed Sheikh Lutfallah mosque. The madrasa (theological college), the pivotal building at the southern end of the institutional ensemble, is built around an idyllic central courtyard within which the ancient Farshadi Canal is reinterpreted as a central waterway lined with cypress and fir trees.

The early Safavid tradition of decorative color and tile work can be traced through the yellow and blue tiles in the madrasa's main portal iwan and the enameled tile work on its flanking minarets. The southern dome adorned with intertwining black, yellow and white arabesques on a turquoise blue base also has the traditional alternate bands of lapis lazuli blue and calligraphic inscriptions on its drum. The previous emphasis on floral patterns is reduced in favor of a simpler contrast between yellow brickwork and inset tile mosaic panels. Greatly restored in recent decades, the caravanserai and bazaar of Shah Husain I still provide calm oases in the center of Isfahan (Blake, 1999; Blunt, 2009; Hillenbrand, 2004; Hutt & Harrow, 1978).

Sample 58   Chahar Bagh School building information.

Name	Madrasa Chahar Bagh
Variant Names	Madrese–ye Madar–e Shah, Madrasa–i Madar–i Shah, Madar–i Shah Madrasa, Madrasah–i Madar–i Shah, Madrasa Mader–i Shah, Madar–e Shah Madrasa, College of the Mother of the Shah, Madrasah–i va Masjid–i Chahar Bagh, Chahar Bagh Mosque and Madrasa, Shah Husain I Madrasa and Caravanserai, Ribat–i va Bazar–i Shah Husain I, Ribat va Bazar Shah Husain I, Shah Husayn I Madrasa, Caravanserai and Bazaar of Shah Husain I.
Location	Isfahan, Iran
Position	Entrance
Date	1706–1714/1117–25 AH
Period	Safavid
Century	18th
Building Type	Commercial, educational, landscape
Building Usage	Caravanserai, madrasa, garden
Copyright	Hamidreza Kazempour

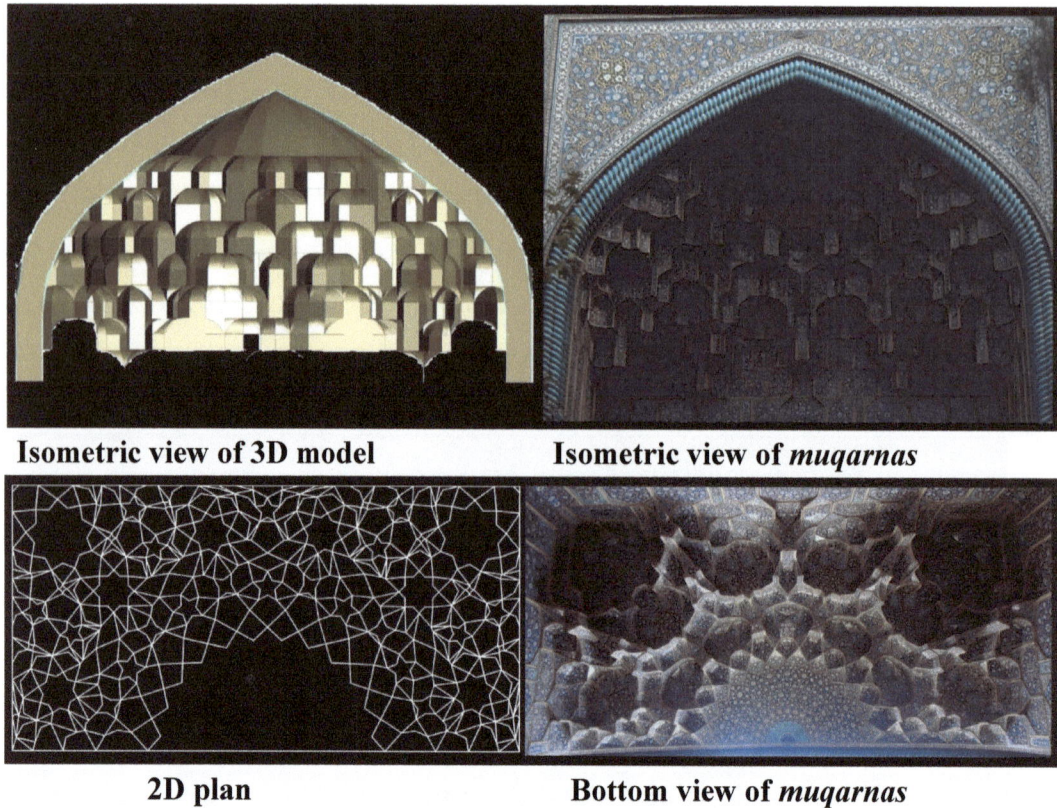

Isometric view of 3D model   **Isometric view of *muqarnas***

2D plan   **Bottom view of *muqarnas***

Figure S. 58   View of *muqarnas* in Chahar Bagh School.

A.29. Ali Gholi Agha Mosque, 1710 CE

Ali Gholi Agha, one of the high ranking court officials at the last Safavid king Shah Sultan Hussein, built a bath house, bazaar and mosque. The mosque and its portal decorated with beautiful geometrical mosaic tile work were built in 1710. The small courtyard also has geometrical tile work but is quite different from the other Safavid monument. The bazaar with a junction under a 12–meter in diameter dome is opposite this mosque.

The Mosque contains a winter prayer gallery, or nave, which has been heavily re–decorated, even raising the level of the old floor. There is a small mihrab on the southern side of the courtyard which contains some unusually dark red coloured tiles and has a dome above it whose decoration appears to derive from the dome at the entrance to the Chahar Bagh Theological College. Looking back to the entrance you can see the upper storey of the mosque, which is given over to quarters for some 30 theological students, however there is no longer any teaching within the mosque (Jenab, 2007).

Sample 59   Ali Gholi Agha Mosque building information.

Name	Masjid Ali gholi Agha
Variant Names	Masjid Imam, Masjid–e Shah, Masjid–e Sultani, Imam Mosque, Imam Khomeini Mosque
Location	Isfahan, Iran
Position	Entrance
Date	1710 / 1122 AH
Period	Safavid
Century	17th
Building Type	Religious
Building Usage	Mosque
Copyright	Hamidreza Kazempour

Isometric view of 3D model   **Isometric view of *muqarnas***

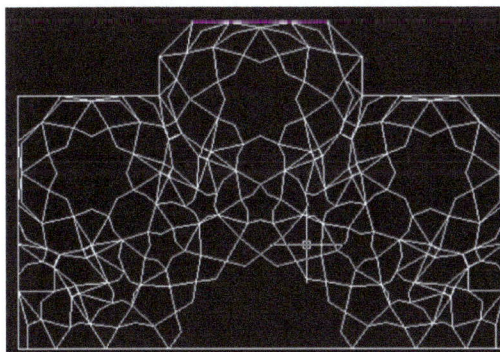

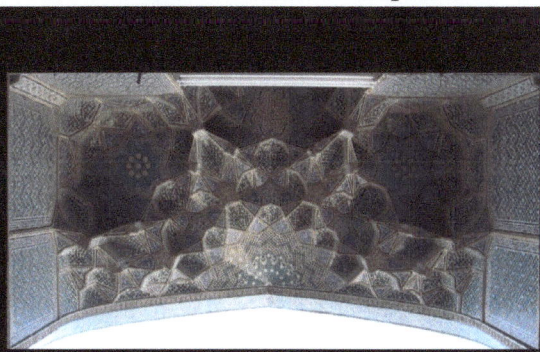

2D plan   **Bottom view of *muqarnas***

Figure S. 59   View of *muqarnas* in Ali Gholi Agha Mosque.

A.30. Bagh Nazar Garden, 1749 CE

The Pars museum, where Zandieh Museum is located, was originally a pavilion in Bagh–i Nazar, a garden in Shiraz built under the Zand ruler Karim khan (reg. 1749–1779). The design of the garden followed the quadripartite or chahar bagh plan, with the pavilion situated at the center and four axial pools. In the past, the building was used for hosting foreign ambassadors or official ceremonies.

In 1936, when the northern part of the Nazar garden was destroyed to make way for a street, the pavilion was renovated to serve as a museum, i.e. Pars Museum. The exterior of the building is decorated with pretty tileworks, the patterns of which consist of flowers, plants, birds, hunting and the scenes of coronation of Soleiman (AS). The building has a central domed hall and four royal chambers located in the north, south, east and west directions. In the middle of the hall is a beautiful marble pond. The ceiling of the building is decorated with attractive embellishments.

The tomb of Karim Khan Zand is situated in the eastern royal chamber. Sections of this hall exhibit 11 paintings by famous painters of Zandieh era, the most famous of whom is Aqa Sadeq (Behehsti & Bidhendi, 2009).

Sample 60   Bagh Nazar building information, Pavilion 1.

Name	Bagh Nazar
Variant Names	Bagh Nazar, Pars Museum, Bagh–i Nazar, Bagh–e Nazar, Nazar Garden, Pars Pavilion
Location	Shiraz, Iran
Position	Pavilion 1
Date	1749–1779/1930
Period	Zand
Century	18th, 20th
Building Type	Landscape, public/cultural, public/cultural
Building Usage	Garden, pavilion, museum
Copyright	Hamidreza Kazempour

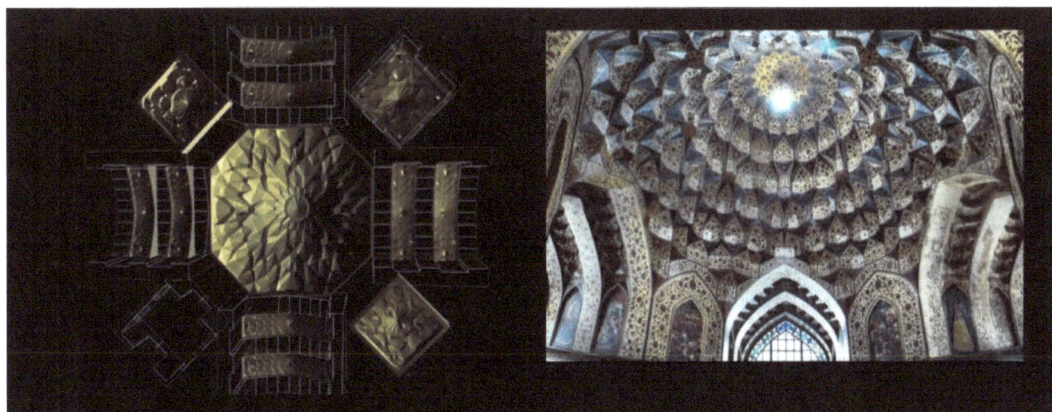

Isometric view of 3D model   **Isometric view of *muqarnas***

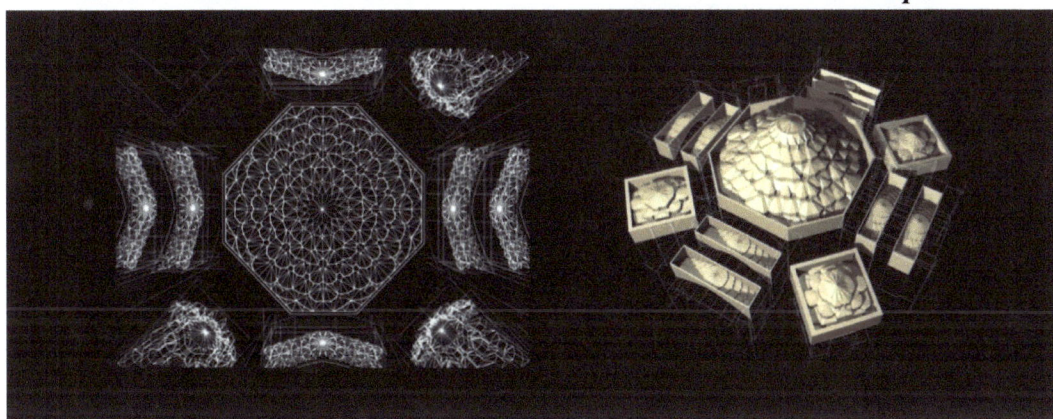

2D plan   **Bottom view of *muqarnas***

Figure S. 60   View of *muqarnas* in Bagh Nazar, Pavilion 1.

Sample 61   Bagh Nazar building information, Pavilion 2.

Bagh Nazar	**Position:** Pavilion 2	**Date:** 1749

Sample 62   Bagh Nazar building information, Pavilion 3.

Bagh Nazar	**Position:** Pavilion 3	**Date:** 1749

Sample 63   Bagh Nazar building information, Pavilion 4.

Bagh Nazar	**Position:** Pavilion 4	**Date:** 1749

Sample 64   Bagh Nazar building information, Pavilion 5.

Bagh Nazar	**Position:** Pavilion 5	**Date:** 1749

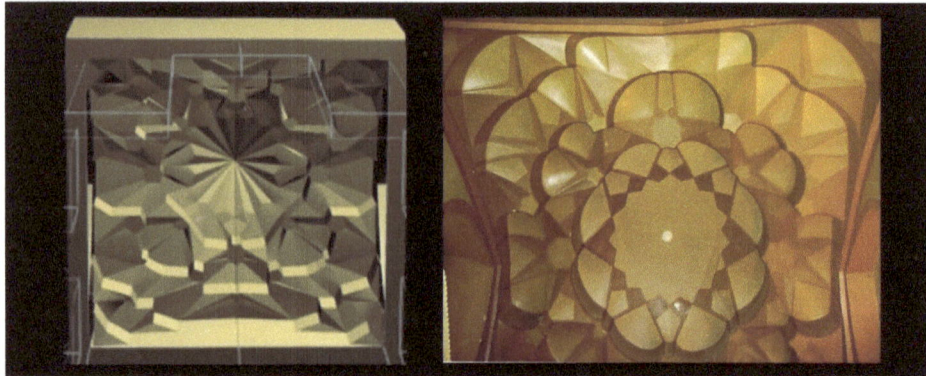

Isometric view of 3D model   **Isometric view of *muqarnas***

2D plan   **Bottom view of *muqarnas***

Figure S. 61 View of *muqarnas* in Bagh Nazar, Pavilion 2.

Isometric view of 3D model   **Isometric view of *muqarnas***

2D plan   **Bottom view of *muqarnas***

Figure S. 62 View of *muqarnas* in Bagh Nazar, Pavilion 3.

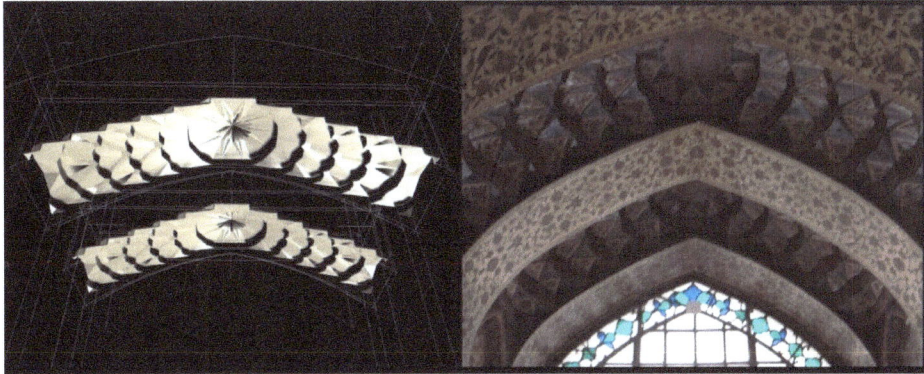

Isometric view of 3D model **Isometric view of *muqarnas***

2D plan **Bottom view of *muqarnas***

Figure S. 63 View of *muqarnas* in Bagh Nazar, Pavilion 4.

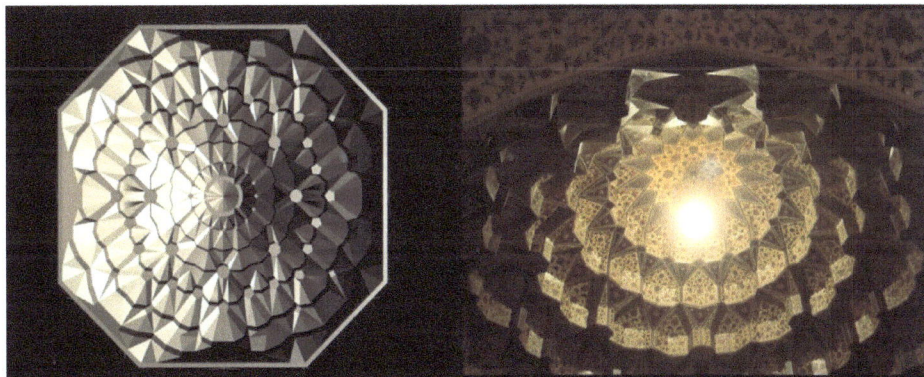

Isometric view of 3D model **Isometric view of *muqarnas***

2D plan **Bottom view of *muqarnas***

Figure S. 64 View of *muqarnas* in Bagh Nazar, Pavilion 5.

A.31. Nabi Mosque, 1787 CE

The al–Nabi Mosque Masjed–e Soltani is a famous mosque in Qazvin. The mosque has an area of about 14,000m², and bears inscriptions indicating that Fath Ali Shah of the Qajar dynasty was the founder of the mosque. Other sources however indicate that the mosque has been in existence since the Safavid period. It is now believed that the architect of the structure was Ustad Mirza Shirazi with the date of construction being 1787. The mosque has four iwans in its courtyard. The portal contains an inscription in nastaliq calligraphy dated 1787 CE. Similar to Qazvin's Masjed e Jame, this mosque has a nave that is now used as a library (Bozorgmehri, 2002).

Sample 65 Nabi Mosque building information, Entrance Portal.

Name	Masjid Nabi Qazvin
Variant Names	Masjed–e Soltani, Masjid Nabi, Al–Nabi Mosque
Location	Qazvin, Iran
Position	Entrance
Date	1787 / 1131 AH
Period	Qajar
Century	16th
Building Type	Religious
Building Usage	Mosque
Copyright	Hamidreza Kazempour

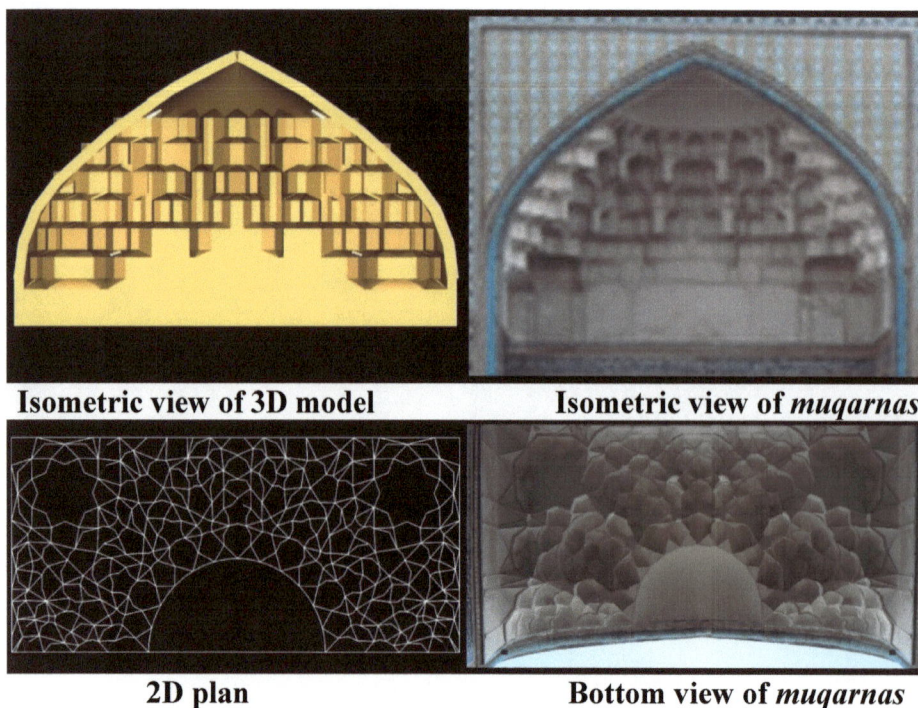

Isometric view of 3D model   **Isometric view of *muqarnas***

2D plan   **Bottom view of *muqarnas***

Figure S. 65 View of *muqarnas* in Nabi Mosque, Entrance.

313

Sample 66   Nabi Mosque building information, Mihrab I

Masjid Nabi Qazvin	Position: Mihrab 1	Date: 1787

Isometric view of 3D model   **Isometric view of *muqarnas***

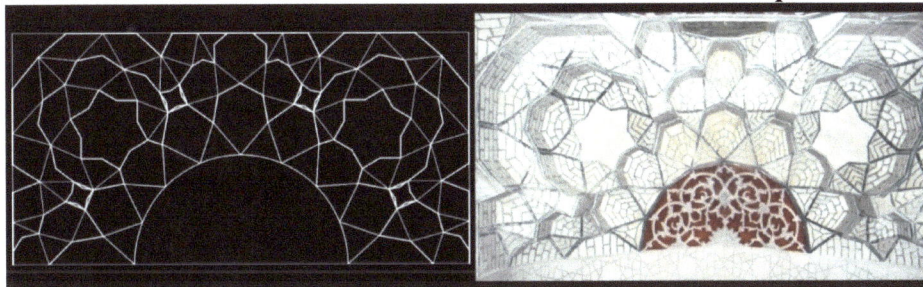

2D plan   **Bottom view of *muqarnas***

Figure S. 66   View of *muqarnas* in Nabi Mosque, Mihrab I.

Sample 67   Nabi Mosque building information, Mihrab II

Masjid Nabi Qazvin	Position: Mihrab 2	Date: 1787

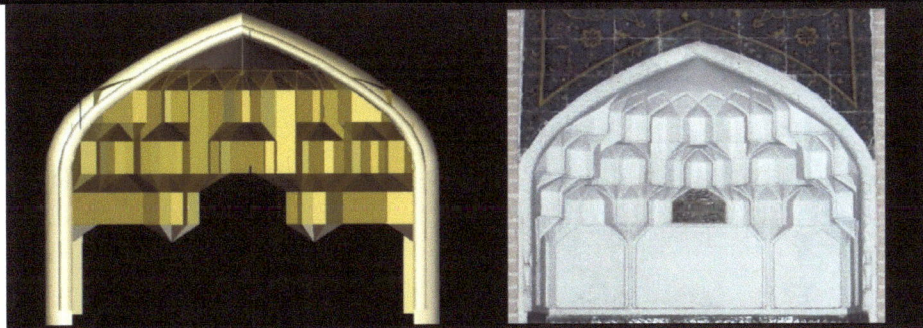

Isometric view of 3D model   **Isometric view of *muqarnas***

2D plan   **Bottom view of *muqarnas***

Figure S. 67   View of *muqarnas* in Nabi Mosque, Mihrab II.

A.32. Shiraz Jame' Mosque, 1793 CE

The Friday Mosque of Shiraz, also known as the Masjid–i Atiq, was first built in 875 during the reign of Safavid ruler Amr b. al–Layth (878–900). It was rebuilt, restored, and expanded various times thereafter. Most of the present day structure, i.e. a four–iwan courtyard mosque, dates back to the 18th century (Wilber, 1955). Damaged by numerous earthquakes, it was repaired and restored extensively after 1935. The center of its courtyard is occupied by the Khuda Khane (House of God). Commissioned by Inju'id ruler Mahmud Shah (1325–1336) in 1351 for the storage of Qur'ans, this small kiosk is also known as Bayt al–Mashaf (House of Qur'ans, or House of Books).

Khuda Khane consists of a rectangular core, with a loggia of three arched bays on each side, with solid circular towers projecting at the outer corners. Only the towers, the platform and ruined inner walls were remaining of the original structure in the early twentieth century. The rectangular core of the Khuda Khane contains a square hall with four central doorways. The four doorways on its exterior – three for the square hall and one leading into the vestibule – are flanked by engaged columns and topped with flat *muqarnas* crowns.

The twelve pointed arches of the veranda are carried on marble columns with bulbous bases and *muqarnas* capitals and square columns built into the corner towers. The arches of the east and west loggias are slightly wider; they are separated with rectangular panels resting on two pairs of columns at the center. A *muqarnas* cornice wraps around the loggia arcades and the corner towers below the flat roof.The Khuda Khane is adorned with a wide tile band below the *muqarnas* cornice containing a white thuluth inscription on a blue background with floral arabesques. The date of construction 1935 / 752 A.H. is seen on the southeast tower (Bruke et al., 2004; Byron, 2007; Golombek & Wilber, 1988; Loveday, 1993; O'Kane, 1996; Pope, 1934; Pourjavady & Booth-Clibborn, 2001; Wilber, 1972; Wilber, 1955).

Sample 68   Shiraz Jame' Mosque building information, Entrance Portal I.

Name	Masjid–i Atiq–i Shiraz
Variant Names	Friday Mosque of Shiraz, Masjid–i 'Atiq–i, Masjid–e Jame, Congregational Mosque, Bayt al–Mushaf, Masjid–i Jami', Masjid–i Atik, Jame' Atigh Mosque, Masjed–e Atiq, Jame' Atigh, Jameh–ye Atigh, Old Mosque of Shiraz, Khuda Khaneh, Huda Khana, Huda Khane, Bayt al–Mashaf, Beit al–Moshaf, Mushafkhana, Masjid–i Jami': Khuda Khaneh
Location	Shiraz, Iran
Position	Entrance
Date	1793/1193 AH – 1351/751–52 AH – 894/280–81 AH
Period	Inju'id, Safavid, Timurid
Century	9th, 14th
Building Type	Religious
Building Usage	Mosque
Copyright	Hamidreza Kazempour

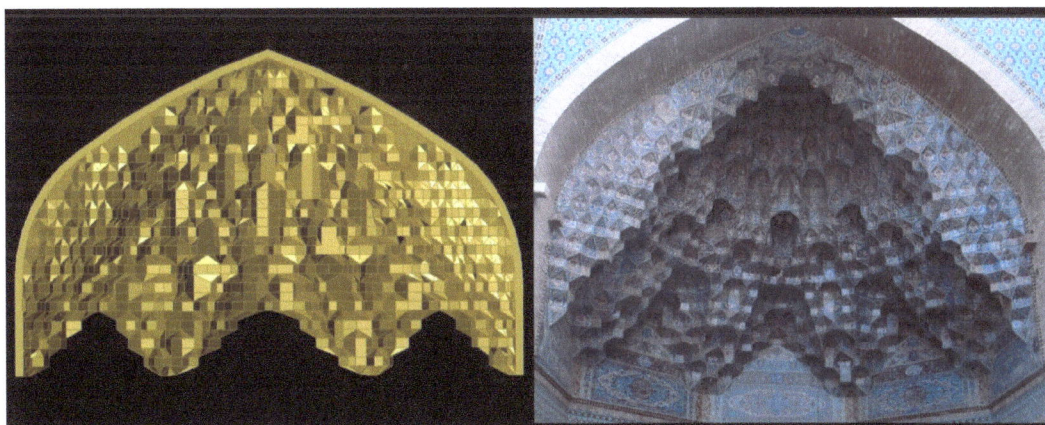

Isometric view of 3D model   **Isometric view of _muqarnas_**

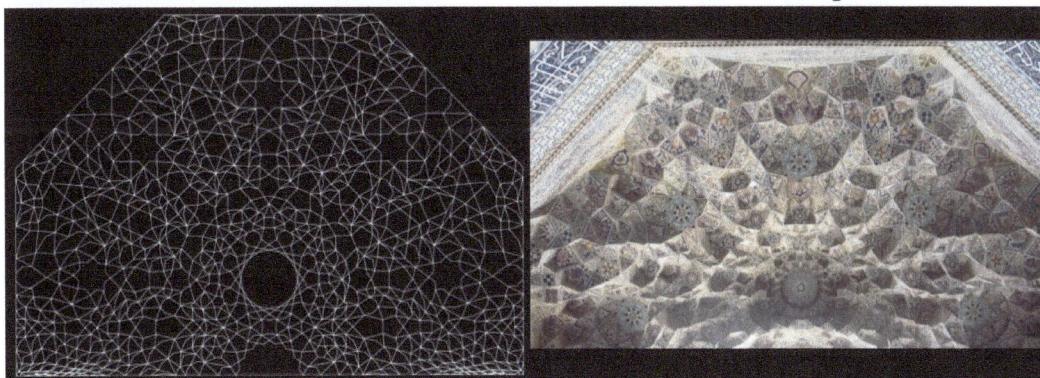

2D plan   **Bottom view of _muqarnas_**

Figure S. 68   View of _muqarnas_ in Shiraz Jame' Mosque, Entrance I.

Sample 69   Shiraz Jame' Mosque building information, Entrance Portal II.

Name	Masjid–i Jami'–i Shiraz
Variant Names	Friday Mosque of Shiraz, Masjid–i 'Atiq–i, Masjid–e Jame, Congregational Mosque, Bayt al–Mushaf, Masjid–i Jami', Masjid–i Atik, Jame' Atigh Mosque, Masjed–e Atiq, Jame' Atigh, Jameh–ye Atigh, Old Mosque of Shiraz, Khuda Khaneh, Huda Khana, Huda Khane, Bayt al–Mashaf, Beit al–Moshaf, Mushafkhana, Masjid–i Jami': Khuda Khaneh
Location	Shiraz, Iran
Position	Entrance Tagh Morvarid
Date	1793/1193 AH – 1351/751–52 AH – 894/280–81 AH
Period	Inju'id, Safavid, Timurid
Century	9th, 14th
Building Type	Religious
Building Usage	Mosque
Copyright	Hamidreza Kazempour

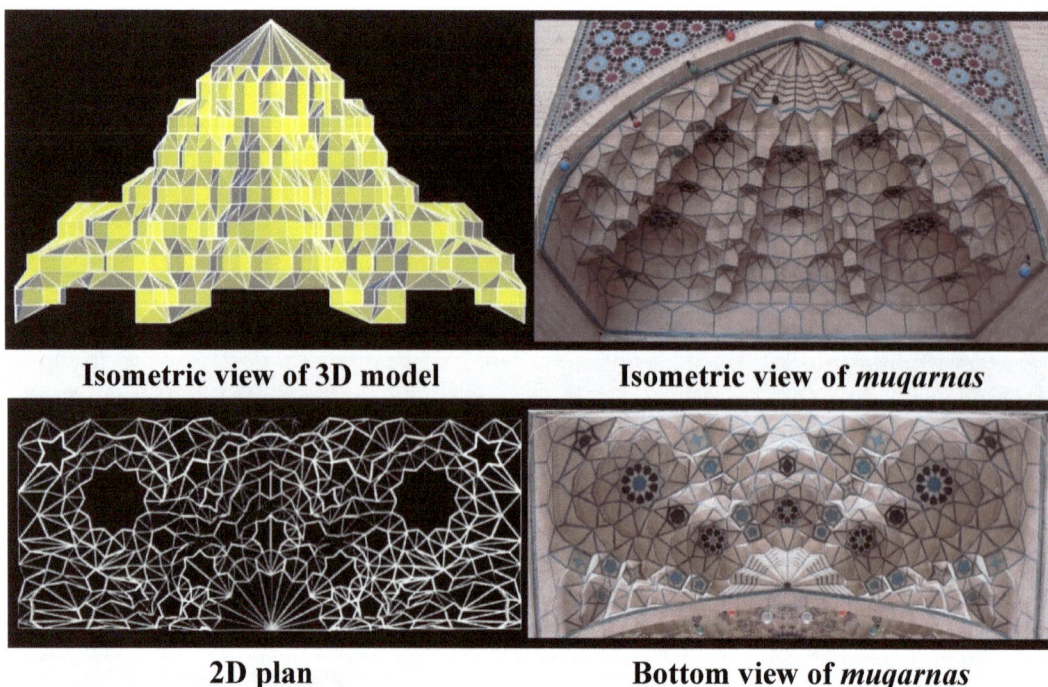

Isometric view of 3D model   Isometric view of *muqarnas*

2D plan   Bottom view of *muqarnas*

Figure S. 69   View of *muqarnas* in Shiraz Jame' Mosque, Entrance II.

A.33. Sepahsalar Mosque, 1804 CE

The Sepahsalar or Motahari Mosque and School complex has one of the largest mosques in Tehran. With its distinctive ten minarets, it is an important Qajar era landmark. The Mosque was built through the largess of Nasser Al–din Shah's Grand Minister for Foreign Affairs and Commander–in–Chief (Sepahsalar) Mirza Hossein Khan. The mosque, a school that has the same name in Tehran is located in the National Assembly (Baharestan Square). Sepahsalar mosque and school are of the first and largest ones in Tehran that are Iran's closest integration with the architecture of mosques in Istanbul. Sepahsalar has a special dome and 10 minarets of common size that resembles the dome–style architecture of the mosques in Turkey.

Because of structural and nonstructural decorating scheme because Yazdi and *Muqarnas* and bowl at the mosque used to tile work and architecture of Qajar art is not only one of the most beautiful mosques, but also a school. Important types of tile work in the mosque, seven color mosaic or brick and can be named as the latter two arabesques and Chinese knots and prominent mosaic over the entrance is unique in its kind. Stone and stone decorations used in this series and move to other buildings of the Qajar era architecture is different and is similar Zandieh (Teimouri, 2010).

Sample 70   Sepahsalar Mosque building information, Iwan I.

Name	Masjid Sepahsalar
Variant Names	Masjid Motahari, Sepahsalar Mosque
Location	Tehran, Iran
Position	Western Iwan
Date	1804 /1215 AH
Period	Qajar
Century	18th
Building Type	Educational, religious
Building Usage	Madrasa, Mosque
Copyright	Hamidreza Kazempour

Sample 71   Sepahsalar Mosque building information, Iwan II.

Masjid Sepahsalar	Position: Iwan 2	Date: 1804

Isometric view of 3D model   **Isometric view of *muqarnas***

2D plan   **Bottom view of *muqarnas***

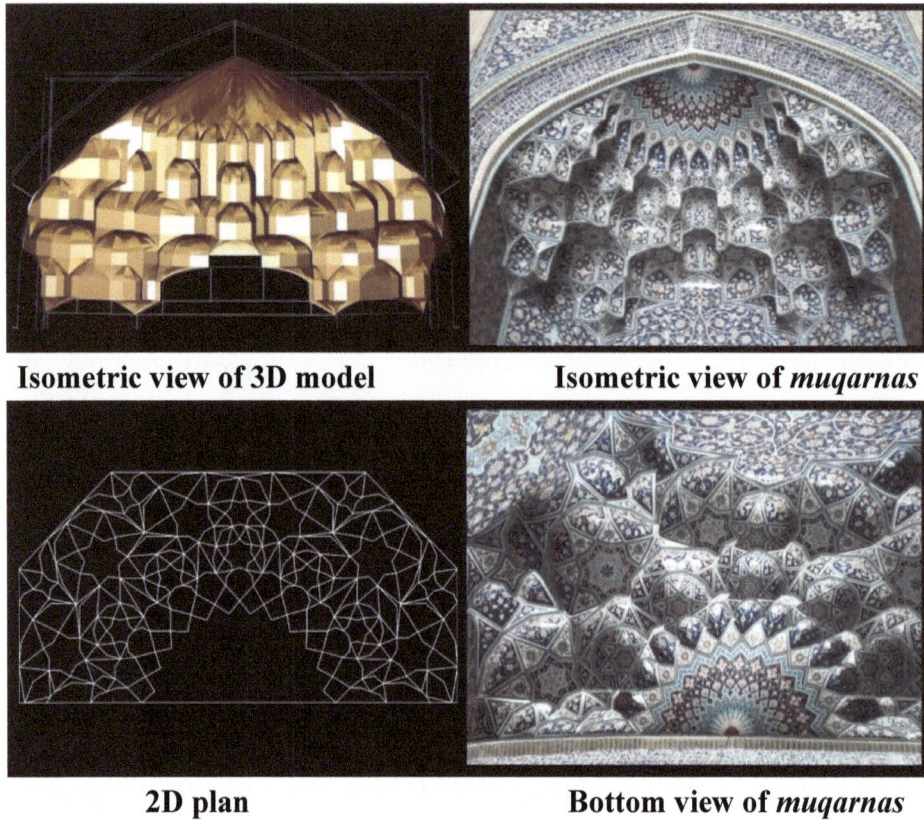

Figure S. 70   View of *muqarnas* in Sepahsalar Mosque, Iwan I.

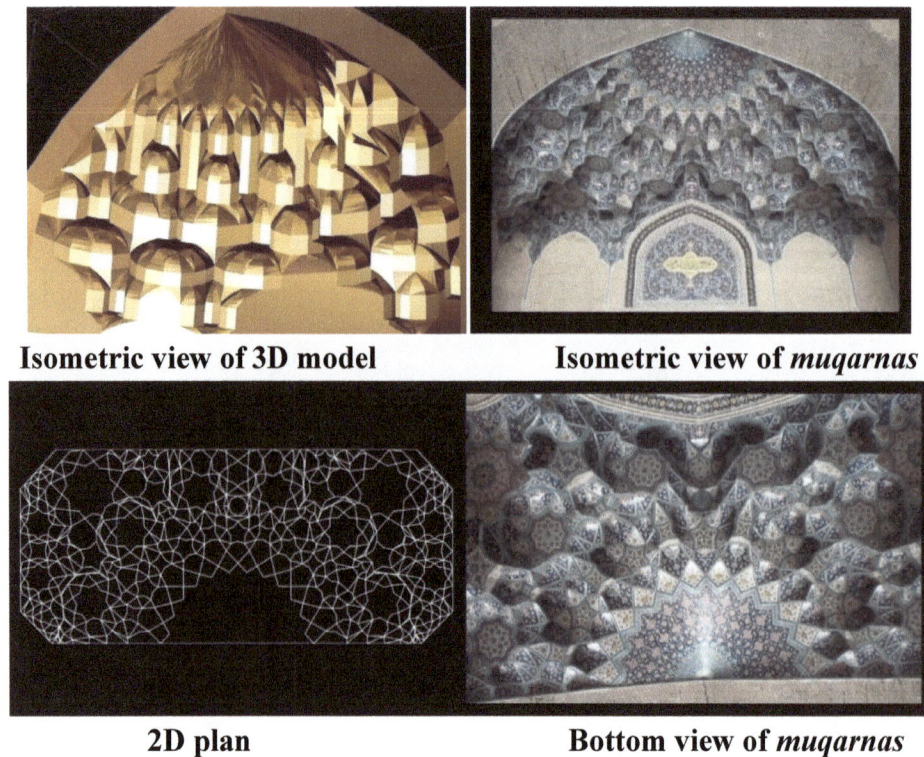

Isometric view of 3D model   **Isometric view of *muqarnas***

2D plan   **Bottom view of *muqarnas***

Figure S. 71   View of *muqarnas* in Sepahsalar Mosque, Iwan II.

A.34. Imam Mosque of Tehran, 1810 CE

This mosque is located at the end of Naser Khosrow avenue near the Tehran Bazaar. The Shah or Sultani Mosque dates back to the reign of Fathali Shah Qajar and during the reign of Naseredin Shah the two current minarets were added to the structure. This mosque proves to be a great piece of architecture and beauty due its vast courtyard, beautiful Naves (or nocturnal areas), the giant tile worked dome and interesting arches and entrances.

The Imam Khomeini Mosque is right inside the Tehran bazaar and is very much a working mosque and one of the largest and busiest in Tehran. The building itself dates from the early 18th century. The courtyard is accessed from several parts of the bazaar and hundreds of people pass through here, so it's usually possible for non–Muslims to stand and watch the faithful performing their ablutions and praying (Shahidi Mazandarani, 2004; Mostafavi, 1996).

Sample 72   Imam Mosque Tehran building information.

Name	Masjid Imam Tehran
Variant Names	Masjid Imam, Masjid–e Shah, Masjid–e Sultani, Imam Mosque, Imam Khomeini Mosque
Location	Tehran, Iran
Position	Entrance
Date	1810–1825 / 1225–1240 AH
Period	Qajar
Century	18th
Building Type	Religious
Building Usage	Mosque
Copyright	Hamidreza Kazempour

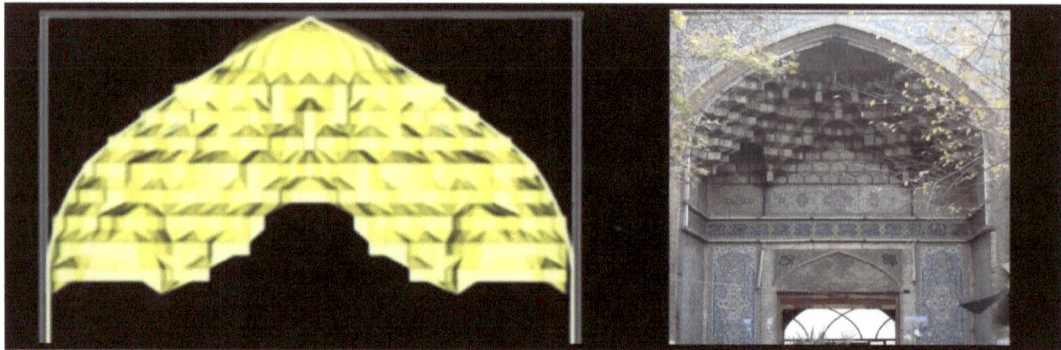

Isometric view of 3D model

Isometric view of *muqarnas*

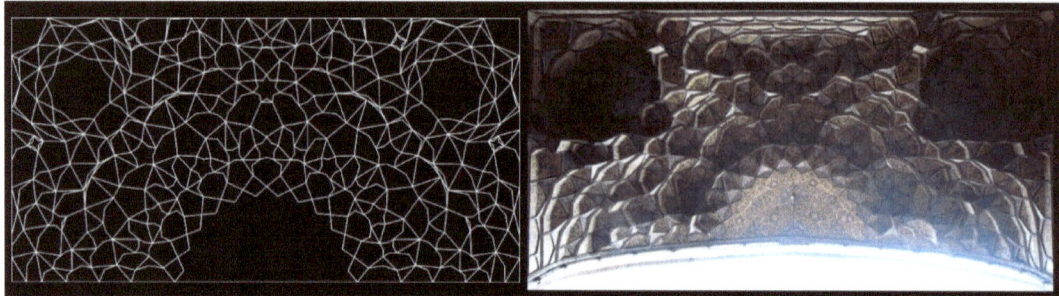

2D plan

Bottom view of *muqarnas*

Figure S. 72   View of *muqarnas* in Imam Mosque Tehran.

A.35. Seyyed Mosque, 1815 CE

Seyyed mosque is the biggest and the most famous mosque from the Qajar era in Isfahan. It was founded by Seyyed Mohammad BagherShafti, one of the most famous clergymen in Isfahan. It was founded in the middle of the 19th century, but its tiling lasted until the end of this century.

It is the largest mosque which was built during Qajar time in Isfahan. This beautiful monument was built at the order of Hojat al–Islam Seyed Mohannad Bagher Shafti. Donations of faithful Moslems and religious payment afforded all necessary expenses of the Seyed Mosque.

The mosque is so large which was not completed at the time of Shafti. After his death, his son and grandson continued the work and completed the mosque.

The Seyed Mosque has a four Iwan plan too. There are entrances on the north and east, but the northern portal is the most magnificent entrance, which opens to the Seyed Street.

Some parts of this mosque are similar to the Imam Mosque of the Naghsh–e Jahan Square. There is a pool in the center of the courtyard, and arcades which has been built between four Iwans. The most important part of the mosque is the south Iwan, and the southern dome of the sanctuary. The main Mihrab is located in the same sanctuary. Tiles and tile mosaics show bright colors of Qajar period. Plaster work, stucco moldings and paintings are other types of ornamentation used in the mosque (Yaghoubi & Beheshti, 2004; Haji Ghasemi, 1996).

Sample 73   Seyyed Mosque building information, Entrance I.

Name	Seyyed Mosque
Variant Names	Seyyed Mosque
Location	Isfahan, Iran
Position	Bazaar Entrance
Date	1815/1215 AH
Period	Qajar
Century	19th
Building Type	Religious
Building Usage	Mosque
Copyright	HamidrezaKazempour

Sample 74   Seyyed Mosque building information, Mihrab.

Seyyed Mosque	Position: Mihrab	Date: 1815

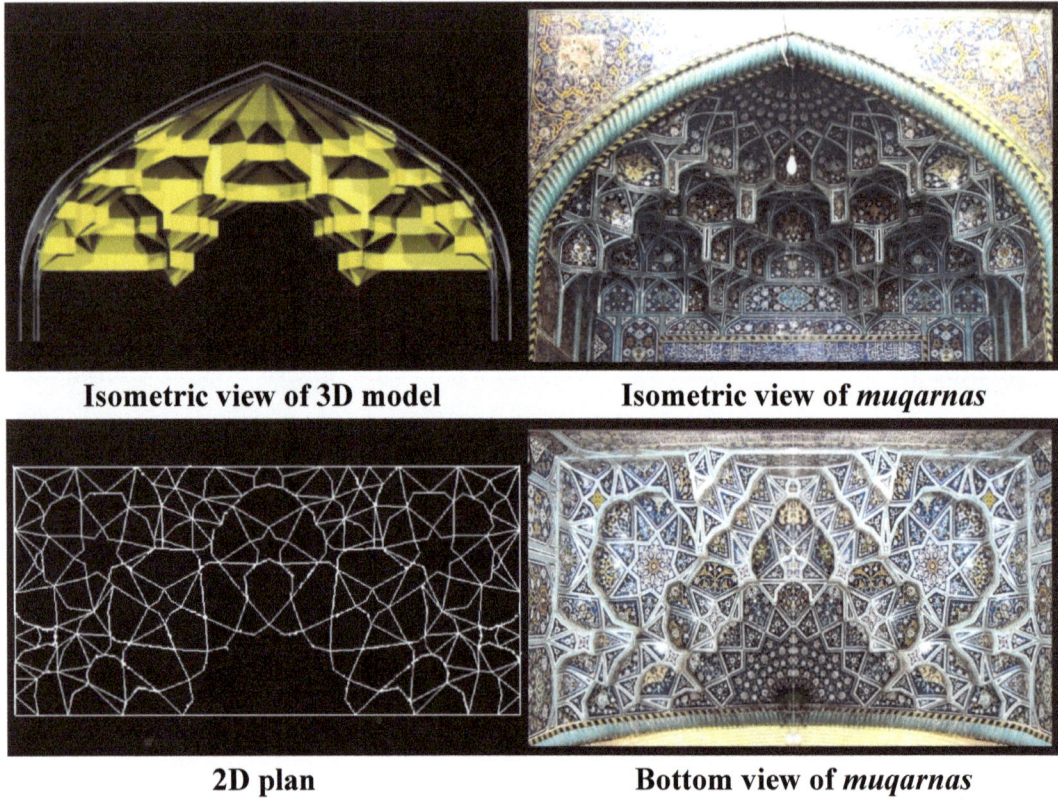

Isometric view of 3D model

Isometric view of *muqarnas*

2D plan

Bottom view of *muqarnas*

Figure S. 73 View of *muqarnas* in Seyyed Mosque, Entrance I.

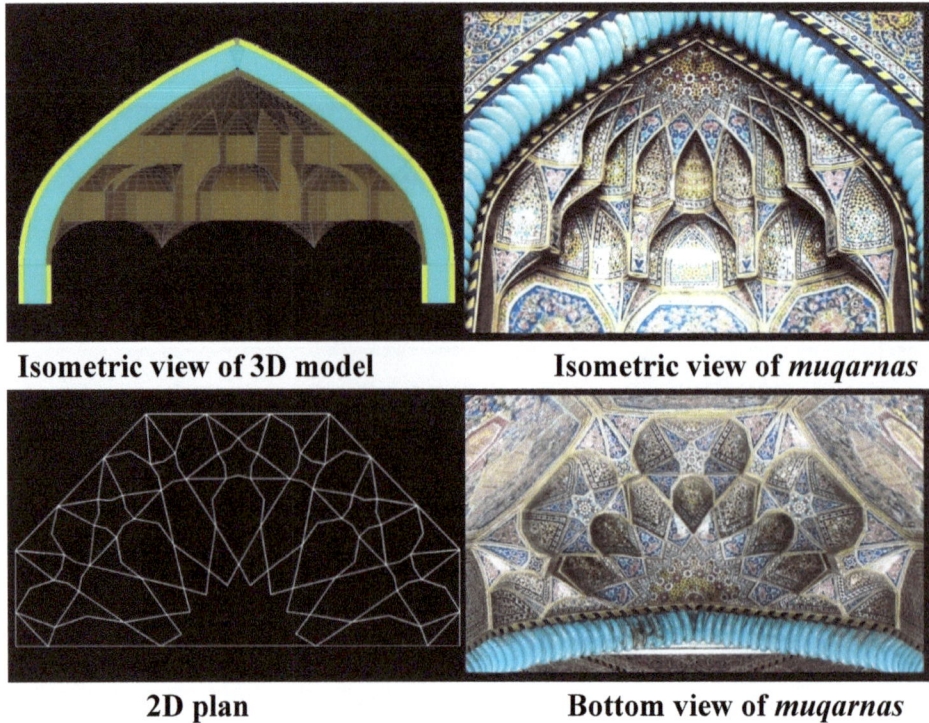

Isometric view of 3D model

Isometric view of *muqarnas*

2D plan

Bottom view of *muqarnas*

Figure S. 74 View of *muqarnas* in Seyyed Mosque, Mihrab.

Sample 75   Seyyed Mosque building information, Entrance II.

Seyyed Mosque	**Position:** Mihrab	**Date:** 1815

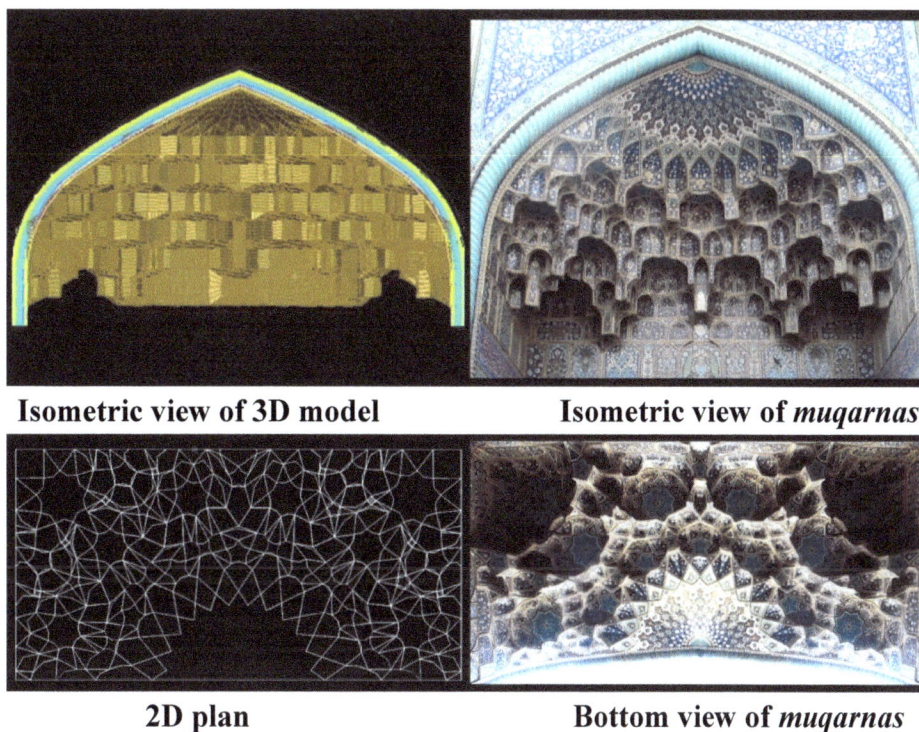

Isometric view of 3D model   **Isometric view of *muqarnas***

2D plan   **Bottom view of *muqarnas***

Figure S. 75   View of *muqarnas* in Seyyed Mosque, Entrance II.

A.36. Mulla Hasan Kashi Mausoleum, 1821 CE

The Molla Hassan Kāshi Mausoleum is a free–standing isolated edifice located 2.5 km to the south of Soltaniyeh, Iran. This 16th century mausoleum was built during Shah Tahmasp I, to honor Molla Hassan Kāshi, a 14th century mystic whose recasting of Islam's historical sagas as Persian poetic epics unwittingly had a vast influence over Shia Islam's future direction.

The monument is composed of a small esplanade serving as an entrance, and the mausoleum itself. The mausoleum displays an octagonal plan from the exterior though it is, in reality, a 6x6 meter square hall with additional galleries at the corners giving the aspect of an octagon. The mausoleum's octagonal exterior is uniquely shaped with four sides measuring 80.3 meters in length, and the other four sides measuring 5.75 meters in length.

In each gallery, there is a narrow staircase leading to an upper storey. The square building is roofed with a double–shelled dome, which has another layer of blue glazed bricks consisting of repeating geometric designs, as well as repeating Kufic calligraphy. The interior decoration of stucco stalactites was done at the time of Fath–Ali Shah, a Qajar king, in the early 19th century (Godard, 1954; Burke et al., 2004).

Sample 76   Molla Hasan Kashi Mausoleum's building information.

Name	Mullah Hassan Kashi Mausoleum
Variant Names	Shrine of Mulla Hasan–i–Kashan, Tomb of Mulla (Mawlana) HashaKashi
Location	Zanjan, Iran
Position	Under Dome
Date	1821 /1221 AH
Period	Qajar
Century	19th
Building Type	Religious
Building Usage	Shrine
Copyright	HamidrezaKazempour

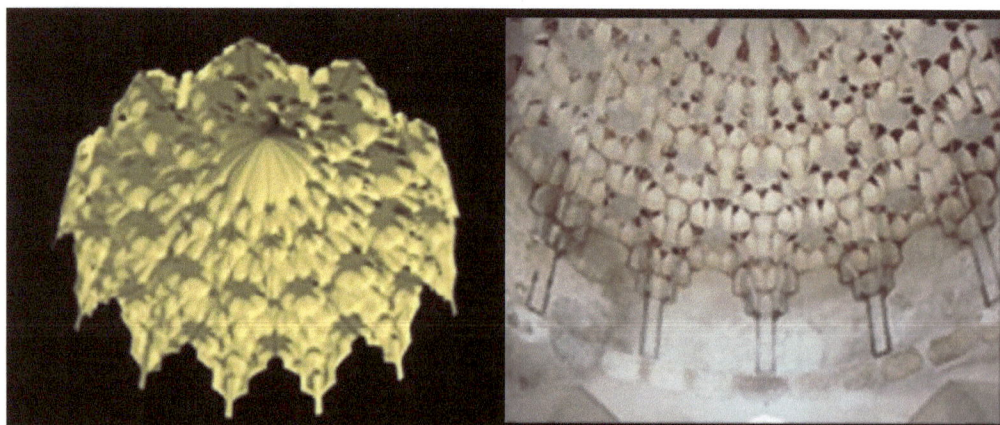

Isometric view of 3D model   Isometric view of *muqarnas*

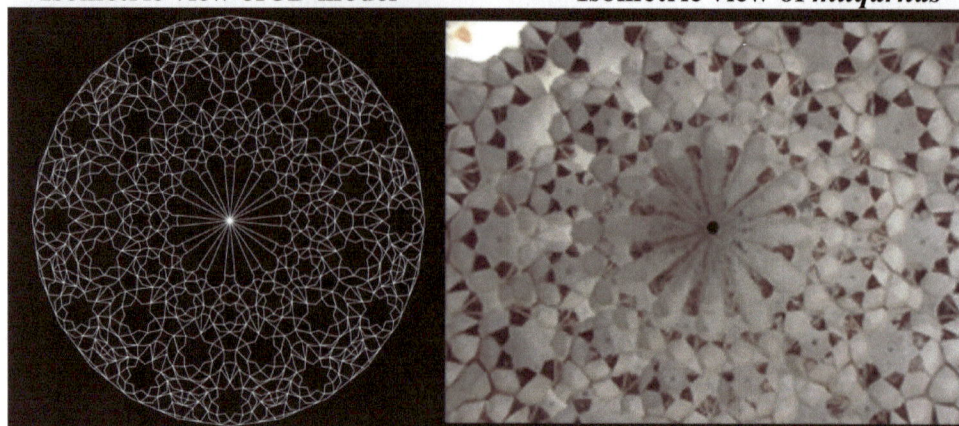

2D plan   Bottom view of *muqarnas*

Figure S. 76   View of *Muqarnas* in Molla Hasan Kashi Mausoleum.

A.37. Agha Bozorg Mosque, 1832 CE

Āghā Bozorg Mosque (Masjed–e Āghā Bozorg) is a historical mosque in Kāshān, Iran. The mosque was built in the late 18th century by master–mimar Ustad Haj Sa'banali, the mosque and theological school (madrasah) is located in the center of Kāshān. Agha Bozorgh Mosque was constructed for prayers, preaching and teaching sessions held by Molla Mahdi Naraghi II, known as Āghā Bozorgh.

The mosque consists of two large iwans, one in front of the mihrab and the other by the entrance. The courtyard has a second court in the middle which comprises a garden with trees and a fountain. Surrounding the courtyard are arcade. The iwan in front of mihrab has two minarets with a brick dome. The colors of arcades and eivan are restricted to blue, red, or yellow against a brick ground. It was here where Ustad Ali Maryam as a pupil started his career as a brilliant architect (Godard, 1965; Hatim, 2000; Pope, 1965; Pope, 1997a; Sourdel-Thomine & Wilber, 1974; Farhangi, 2003).

Sample 77   Agha Bozorg Mosque building information, Iwan.

Name	Masjid Agha Bozorg
Variant Names	Agha Bozorg Mosque and Madrasa, Masjid–i Agha Bozorg, Madaras–i Agha Bozorg, Agha Bozorg Mosque and Madrasa
Location	Kashan, Iran
Position	Iwan
Date	1832–33/1248 AH
Period	Qajar
Century	19th
Building Type	educational, religious
Building Usage	madrasa, mosque
Copyright	Hamidreza Kazempour

Sample 78   Agha Bozorg Mosque building information, Minaret.

Agha Bozorg Mosque	Position: Minaret	Date: 1832

326

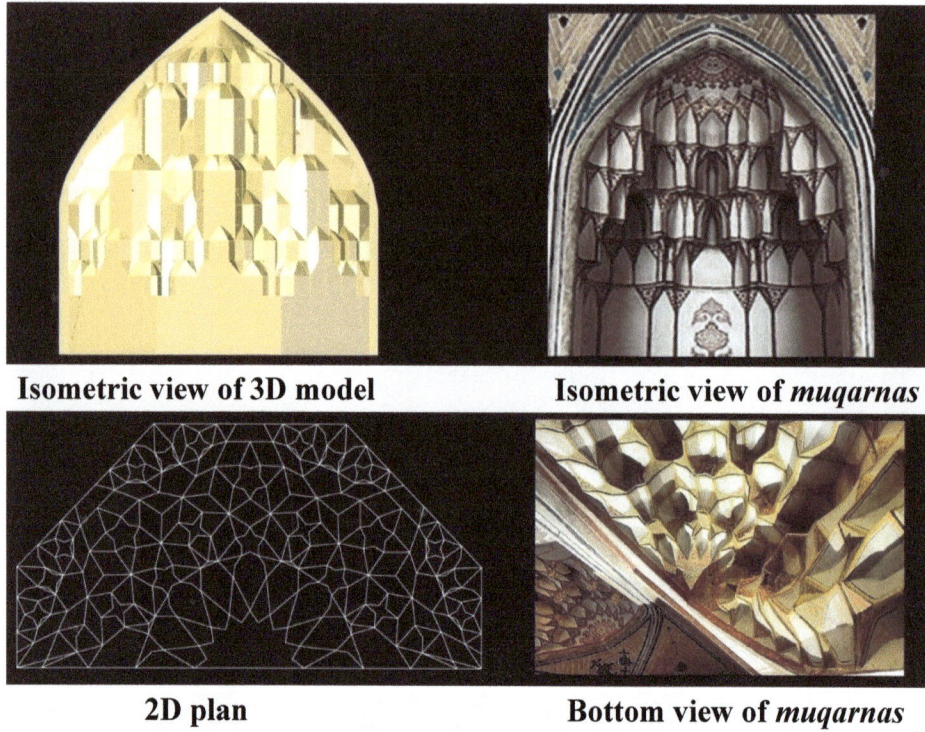

Isometric view of 3D model **Isometric view of *muqarnas***

2D plan **Bottom view of *muqarnas***

Figure S. 77 View of *muqarnas* in Agha Bozorg Mosque, Iwan.

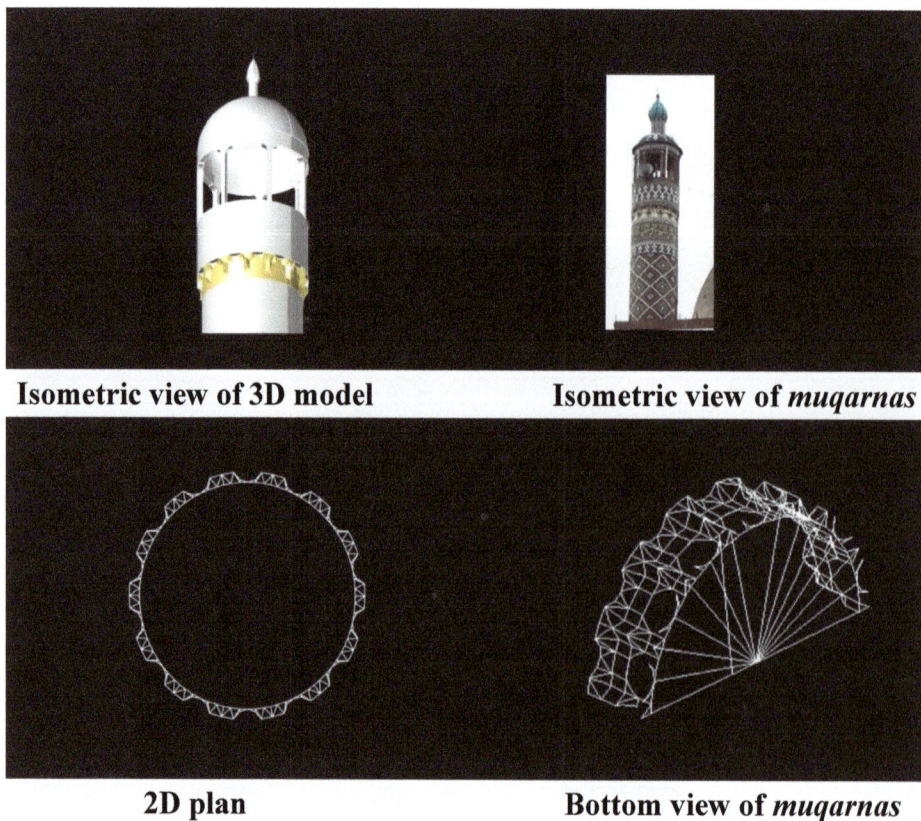

Isometric view of 3D model **Isometric view of *muqarnas***

2D plan **Bottom view of *muqarnas***

Figure S. 78 View of *muqarnas* in Agha Bozorg Mosque, Minaret.

A.38. Nasir Al–Mulk Mosque, 1876 CE

The Nasir al–Mulk Mosque or Pink Mosque is a traditional mosque in Shiraz, Iran, located in Goade–e–Araban place (near the famous Shah Cheragh mosque). The mosque was built during the Qājār era, and is still in use under protection by Nasir al Mulk's Endowment Foundation. It was built by the order of Mirza Hasan Ali Nasir al Molk, one of the lords of the Qajar Dynasty, in 1876 and was finished in 1888. The designers were Muhammad Hasan–e–Memar and Muhammad Reza Kashi Paz–e–Shirazi.

The mosque extensively uses colored glass in its facade, and displays other traditional elements such as panjkāseh–i (five concaves) in its design; it is also named in popular culture as Pink Mosque due to the usage of beautiful pink color tiles for its interior design.

Restoration, protection and maintenance of this valuable monument, observing all international standards on historical places, has duly begun and is continuing by The Nasir Al–Molk Endowment Foundation. This foundation is one of the largest in Fars Province and is under the administration of Mr. Mahmood Ghavam. The foundation actively devotes most of its yearly income towards feeding the poor during and after the religious ceremonies of Muharram and Safar, and towards the renovation and up–keep of this precious mosque (Fisher et al., 1991; Getlein, 2005).

Sample 79   Nasir Al–Molk Mosque building information.

Name	Masjid Naseer–Al–Molk
Variant Names	Nasr al–Mulk Mosque, Masjid–i Nasir al–Mulk, Masjid of Nasr al–Mulk, Nasir al–Mulk Mosque, Masjid of Nasir al–Mulk
Location	Shiraz, Iran
Position	Entrance Portal
Date	1876–88/1292–1305 AH
Period	Qajar
Century	19th
Building Type	Religious
Building Usage	Mosque
Copyright	HamidrezaKazempour

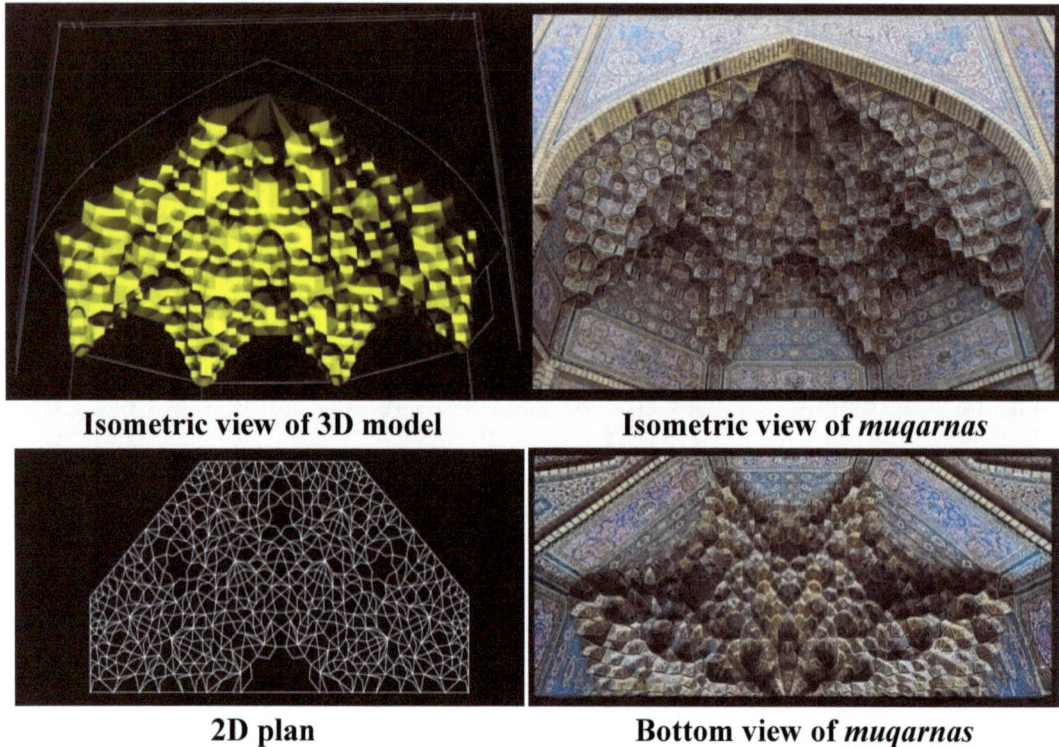

Isometric view of 3D model   **Isometric view of *muqarnas***

2D plan   **Bottom view of *muqarnas***

Figure S. 79   View of *muqarnas* in Nasir Al–Molk Mosque.

A.39. Imam Mosque of Semnan, 1931 CE

Imam Mosque (Soltani Mosque) built under the Qajar dynasty; it is located in the city center of Semnan. This mosque is a rare four–terrace mosque. This mosque has four large entrances on four sides. The doors to the northern, southern and eastern sides have a vestibule (usually octagonal in shape) and corridors. The design of the Imam Mosque utilized the expertise of Iranian architecture of the time, providing all sectors of the complex with equal acoustic sound systems. The upper portion of the northern and eastern doors are arched and artistically worked in plaster and tile. The ceilings of the vestibules, on the north, south and eastern sides are constructed of bricks, and are domed shaped with numerous arches and designed with tiles. There are four porches on four sides along with an area for nocturnal prayers. Behind the western porch is an inscription revealing the final date of construction (1342 AH).

There is a marble pulpit besides the altar. On the top of western porch is a small dome adorned with blue tiles. There is also a tiled inscription from the Holy Quran, all around the western porch worked in white tiles on a blue (tiled) back ground (Godard, 1965; Hillenbrand, 2004; Pope, 1965; Pope, 1997a).

Sample 80 Imam Mosque Semnan building information.

Name	**Masjid Imam**
Variant Names	Masjid Imam, Masjid–e Shah, Masjid–e Sultani, Imam Mosque, Imam Khomeini Mosque, Semnan Royal Mosque
Location	Semnan, Iran
Position	Entrance
Date	1931 /1342 AH
Period	Qajar
Century	19th
Building Type	Religious
Building Usage	Mosque
Copyright	Hamidreza Kazempour

Isometric view of 3D model   **Isometric view of *muqarnas***

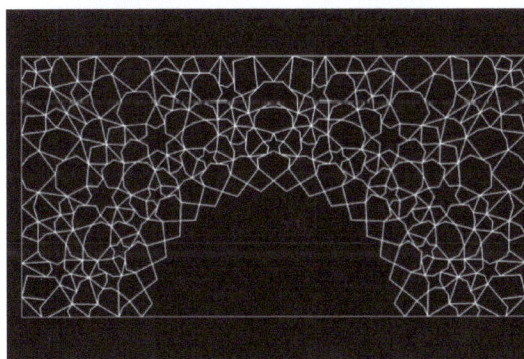

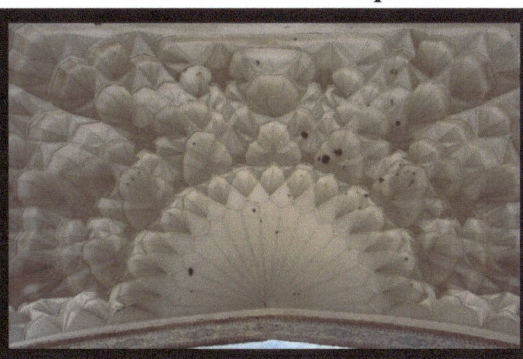

2D plan   **Bottom view of *muqarnas***

Figure S.80 View of *muqarnas* in Imam Mosque Semnan.

Appendix B

As mentioned in Chapter 3, most of the data collection procedure of this manuscript was accomplished in Author's class in two universities, namely Iran University of Science and Technology (IUST), and Shahid Rajaee Teacher Training University (SRTTU). The data collection was done by about 70 students as a part of their undergraduate courses "Islamic Architecture" and "Measurement of Historical Buildings" and the reports are available for further reference in the mentioned universities.

Although the whole procedure was accomplished under direct supervision of the author and Dr Memarian, and the building were chosen based on the final goal of this manuscript, the whole set of reports were reviewed again for obtaining the final set of 100 samples. Selected reports were finally double-checked by author for accuracy of the plan and 3Ds before being used here as initial data and some other recording was accomplished later again by author. Tables A and B shows the list of the names of the students involved in data gathering in IUST and SRTTU, respectively, whose efforts are really appreciated by the author. Among the names, some are highlighted as those whose plans and 3Ds were used in the main body of the manuscript, as part of selected 20 conducting samples. Furthermore, Figures A shows some photos of data collection days with students of the two mentioned universities.

Table A   List of the name of undergraduate students who took part in data collection procedure in IUST, Tehran.

1	Behzad Hajibeiklo	2	Shoresh Karimi	3	Kave Baghbeh
4	Fatemeh Gharoni	5	Zahra Bidarbakht	6	Neda Najafi
7	Mostafa Zarodi	8	Mitra Hashemi	9	Somayeh Taghavi
10	Kambiz Mozafari	11	Samaneh Abdi	12	Hakimeh Molazadeh
13	Nima Dehghani	14	Maryamsadat Banayi	15	Pegah Jafari
16	Ali Rastegar	17	Mansor Mohammadi	18	Elham Morovati
19	Fatemeh Nasrolahi	20	Mokhtar Naghavian	21	Mina Ghasemi
22	Bahare JahanBakhsh	23	Zahra Bakhshayi	24	Bita Langari
25	Negin Sheikhdavodi	26	Bita Ahmadi	27	Roya Diyanati
28	Maryam Abdolahi	29	Sara Zahedi	30	Mohammad Arefi
31	Shima Shahrjerdi	32	Nasrin Shishegar	33	Alireza Savadkohi
34	Hajarsadat Yadanfar	35	Zahrasadat Rajayi	36	Yasamin Bagheri
37	Parisa Legzi				

Table B   List of the name of undergraduate students who took part in data collection procedure in SRTTU, Tehran.

1	Elahe Faregh	2	Reza Soltani	3	Ahmad Khodakaram
4	Neda Rahmani	5	Saeed Rahanjam	6	Shima Shabani
7	Javaneh Abaspour	8	Raheleh Hoseini	9	Najmeh Etemadi
10	Elham Jamshidi	11	Hoda Bakhshi	12	Ameneh Ebrahimi
13	Arta Panosian	14	Shida Ramzani	15	Narges Yazdani
16	Farideh Kalhor	17	Elham Shams	18	Zeinab Bayat
19	Fatmeh Ghotbzade	20	Maryame Gholami	21	Samineh Jorbandian
22	Narges Aghabozorg	23	Mostafa Mokhtari	24	Rezvaneh Rezayi
25	Sahar Fathiazar	26	Alireza Ghafori	27	Masomeh Najafi
28	Sara Nemati	29	Naser Naghavi	30	Neda Rahimi
31	Ghazal Tahami	32	Seyed Hasan Alavizadeh	33	Seyed Hosein Hoseini
34	Noshin Bornayi				

Figure A   Some images from data collection procedure in Iran.